STP 1425

Fretting Fatigue: Advances in Basic Understanding and Applications

Y. Mutoh, S. E. Kinyon, and D. W. Hoeppner, editors

ASTM Stock Number: STP1425

ASTM International
100 Barr Harbor Drive
PO Box C700
West Conshohocken, PA 19428-2959

Printed in the U.S.A.

Library of Congress Cataloging-in-Publication Data
ISBN:

Fretting fatigue : advances in basic understanding and applications / Y. Mutoh, S.E.
Kinyon, and D.W. Hoeppner, editors.
   p. cm.—(STP ; 1425)
   "ASTM stock number: STP1425."
   Includes bibliographical references and index.
   ISBN 0-8031-3456-8
   1. Metals—Fatigue—Congresses. 2. Fretting corrosion—Congresses. 3. Contact
mechanics—Congresses. I. Mutoh, Y. (Yoshiharu), 1948– II. Kinyon, S. E. (Steven E.),
1966– III. Hoeppner, David W. IV. ASTM special technical publication; 1425

TA460.F72 2003
620.1′66—dc21

          2003041827

Photocopy Rights

Peer Review Policy

Each paper published in this volume was evaluated by two peer reviewers and at least one editor. The authors addressed all of the reviewers' comments to the satisfaction of both the technical
editor(s) and the ASTM International Committee on Publications.

To make technical information available as quickly as possible, the peer-reviewed papers in this
publication were prepared "camera-ready" as submitted by the authors.

The quality of the papers in this publication reflects not only the obvious efforts of the authors
and the technical editor(s), but also the work of the peer reviewers. In keeping with long-standing
publication practices, ASTM International maintains the anonymity of the peer reviewers. The ASTM
International Committee on Publications acknowledges with appreciation their dedication and contribution of time and effort on behalf of ASTM International.

Printed in Bridgeport, NJ
March 2003

Foreword

This publication, Fretting Fatigue: Advances in Basic Understanding and Applications, contains papers presented at the symposium of the same name held in Nagaoka, Japan, on 15-18 May 2001. The symposium was sponsored by ASTM Committee E08 on Fatigue and Fracture. The symposium co-chairpersons were Yoshiharu Mutoh, Nagaoka University of Technology, David Hoeppner, The University of Utah, Leo Vincent, Ecole Centrale de Lyon, Toshio Hattori, Hitachi LTD., Trevor Lindley, Imperial College of Science, and Helmi Attia, McMaster University.

Contents

Overview

The Third International Symposium on Fretting Fatigue was held in Nagaoka, Japan on May 15–18, 2001. This symposium is a follow-up to the First International Symposium on Fretting Fatigue held at the University of Sheffield in April 1993 (see Fretting Fatigue, ESIS Publication 18, edited by Waterhouse and Lindley, 1994) and the Second International Symposium on Fretting Fatigue held at the University of Utah on August 31, 1998 (see Fretting Fatigue: Current Technology and Practices, ASTM STP 1367, edited by Hoeppner, Chandrasekaran and Elliott, 2000). Fretting is well known to degrade fatigue strength significantly. Fretting fatigue failure has been increasingly disclosed in service components because those components have suffered more severe loading conditions than before due to the demands of save-energy and environment-preservation. One of major magic behaviors in fretting fatigue problems will be that a micro-slip between two combined components occurs under service loading, while such a slip is never expected at the designing stage. Great efforts have been devoted for understanding the fretting fatigue phenomenon and for developing the fretting fatigue design. This symposium was organized to focus on the progress in basic understanding and application and to continue the extensive interchange of ideas that has recently occurred.

Fifty-seven delegates from seven countries attended the symposium to present papers and participate in lively discussions on the subject of fretting fatigue. Dr. Waterhouse, who did pioneering research since the 1960s and is well known as a father of fretting research, was invited to this symposium. Technical leaders in the area of fretting fatigue were in attendance from most of the leading countries that are currently involved in fretting fatigue research, development, and engineering design related matters, as well as failure analysis and maintenance engineering issues. ASTM International Committee E8 provided the ASTM International organizational support for the symposium. The collection of papers contained in this volume will provide as an update to a great deal of information on fretting fatigue. This volume surely serves engineers that have a need to develop an understanding of fretting fatigue and also serves the fretting fatigue community including both newcomers and those that have been involved for some time.

The Symposium was sponsored by the following organizations as well as ASTM International: 1) Materials and Processing Division of JSME, 2) MTS Systems Corporation, 3) SHIMADZU Corporation, 4) HITACHI Ltd., and 5) JEOL Ltd.

All of the above organizations provided valuable technical assistance as well as financial support. The Symposium was held at Nagaoka Grand Hotel in the center of Nagaoka city, which is famous for fireworks and excellent rice and related products, such as Japanese sake and snacks. Many of the delegates would enjoy them.

The organizing committee members were: Dr. Yoshiharu Mutoh, Chair (Japan), Dr. David Hoeppner (USA), Dr. Leo Vincent (France), Dr. Toshio Hattori (Japan), Dr. Trevor Lindley (UK), and Dr. Helmi Attia (Canada). At the conclusion of the symposium, the organizing committee announced that the next symposium would be held a few years after this symposium in France with Dr. Vincent as coordinator and chair.

Editing and review coordination of the symposium was done with the outstanding coordination of Ms. Maria Langiewicz of ASTM International. The editors are very grateful to her for her extensive effort in assisting in concluding the paper reviews and issuing this volume in a timely manner.

The symposium opened with remarks by the symposium chair. Subsequently, Dr. Robert Waterhouse gave the distinguished invited lecture on Fretting in Steel Ropes and cables. Six keynote lectures were given in the following sessions, which were "Fretting wear and crack initiation", "Fretting fatigue crack and damage", "Life prediction", Fretting fatigue parameter effects", Loading condition and environment", Titanium alloys", "Surface treatment", and "Case studies and applications". Forty-three papers were presented and this volume contains twenty-nine of those papers.

The new knowledge about the process of fretting crack nucleation under fretting wear was provided through both detailed in-situ observations and mechanical models, which included not only fracture mechanics but also interface mechanics. Fretting fatigue crack propagation under mixed mode was discussed based on fracture mechanics approach. However, small crack problems, especially those related to threshold and under mixed mode, are still remained for future efforts. Fretting fatigue life estimations were attempted based on various approaches including fracture mechanics, notch fatigue analysis and multiaxial fatigue parameters. A number of factors are well known to influence on fretting fatigue behavior and strength. Effects of those parameters, which included contact pressure, friction coefficient, contact pad geometry, mating material and so on, were discussed. Effect of loading conditions including block loading, high frequency and service loading was also presented. The knowledge about loading wave effect has been limited until now. Improvements of fretting fatigue strength by using coating techniques were presented. Titanium alloys have been typically used for structural components suffering fretting fatigue, such as turbine components and bio-joints, due to their lightweight as well as excellent corrosion resistance. A lot of works on this material including a review paper were presented to understand fretting fatigue behavior in various conditions. Case studies on electrical cables, dovetail joints, pin joints and rollers were introduced. Methods for bridging between specimen-based research works and case studies are required, when a fretting fatigue test method would be standardized. These topics will be also important future work.

This publication is only one aspect of the symposium. The sessions and discussions contribute greatly to the mission of the symposium. The effort of the co-chairmen of the sessions is acknowledged and appreciated. The editors are thankful to the attendees of the symposium for interesting points and useful comments they made during the discussions that followed the paper presentations. Their enthusiasm to follow up this symposium with the next symposium in France is appreciated and well taken. The editors hope that those concerned with the subject of fretting fatigue will find this publication useful and stimulating.

Y. Mutoh
Nagaoka University of Technology
Nagaoka, Japan
Symposium co-chairman and editor

D. W. Hoeppner
University of Utah
Salt Lake City, UT
Symposium co-chairman and editor

S. E. Kinyon
MTS Systems Co.
Eden Prairie, MN
Editor

INVITED PAPER

Robert B.Waterhouse[1]

Fretting in Steel Ropes and Cables – A Review

REFERENCE: Waterhouse, R. B., "**Fretting in Steel Ropes and Cables—A Review,**" *Fretting Fatigue: Advances in Basic Understanding and Applications, STP 1425,* Y. Mutoh, S. E. Kinyon, D. W. Hoeppner, and, Eds., ASTM International, West Conshohocken, PA, 2003.

Abstract: The numerous inter-wire contacts in ropes and cables are potential sites for nucleation of fatigue cracks by fretting. At larger amplitudes of movement excessive wear can occur leading to decreased cross-section of a wire and increased stress, or if debris accumulates, forcing apart of the wires. In locked coil ropes this can cause local stiffness and ingress of corrodants. The wires themselves have an asymmetric residual stress distribution circumferentially resulting from the drawing process. The position of an inter-wire contact relative to this greatly influences the fretting fatigue strength. Hot dip galvanising reduces these stresses as well as giving cathodic protection. Local friction is reduced by incorporating a lubricant in locked coil ropes. Conventional grease-based lubricants with high shear strength are most effective in reducing friction but recent experiments with a lower viscosity oil containing graphite show promise

Keywords: fretting fatigue, fretting wear, wire ropes

Introduction

The great virtue of ropes and cables fabricated from steel or other metal wires, is their flexibility which allows them to be wound on drums or passed over pulleys and sheaves, combined with their great strength, which under tension, often approaches that of the individual wires. They are therefore widely used in the construction industry as the support in suspension and cable-stayed bridges and also in the roofs of some buildings. The lifting ropes of cranes and elevators are a further application. In marine engineering they are used as hawsers, rigging and as anchor and mooring ropes. The involvement of the author with the latter has been with the mooring of the off-shore oil rigs, which are tethered with four ropes the size of tree trunks at each corner. In mountainous countries ski-lifts and cable cars which cater for the recreational and tourist trades are entirely dependent on steel cables. In some instances they provide the compressive stress in post-tensioned concrete beams [1]. A further use, which has developed in more recent years, is as strengthening in the bead, belt and carcass areas of pneumatic automobile radial tyres. These are very different from the traditional steel rope in that they consist of two or three wires of diameter 0.15 to 0.38mm of strength 2300 to 3000MPa which are brass plated to bond with the rubber [2]. The optimum design for a two-filament cord with superior fatigue resistance is for one to be maintained straight and the other to be wound around it [3].

---

[1] Department of Materials Engineering, University of Nottingham, Nottingham NG7 2RD, UK.

3

In all these applications there are obviously an astronomically high number of inter-wire contacts and therefore the possibility of fretting under operating conditions is absolutely certain. The question is to what extent does fretting prejudice the function of a particular rope? Some evidence is available from the examination of ropes taken from a motorway bridge over the Norderelbe, which had been in service for 25 years. They were tested in static and dynamic loading and showed remarkable remaining fatigue life and carrying capacity [4]. However it was stated that fretting had a dominating influence on damage extension. As failure is initiated by the fracture of an individual wire, the total failure of a rope is usually apparent before it actually occurs particularly if there is some system in force to detect individual wire failures. For this reason catastrophic failures are relatively rare.

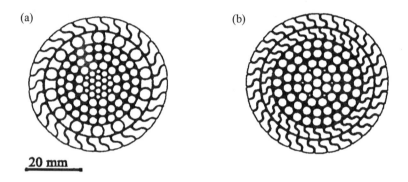

Figure 1 – *Types of locked coil ropes: (a) half lock 1½x, (b) full lock 2x*

Design of ropes

Ropes are divided into two categories, single strand and stranded. The simplest rope is where six wires of circular diameter are wound around a central (king) wire. In a stranded rope such strands are further wound together in combinations of increasing complexity. The author's experience has been entirely with single strand ropes consisting of a much larger number of wires. In the case of the mooring ropes for the off-shore oil rigs the wires were 5mm dia. wound in close packed layers with succeeding layers being obliquely inclined to each other. This results in the inter-wire contact in the same layer being a line contact whereas the contact between wires in adjacent layers being a so-called "trellis" contact. The rope consisted of several hundred wires and was filled with bitumen and encased in a thick polymer coating so that its overall diameter was some 24cm, like a tree-trunk. Since they were operating in seawater with the possibility of outward damage and penetration of seawater, the wires were hot-dipped galvanised. In other ropes particularly lift ropes and those for cable cars, they were of the locked coil type where the wires in the external layers were shaped to fit together to form almost solid sheets. Figure 1 shows some examples. With such ropes a lubricant can be incorporated within the rope and this is an important factor in the prevention of fretting. Power cables consist of a central steel strand surrounded by three layers of aluminium wires (Figure 2). In some cases the central strand may also be of aluminium. Another design of rope which is favoured in Japan, is the parallel wire strand rope which consists of a bundle of parallel wires which is given a certain amount of twist. This has the advantage of avoiding trellis inter-wire contacts (Figure 3).

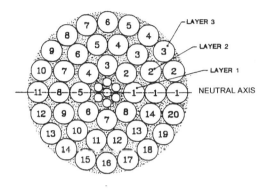

Figure 2 – *Power line conductor*

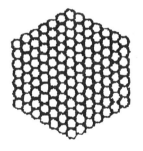

Figure 3 – *Parallel wire rope*

Although fretting on line contacts will have an effect on the fatigue behaviour of a rope, the effect of trellis contacts is much greater because they are localised and the contact stresses are consequently much higher than on line contacts. Hobbs and Raoof [5] and Raoof [6] separately have conducted detailed analyses into the conditions at such contacts when the rope is subjected to tensile, torsional and bending stresses. Applying this to a single strand mooring rope, the theory indicates that internal wire fractures occur prior to external fractures and that they are located at trellis point contacts between an outermost wire and the layer immediately below. This has been found to be the case in locked coil ropes [7,8]. Figure 4 shows fretting marks on a wire from a single strand rope and Fig 5 shows fatigue cracks generated at such marks in a laboratory test. Since the amplitude of the fretting movement is extremely small the situation is that of partial slip with the cracks developing at or towards the edge of the contact spot as in his example. If however the movement becomes larger, fretting wear may occur and failure may result from increased stresses at the reduced cross-section, Fig 7. Fatigue tests on ropes confirm that fractures occur at trellis points [9].

Figure 4 – *Wire from rope with fretting marks from a trellis contact*

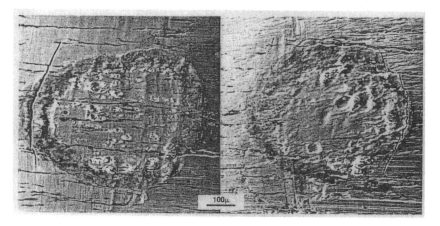

Figure 5 – *Fretting marks with cracks*

One of the problems in fatigue testing ropes and individual wires is gripping the ends of the specimen in such a way as to prevent fretting failure in the anchorage or grips. Related to this fact is the observation that failures in power cables are usually located in the vicinity of clamps near pylons [10]. Anchorages are obviously important in the construction of suspension bridges and the topic has been dealt with in detail by Chaplin [11]. The development of zero-loss anchorages in which either the relative Displacements or atmospheric oxygen is eliminated such that fretting corrosion is reduced to an insignificant level results in the fatigue strength not being diminished [12]. The design of a rope also influences the fatigue behavior in service. In locked coil ropes the 2x arrangement attains 80% higher service life than the 1 1/2x (Fig 1). This is attributed to the larger contact areas in the former [7].

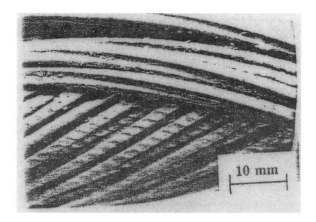

Figure 6 – *Fretting marks on second layer of locked coil rope*

Figure 7 – *Wire failure due to wear*

The Wires

In the construction of a rope or cable the wires themselves are the vital component. In steel ropes the wires are high strength eutectoid or hypereutectoid steel which is drawn down through twelve or thirteen dies. Solid lubricants are applied during the drawing process. The original steel rod is passed through a zinc phosphate bath which deposits an iron-zinc conversion coating and then through a 20% sodium tetraborate bath to neutralize the acid from the phosphate bath and also acts as a sealant and finally powdered sodium stearate is applied as a solid lubricant. Remnants of this treatment on the surface of the wire have some effect in the eventual fretting contacts. The wire is drawn at an angle to the final die so that the wire forms a coil. It is the residual stress thus introduced that has a major influence on the properties of the wire. The level of the residual stress is also governed by the degree of reduction in the final die. A further metallurgical complication which may occur is decarburization often traceable back to the original billet.

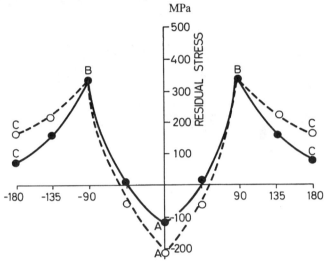

Figure 8 – *Stress distribution around circumference of wire*

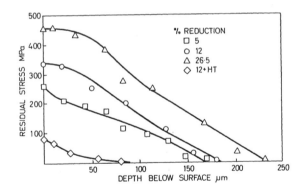

Figure 9 – *Effect of reduction in final die on residual stress*

Figure 8 shows the residual stress pattern round the circumference of a 5 mm dia wire. The residual stress has a maximum tensile value at the ends of a diameter perpendicular to the plane of the coil of the wire (B position). On the convex surface the stress is compressive (A position) and a lower tensile stress on the concave surface. On straightening the wire the latter two stresses both increase as indicated in Figure 8. The level of these stresses also depends on the degree of reduction in the final die, which is shown in Figure 9 for the B position. Hot dip galvanizing greatly reduces the stresses and this is included in Figure 9 where the wire has been given a heat treatment comparable with the galvanizing process.

The fretting fatigue behavior of individual wires depends on the location of interwire contact points in relation to the residual stress pattern, which is shown in Figure 10. The higher tensile stress at position B is where a crack will initiate leading to a lower fatigue strength. The level of stress also has its expected effect Figure 11. The beneficial effect of the galvanizing heat treatment should be noted.

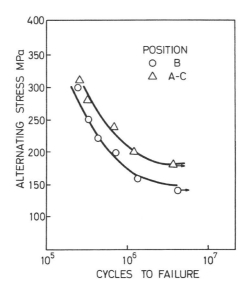

Figure 10 – *Fretting fatigue curves in seawater for B and A-C contact*

Environment

Most ropes are operating in a corrosive atmosphere particularly those in marine conditions such as the moorings of ships and off-shore oil rigs as well as cranes and haulage ropes in coastal dockyards. Although every effort is made to protect the rope in the case of moorings by filling the cable with bitumen and surrounding it with a polymer coat, there is the possibility of this becoming damaged and seawater getting into the rope. Figure 12 shows the further reduction in fatigue strength caused by fretting in seawater. Zinc galvanizing is applied as further protection as illustrated. The cathodic protection conferred by the zinc coating results in a reduction in the crack growth rate as shown in Figure 13 [*14*].

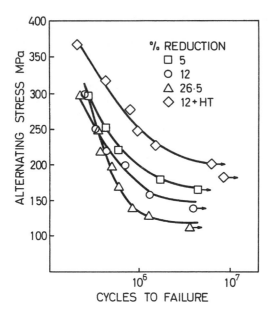

Figure 11 - *Effect of residual stress on fretting fatigue curves*

Lubricants for locked coil ropes

One of the dangers of fretting in ropes besides that of initiating fatigue cracks, is that fretting wear may occur producing oxide debris which is of greater volume than the metal consumed in its creation. This can cause the wires to be forced apart so that ingress of the environment is more likely and the increase in coefficient of friction between the wires leads to local stiffness in the rope which becomes apparent if it is passing over a pulley or sheave. Lubricants are incorporated to ensure inter-wire friction remains low. Those in use are proprietary products whose composition is not divulged. However, a study by the author of five of these materials allowed them to be ranked in their ability to improve resistance to fretting fatigue failure. The increase in the fretting fatigue strength of the wires by their application ranged from 7.5 to 32%. The properties of the best lubricant were a high drop point, low unworked penetration and a high shear strength and resistance to shear i.e. a low increase in penetration on working, and a high value of shear strength in an extrusion test. Zhou et al have investigated the effect of a proprietary reversible protective grease on the fretting behaviour of an electrical conductor in cyclic bending and found a decrease in contact wear and particle oxidation together with shorter fretting cracks and a higher service life [*15*] .

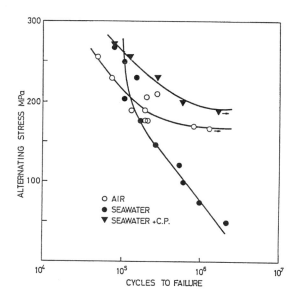

Figure 12 – *Effect of galvanising of fretting fatigue curves*

Since the main purpose of lubrication is to reduce the coefficient of friction, it was decided to measure friction in fretting wear tests rather than fatigue testing. Tests on the as received 2.8 mm dia wire, so-called "bright wire" with the remnants of the drawing lubricant on the surface showed that the latter resulted in low friction of 0.15 for the initial 3000 to 5000 cycles after which it rose to 0.65. When the wire was abraded the friction was 0.65 from the start. After 100 000 cycles the friction rose steadily to 0.9, Figure 14. Contact resistance measurements indicated that this was due to the accumulation of debris. A suspension of graphite in a low viscosity paraffinic oil smeared on to the wire surfaces maintained a friction of 0.15 for the duration of the test, 300 000 cycles, Figure 15 [*16*]. Further experiments with chopped carbon fibre and Kevlar in a paraffinic oil were less successful. The Kevlar actually produced surface damage by indenting the hard steel surface.

Surface Treatment

The standard and most effective method of dealing with fretting problems is to apply some form of surface treatment such as shot-peening, ion implantation or coatings. To employ such methods in the case of wire ropes is difficult practically and not viable economically. However the problem has been tackled by a group in Australia with a proposal to apply coatings using contact resistance heating between a roller and the wire and feeding in a powder of the coating material [18]. Tests have shown that a promising coating is molybdenum which when applied by PVD provided useful wear resistance

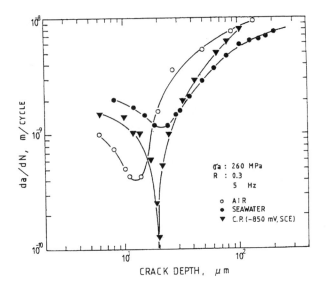

Figure 13 – *Effect of cathodic protection on fretting crack growth*

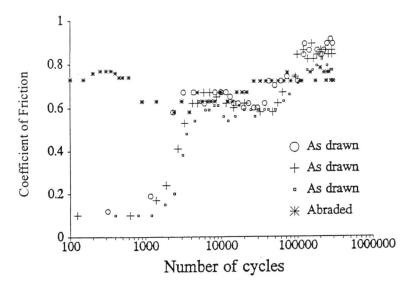

Figure 14 – *Coefficient of friction of as-received wire and after abrasion*

although the friction remained high at 0.7. Whether this would have any beneficial effect on fatigue has yet to be demonstrated.

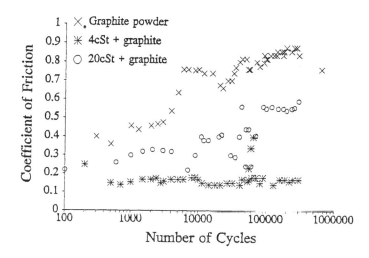

Figure 15 – *Effect of graphite in oil on coefficient of friction*

Conclusion

In a survey of the future development of cables and ropes particularly for bridges Gourmelon has suggested that materials which are insensitive to fretting should be explored [*18*]. Aramid fibre cables have the required tensile strength comparable with steel i.e. 2000 MPa, with the advantage of low density and resistance to corrosion but nevertheless steel still has the competitive advantage. One of the disturbing features of steel wire is the residual tensile stress which exists in 75% of the circumference of the wire which is not ideal from the fatigue point of view. The way forward would seem to be exploring new coatings, including improvements in galvanising and the search for more efficient lubricants.

References

[1] Oertle, J. and Thurlimann, B., "Reib Ermudung einbetonierter Spannkabel" *Schweizer Ingenieur und Architekt*, 5, 1987, pp. 295-300.
[2] Prakash, A., Shemenski, R.M. and Kim, D.K., "Life Prediction of Steel Tire Cords" *Rubber World,* May 1987, pp. 36-43.
[3] Cipparone, M. and Doujak, S., "Steel Cords with Improved Fatigue Resistance" *Tire Technology International: Annual Review of Tire Materials and Tire Manufacturing Technology,* 1999, pp. 45-46.

[4] Harre,W., "Erkentnisse aus der Prufung baupraktisch vorbelasteter voll-
     verschlossenner Bruckenseile der Autobahnbrucke uber die Norderelbe,"
     Bauingenieur, 67, 1992, pp. 91-99.
[5] Hobbs, R.E. and Raoof, M., "Mechanism of Fretting Fatigue in Steel Cables,"
     International Journal of Fatigue, 16, 1994, pp. 273-280.
[6] Raoof, M., "Prediction of Axial Hysteresis in Locked Coil Ropes," *Journal of Strain
     Analysis,* 31, 1996, pp. 341-351.
[7] Woodtli, J.,"Microscopical Identification of Fretting and Wear Damage due to
     Fatigue in Locked Coil Wire Ropes," *Fretting Fatigue,* R.B.Waterhouse and
     T.C.Lindley, Eds., MEP, London, 1994, pp. 297-306.
[8] Harris, S.J.,Waterhouse, R.B. and McColl, I.R., "Fretting Damage in Locked Coil
     Steel Ropes," *Wear,* 170, 1993, pp. 63-70.
[9] Suh, J.-I., and Chang, S.P., "Experimental Study on Fatigue Behaviour of Wire
     Ropes," *International Journal of Fatigue,* 22, 2000, pp. 339-347.
[10] Cardou, A., Leblanc, A., Goudreau, S. and Cloutier,L., "Electrical Conductor
     Bending Fatigue at Suspension Clamp: a Fretting Fatigue Problem," *Fretting
     Fatigue,* R.B.Waterhouse and T.C.Lindley, Eds, MEP, London, 1994, pp. 257-
     266.
[11] Chaplin, C.R., Rebel, G. and Ridge, I.M.L. "Tension-Torsion Fatigue Effects in
     Wire Ropes," *Tecnologie del Filo,* 6, 1999, pp. 56-61.
[12] Kohler, W., and Nurnberger, U., "Verbesserung des Schwing-festigkeitsverhaltens
     von Spannkabel- und Seilveranker-ungen," *International Association for Bridge
     and Structural Engineering,* 1982, pp. 673-680.
[13] Smallwood, R. and Waterhouse, R.B., "Residual Stress Patterns in Cold Drawn
     Steel Wires and Their Effect on Fretting-Corrosion Fatigue Behaviour in
     Seawater," *Applied Stress Analysis*, T.H. Hyde and E. Ollerton, Eds, Elsevier
     Applied Science, London and New York, 1990, pp. 82-90.
[14] Takeuchi, M. and Waterhouse, R.B., "The Initiation and Propagation of Fatigue
     Cracks under the Influence of Fretting in 0.64C Roping Steel Wires in Air and
     Seawater," *Environment Assisted Fatigue,* P. Scott and R.A. Cottis, Eds, MEP,
     London, 1990, pp. 367-379.
[15] Zhou, Z.R., Fiset, M., Cardou, A., Cloutier, L. and Goudreau, S. "Effect of
     Lubricant in Electrical Conductor Fretting Fatigue," *Wear,* 189, 1995, pp. 51-57.
[16] McColl, I.R., Waterhouse, R.B., Harris, S.J. and Tsujikawa, M., "Lubricated
     Fretting Wear of a High-Strength Eutectoid Steel Rope Wire," *Wear,* 185, 1995,
     pp. 203-212.
[17] Batchelor, A. and Stachowiak, G. "Developments in Advanced Coatings for Roping
     Wire," *Wire Industry,* 63, 1996, pp. 275-278.
[18] Gourmelon, J.P., "Cables for Cable-Stayed Bridges – Which Material for
     Tomorrow?," *Revue de Metallurgie, Cahiers d'Information Techniques*, 95,
     1998, pp. 553-562.

FRETTING WEAR AND CRACK INITIATION

Siegfried Fouvry,[1] Philippe Kapsa,[1] and Léo Vincent [2]

A Global Methodology to Quantify Fretting Damages

REFERENCE: Fouvry, S., Kapsa, Ph., and Vincent, L., " **A Global Methodology to Quantify Fretting Damages,**" *Fretting Fatigue: Advances in Basic Understanding and Applications, ASTM STP 1425,* Y. Mutoh, S. E. Kinyon, and D. W. Hoeppner Eds., ASTM International, West Conshohocken, PA, 2003.

ABSTRACT: Fretting wear and fretting fatigue are commonly associated with damage of quasistatic loaded assemblies and with decrease in lifetime. Depending on the sliding condition, wear induced by fretting or cracking induced by fretting can be observed. To quantify such competitive damage phenomena, a fretting map approach has been extensively applied describing the sliding conditions and the damage evolution as a function of the normal force and the displacement amplitude. This approach, considered as a useful methodology to analyze tribo-systems, nevertheless presents the limitation of not allowing a direct comparison between tribo-systems. To rationalize this experimental approach and facilitate the comparison between tribo-systems, normalized sliding condition and crack nucleation fretting maps are introduced. Based on contact mechanics, the sliding transition is quantified using a fretting sliding criterion, and a specific formulation is provided to identify the local friction coefficient under partial slip condition. Cracking, which is mainly observed under stabilized partial slip condition, is analyzed by applying multiaxial criteria and taking into account the size effect. Wear, which is favored under gross slip condition, is quantified through an energy approach. Finally a global methodology is developed by which the sliding condition, the crack nucleation under partial slip condition and the wear kinetics under gross slip regime may be quantified.

KEYWORDS: Fretting, Fretting map, sliding condition, crack nucleation, size effect, wear, dissipated energy.

Introduction

Fretting is a small amplitude oscillatory movement, which may occurs between contacting surfaces which are subjected to vibration or cyclic stress. Considered to be a plague for modern industry, fretting is encountered in all quasi-static loadings submitted

[1] UMR CNRS 5513, Ecole Centrale de Lyon, BP 163, 69131, Ecully, France.
[2] UMR CNRS 5621, Ecole Centrale de Lyon, BP 163, 69131, Ecully, France.

17

to vibration and thus concerns many industrial branches (helicopters, aircraft, trains, ships, trucks, electrical connectors) [1-3] (Figure 1a). Fretting is a very complex problem involving numerous aspects such as tribology, contact mechanics, fatigue mechanics and material science, but also corrosion science. To reproduce such phenomena different fretting tests have been developed which permit the control of the normal loading and displacement amplitudes [4-5]. The relative displacement induces tangential loading, which can be described by the fretting loop: $Q(t)=f(\delta(t))$ (Figure 1b).

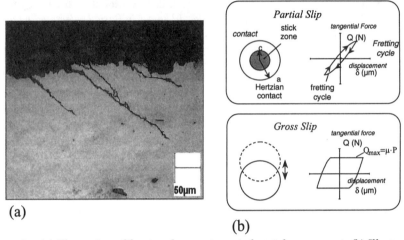

(a)

(b)

Figure 1 – *(a) Illustration of fretting damage in an industrial component, (b) Illustration of the fretting sliding conditions.*

Two fretting conditions are identified : partial slip, which is characterized by a closed elliptical fretting loop, associated with a composite contact of the sliding and sticking zone. The gross slip condition which is identified by a quadratic dissipative fretting loop, is related to full sliding occurring over the entire interface.

The friction coefficient changes during loading and this can lead to a change of sliding condition. It is then possible to define various fretting regimes : Partial slip Regime (P.S.R) : when the partial slip condition is maintained during the test; Mixed Fretting Regime (MFR) : when there is a transition from one condition to another; Gross Slip Fretting Regime (GSR): when the gross slip condition is maintained. Waterhouse et al [1] first indicated a correlation between the sliding regime and damage evolution (Figure 2).

Cracking is mainly encountered in partial slip regimes and mixed fretting regimes stabilized under partial slip condition, whereas wear is observed for larger amplitudes in gross slip regimes. The experimental mapping of the material response (MRFM : Material response fretting mapping) was originally introduced by P. Blanchard et al [6]. A major defect of such a mapping approach is the use of mechanical variables like normal force and displacement amplitude which, depending on the geometry and mechanical properties studied, cannot permit a direct correlation between fretting situations. This paper will develop a quantitative approach defining pertinent variables, which allow the damage evolution to be quantified (Figure 3).

It consists in quantifying the sliding transition, transposing multiaxial fatigue approaches to better predict the crack nucleation under stabilized partial slip conditions and analyzing the wear extent through an energy approach.

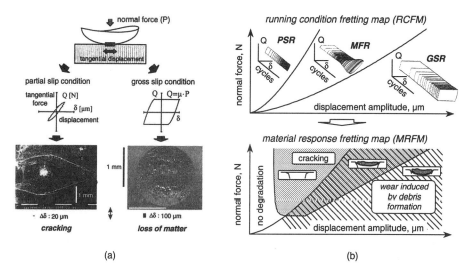

(a)   (b)

Figure 2 – *(a) Damage evolution as a function of the fretting loading condition (sphere/plane contact). (b) Representation of the fretting chart which combines the fretting regime analysis (RCFM) with the material response (MRFM).*

Normalized criterion to determine the sliding transition

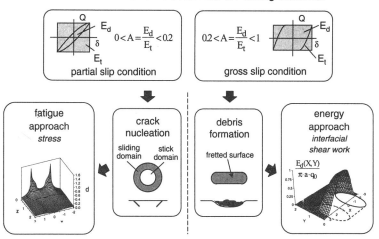

Figure 3 – *Illustration of the methodology for the fretting damage analysis (after [6]).*

Quantification of the sliding transition

The present sliding analysis is developed from the Mindlin formulation of a sphere/plane configuration [7]. The experimental system is a fretting wear test already described elsewhere [5, 6]. During a test, the normal force (P), tangential force (Q) and displacement (δ) are recorded. The magnitude of the tangential force versus the displacement is recorded for each fretting cycle (Figure 4). The surface displacement δ is deduced from the measured value δ_m taking into account the tangential apparatus compliance Cs. This permits an estimation of the contact displacement through relation [6]:

$$\delta(t) = \delta_m(t) - C_S \cdot Q(t) \tag{1}$$

As mentioned before, the sliding transition is classically defined from the fretting loop shape. One major difficulty of such an analysis is the subjectivity of the observation, particularly close to the transition between partial and gross slip.

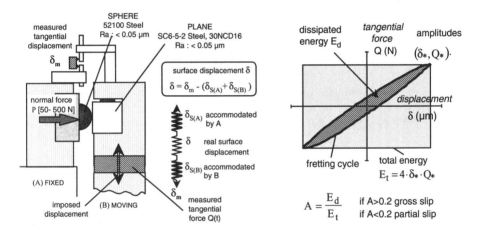

Figure 4 – *Fretting test and quantitative analysis of the fretting loop.*

To overcome this difficulty, several sliding criteria have been introduced to quantify the sliding analysis (Figure 4) [10, 12]. An energy sliding criterion has been introduced which corresponds to the ratio between the dissipated energy (i.e. area of the hysteresis loop) and an equivalent total energy which could be defined as the total energy dissipated by the contact system if it presents infinite tangential compliance.

$$A = \frac{E_d}{E_t} = \frac{E_d}{4 \cdot Q* \cdot \delta*} \tag{2}$$

It has been demonstrated for the elastic sphere/plane configuration that this variable presents a constant value ($A = 0.2$) at the sliding transition (Figure 5) [6].

This value is independent of the contact dimension and the elastic properties of the tribosystems. Therefore, the experimental sliding analysis can be quantified and the sliding regime formalized. Finally, the running condition fretting map is more easily established. Figure 6a clearly illustrates that the sliding transition (δ_t) is mainly controlled by the friction coefficient. The Mindlin analysis confirms this proportional relation.

$$\delta_t = \frac{K_1 \cdot \mu \cdot P}{a} \qquad \text{, with } K_1 = \frac{3}{16} \cdot \left(\frac{2-v_1}{G_1} + \frac{2-v_2}{G_2} \right) \qquad (3)$$

with
v_1, v_2 : Poisson coefficients of the plane (1) and the sphere (2),
G_1, G_2: Shear elastic moduli of the plane (1) and the sphere (2),
a   : Hertzian contact radius.

Regarding the Mindlin formulation, the transition amplitude can also be expressed as a function of the maximum Hertzian pressure (p_0) (or shear stress q_0) and the contact Hertzian radius "a" (Figure 6b) [8].

$$\frac{\pi}{8} \cdot \frac{\mu \cdot p_0}{G^*} = \frac{q_0}{(8 \cdot G^* / \pi)} = \frac{\delta_t}{a} \qquad (4)$$

and G^* (MPa): Equivalent shear modulus expressed by $\dfrac{1}{G^*} = \dfrac{2-v_1}{G_1} + \dfrac{2-v_2}{G_2}$.   (5)

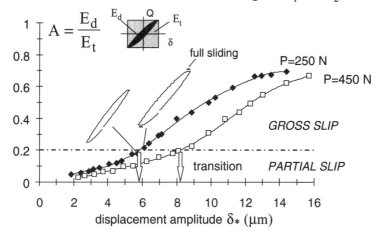

Figure 5 - *Variation of the sliding criteria "A" versus the displacement amplitude SC652/100C6 (R=12.7mm) : C_S=0.0075 μm/N; ◆ : P = 250 N □: P = 450 N.*

A similar analysis has been introduced by Chateauminois et al. [9] where the transition amplitude is normalized by the friction value and the contact radius and plotted versus the mean pressure. The advantage of the present representation is that each part of the equation is non dimensional, which permits a normalized fretting chart to be defined (Figure 6b) with a constant slope equal to one.

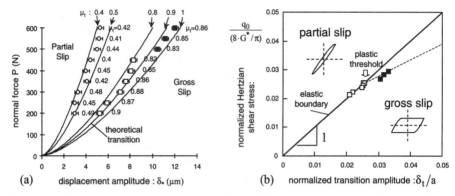

Figure 6 - *Sliding analysis, determination of the sliding transition (counterbody :12.7 mm radius 100C6 ball): SC652 : □: elastic loading, ■ plastic loading; ○ : TiN-SC652 (elastic loading); (a) Conventional Running Condition Fretting Map (RCFM); (b) Normalized Sliding Fretting Map (NSFM) [8].*

The results obtained from different contact dimensions and various pairs of materials are directly compared.

A comparison of the different tribosystems and loading situations confirms the good prediction obtained by the Mindlin formalism as long as the loading remains elastic. For softer materials and higher loading situations, a shift of the transition towards larger amplitudes can be observed. The threshold loading marking the transition toward a plastic accommodation through the contact is currently being studied, with both yield stress and the cyclic hardening behavior of the materials being considered.

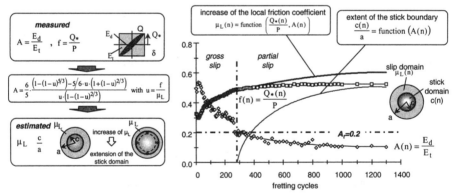

Figure 7 - *Evolution of the mixed regime deduced from the analysis of the "A" and "f" variables. (SC652/52100, R=12.7mm, C_S=0.006 μm/N; P = 300 N; δ* = 4.2 μm).*

As well as analyzing the sliding condition, it is fundamental to correctly identify the local friction coefficient operating through the sliding domain. This latter variable is essential to estimate properly the stress. While determination of the local friction coefficient under gross slip is obvious, it is more complex under partial slip. Indeed, the friction coefficient is not equal to the ratio between the tangential force amplitude and the normal force, depending on whether part of the tangential loading is related to the elastic accommodation through the sticking domain.

Nevertheless, considering elastic condition and applying the Mindlin formalism, the local friction coefficient can be estimated by combining the two measured variables which are the energy ratio A and the force ratio f (Figure 7) [10].

Quantification of the crack nucleation

The stress loading path can correctly be evaluated and the crack nucleation predicted by transposing a multiaxial fatigue analysis [11-15], provided there is a precise description of the sliding condition and a precise definition of the friction coefficient operating through the sliding domain. Note that the crack nucleation analysis developed here is related to the first observed very small cracks (usually inferior to 30 μm length). It corresponds to the incipient crack nucleation condition and at this stage it can not be extrapolated if it will propagate or not.

In opposition to classical fatigue analysis, where the maximum stress field concerns a rather large volume of matter, contact loading is characterized by a very sharp stress gradient on and below the surface. Indeed, the maximum stress field concerns volume dimensions which can be inferior to the grain size. To estimate the contact failure from a macroscopic fatigue approach, a size effect must be taken into account [11, 15].

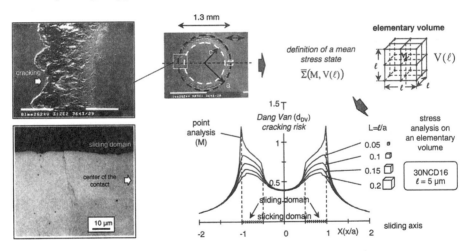

Figure 8 – *Influence of size effect on the cracking risk (30NiMo, c/a=0.5, μ=0.8).*

The local fatigue approach based on a point stress analysis (point M) must be replaced by a non local fatigue description which considers a mean loading state $\overline{\Sigma}(M,t)$ averaged on a micro-volume V(M) surrounding the point on which the fatigue analysis is performed (Figure 8). Such a micro volume is obviously highly dependent on the microstructure. A tempered low alloyed steel (30NCD16) was studied under a fretting wear situation.

When a Dang Van fatigue description was considered, good prediction of the crack nucleation through the contact was found if the contact stress loading path was averaged through a 5-6 µm cubic edge representative micro-volume [11].

This dimension appears smaller than the original austenitic grain size. Therefore, the intrinsic length scale (ℓ) required to conduct an optimized fatigue analysis is very small, grain and even subgrain dimensions need to be considered. This leads to the practical consequence that to predict fretting crack nucleation using FEM modeling, very fine meshing of the fretted surface must be defined. The grain size appears to be a relevant dimension.This multiaxial fatigue description leads us to conclude that the cracking risk under fretting wear can be described through five fundamental variables which are :

- the Hertzian pressure p_0 (normalized by the Hertzian yield pressure under pure indentation $p_{0y} = 1.6\,\sigma_{y0.2}$),
- the fretting sliding condition here expressed by the ratio between the tangential force amplitude and the amplitude at the transition ($Q_t = \mu \cdot P$),
- the friction coefficient operating through the sliding domain under partial slip condition,
- the conventional fatigue properties associated to the shear and bending fatigue limits,
- the intrinsic length scale parameter (ℓ) associated to the studied material.

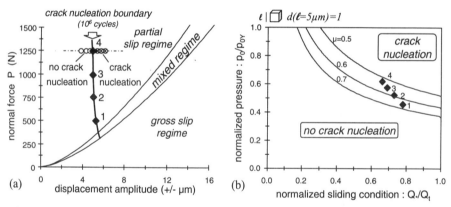

Figure 9 -*Crack nucleation response of the 30NCD16/100C6 (R=50 mm) under partial slip condition. (a) Material Response Fretting Map (MRFM); (b) Corresponding Normalized Crack nucleation Fretting Map (NCFM);* ——— : *Dang Van crack nucleation limits of the 30NCD16 alloy (d(ℓ=5µm)=1) as a function of the friction coefficient :* $p_{0Y} = 1.6 \cdot \sigma_{Y0.2}$, $Q_* / Q_t = Q_* /(\mu_L \cdot P)$ ($\mu_L(1)$=0.66, $\mu_L(2)$=0.6, $\mu_L(3)$=0.56, $\mu_L(4)$=0.53)

By defining the admitted Hertzian pressure as a function of these parameters, the provided approach allows to define a normalized crack nucleation fretting charts (Figure 9). Defining the Dang Van boundary with d=1, it became possible to distinguish contact loading situations inducing crack nucleation. This approach has been extended to different fretting fatigue and fretting on pre-stressed specimen conditions. The comparison between similar tempered low alloyed steels confirms the stability of this approach indicating that for such material the critical length scale parameter (ℓ) defining the present size effect approach is constant ℓ= 5- 6 μm. Then, considering the Dang Van diagram, a normalized representation can be introduced by dividing the critical stress couple ($\hat{\tau}, \hat{p}$) respectively by (τ_d and $\dfrac{\tau_d}{\alpha} = \dfrac{\tau_d \cdot \sigma_d/3}{\tau_d - \sigma_d/2}$).

With : σ_d (MPa)   : Alternating bending fatigue limit (10^6 cycles).

      τ_d (MPa)   : Alternating shear fatigue limit (10^6 cycles).

Thanks to this normalized representation (Figure 10), all the experiments could be compared. The classical result, that crack nucleation is a function of the shear stress amplitude and the mean tensile state, can also be inferred from it.

This size effect approach appears to be a powerful means for the designer to optimize contact geometry and surface treatments, but also to define relevant FEM meshing to compute loading paths and establish cracking risks. This approach has been extended to different materials like aluminium and titanium alloys. The analysis of 2618A aluminium concludes that a significantly longer scale parameter (ℓ>40μm) should be used, confirming the influence of the microstructure on the crack nucleation mechanisms. Current physical EBSD analyses, allowing the local identification of grain orientations, tend to confirm that this length scale parameter is related to the crack arrest phenomenon (i.e. the incipient crack are stopped at the interface of the grain boundary). The calculated length (ℓ =5 μm for the low alloy studied), defined through an average stress approach, can be associated with the maximum size that an incipient crack nucleating through the grain can reach before being stopped by the strongest barrier : the grain boundary.

It verifies the K.J. Miller assumption that "the fatigue limit can be seen to be a function of the maximum non-propagating crack length associated with a particular stress level" [16].

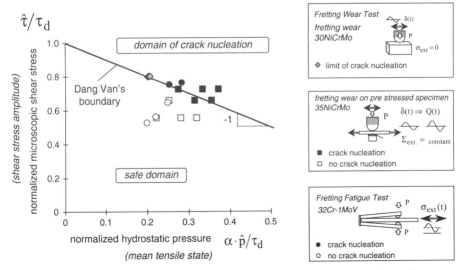

Figure 10 – *Normalized representation of the crack nucleation phenomenon in fretting. (Stress averaging considering intrinsic length parameter : ℓ =5μm)*

Quantification of the wear induced by fretting

Archard approach

The most common wear model proposed in tribology is the Archard model which relates the wear volume to the product of the sliding distance with the normal force. A wear coefficient (K) is usually extrapolated from the following relationships [17]:

$$K = \frac{V}{P \cdot S} \qquad (6)$$

with :
K : Archard wear coefficient,
V : Volume of the fretting scar defined by surface profiles,
P : normal force,
S : sliding distance.

Confirming previous studies, the wear response under gross slip condition shows that for the same material, the K factor strongly depends on the wear mode, the displacement amplitude, the contact geometry and the friction coefficient (Figure 11).
To interpret the various wear behaviors it is fundamental to consider the elasto-plastic response of a metallic material. An elastic-plastic structure such as a fretted contact can respond to cyclic loading in three identifiable ways [18, 19]. Under sufficiently small loads, such that no element of the structure reaches the yield points, the response is perfectly elastic and reversible. For higher loading conditions plastic flow can take place during the first few cycles, but plastic deformation, residual stresses and strain hardening may enable the structure to reach a perfectly cyclic elastic response commonly called "elastic shakedown". The maximum loading for which this evolution is possible is known as the elastic shakedown limit. Above this limit, plastic deformation takes place with each loading cycle.

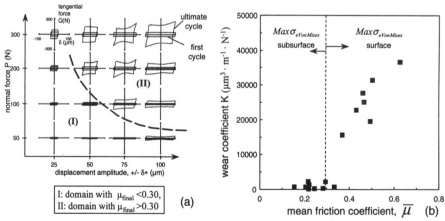

Figure 11 - *Tribological and wear behavior of a sintered steel DC1 displaying a surface porosity containing lubricant (20000 cycles): a – Evolution of the friction coefficient (first and last fretting cycles); b – Wear coefficients versus mean friction values.*

Two behaviors are then observed :
- When a stabilized and closed cycle of plastic strain is reached; this condition is referred to as "plastic shakedown",
- If repetitive accumulation of incremental unidirectional plastic strains is observed, this condition is referred to as "ratchetting".

These different plastic evolutions have been thoroughly analyzed by K.L Johnson and D.A Hills for points [20] and linear contacts [18] under repeated unidirectional sliding and rolling conditions, and by K. Dang Van and co authors for alternating sliding conditions [21]. The elastic shakedown limit is expressed as a function of the maximum Hertzian pressure (p_0) and the friction coefficient μ. From this mechanical point of view, low wear regimes will correspond to the elastic and elastic shakedown domains, whereas the high wear regime will be associated with the repeated plastic flow condition. The wear kinetics significantly increase when the shakedown limit is exceeded (Figure 12). The transition from mild to severe wear kinetics is then related to the elastic shakedown boundary. The analysis of fretting wear under low friction coefficient tends to confirm this tendency [22]. The shakedown boundary highly depends on the friction coefficient. Therefore it can be understood the large dispersion induced by the Archard formalism, which does not take into account such a variable through its formulation.

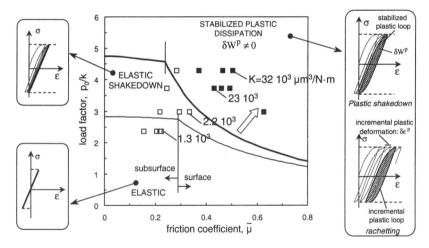

Figure 12 - *Wear map taking into account the shakedown behavior (repeated sphere/plane sliding contact):* ⎯ *elastic limit (Von Mises),* ⎯ *elastic shakedown limit (kinematical hardening, A.D. Hearle et al.), DC1 wear kinetics (Figure 11) reported as a function of the loading parameters* ■ *: high wear regime* $16000 < K(\mu m^3/N.m) < 38000$, □ *: low wear regime,* $K(\mu m^3/N \cdot m) < 3000$.

Energy wear approach

The relation between a high wear regime and a plastic shakedown or a ratchetting condition demonstrates that plastic dissipation is probably one of the most relevant parameters to quantify the wear kinetics of metals. An exact determination of the plastic work requires long and fastidious finite element computation which must be generalized for any test situation. An alternative solution consists in assuming a relationship between the plastic dissipation and the total dissipated energy introduced through the contact. The dissipated energy which is measured from the $Q*-\delta*$ fretting loop activates numerous damage mechanisms such oxidation, debris formation and ejection, as well as material transformation. If the controlling parameter of wear is assumed to be the material's plastic transformation, then a linear relationship between the accumulated dissipated energy and the wear volume could be considered [6,23]. This correlation has been previously observed on high speed and low carbon steels but also for ceramics [23]. For ceramic materials, the low energy wear factors which are identified can be related to the activation of abrasion and oxidation mechanics.

Accordingly, Figure 13 displays the wear volume versus the corresponding accumulated dissipated energy. A linear evolution is observed and an energy wear coefficient α_V (56500 $\mu m^3/J$) can be extrapolated. This expresses the wear volume created by each joule dissipated. The linear approximation is justified by a high regression coefficient ($R^2>0.9$). Comparison with the Archard's analysis (Figure 13) confirms that such a shear work approach is much more relevant to identify the intrinsic wear resistance of a microstructure. The energy approach, which is more stable, can be considered as an extension of the Archard description, which better integrates the influence of the friction coefficient.

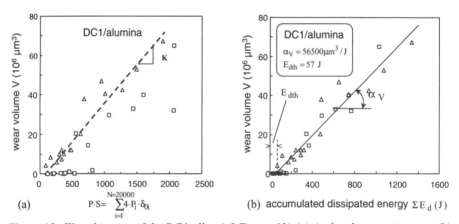

Figure 13- *Wear kinetics of the DC1 alloy (cf. Figure 12) (a) Archard approximation, (b) Energy wear quantification.*

It also permits a more physical description of the damage processes. It can be noted that the linear approximation does not cross the origin but presents a smooth shift along the energy axis (Figure 13b). This was previously observed mainly for steel or metal alloys rather than for ceramics. This energy, called the threshold energy of wear activation (E_{dth}) can be associated with the incipient plastic work first required to transform the material's surface. A layer formed of TTS (tribologically transformed structure) is generated. Hard and fragile, it is continually fractured by the contact loading leading to debris formation [24].

This transient period can also be interpreted with regard to the rachetting phenomenon induced by the contact loading. Indeed, as mentioned by Kapoor, before reaching a stabilized wear evolution, the first layer of the material has to endure a given plastic strain deformation or plastic dissipation [25]. Therefore an initial dissipated energy has to be introduced (E_{dth}) through the contact to transform the material before wear initiates (Figure 14). Therefore the wear extent can be formulated through the following relation :

$$\text{If } \Sigma E_d < E_{dth} \text{ then } V = 0$$
$$\text{If } \Sigma E_d > E_{dth} \text{ then } V = \alpha_V \cdot (\Sigma E_d - E_{dth}) \tag{7}$$

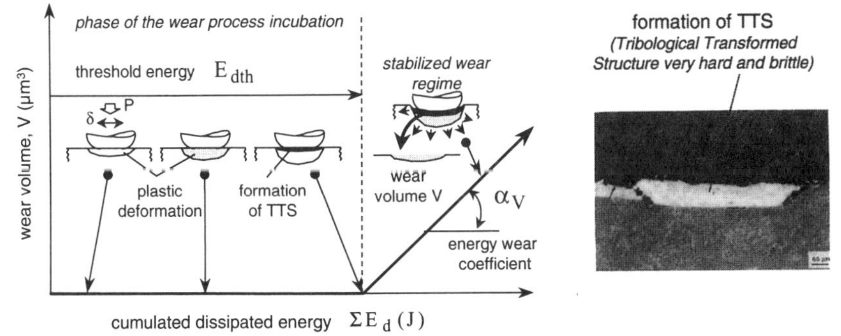

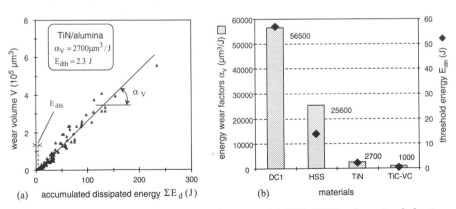

Figure 14 - *Wear process of metallic structure through the energy approach.*

Though it is more stable, and allows more in depth investigations of wear phenomenon, the application of the energy approach nevertheless has to be considered carefully. The wear evolution depends on the third body flowing through the interface. It is a function of the sliding amplitude or the sliding condition (i.e. repeated sliding - like pin on disk - or alternating sliding like fretting).

Figure 15 - *Wear energy analysis of hard coating; (a) PVD TiN/hard coating behavior ($P=50-200N$, $\delta_* = +/- 25 – 200\mu m$, $R=12.7mm$); (b) Comparison for similar loading conditions between metal and hard coatings [23].*

Moreover, the accumulated plastic deformation differs, according to whether it is repeated sliding (pin on disk) or alternating (fretting). Both the determination and the application of the energy wear coefficient are consequently restricted to the studied loading domain (i.e. pressure and sliding condition). Now for given loading range, different materials and coatings can be compared like PVD TiN (Fig. 15a) and modulated TiC-VC multilayer structures. Figure 15b compares the different wear energy coefficients, outlining a significantly higher wear resistance of hard coatings compared to metals [23].

Synthesis and conclusion

The methodology developed here to attempt to quantify fretting damage brings out the following aspects:
- A simple mechanical analysis of the contact loading as a function of the sliding amplitudes shows that cracking will be favored under stabilized partial slip conditions (partial and mixed fretting regimes) whereas wear by debris formation is predominant under gross slip regime. Nevertheless the classical Normal force – Displacement fretting map approach displays the limitation that it is restricted to a given contact geometry ; therefore to extend and formalize this approach, normalized representations have been introduced (Figure 16).

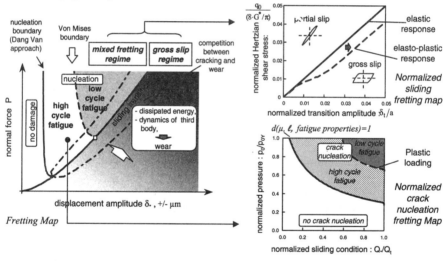

Figure 16 - *Formalization of the fretting responses: Development of the classical fretting map and quantification of the sliding condition and crack nucleation based on normalized representations.*

- To quantify the sliding condition, an energy sliding ratio was introduced. As an online indicator it allows the fretting regime evolution to be controlled. Using a simple formulation it can also be used to identify the local friction coefficient operating under partial slip conditions. A normalized sliding fretting map was introduced to consider plastic accommodation under severe loading conditions.

- Cracking was investigated using a multiaxial approach. It was shown that to correctly predict the crack nucleation risk, fatigue analysis has to consider the size effect related to the very·sharp stress gradient imposed by the contact loading [26]. A critical length scale parameter was considered that seems to be related to the microstructure. Recent investigations tend to associate such a "crack nucleation length variable" to a crack arrest condition. By contrast, the propagation analyses [27,28] assume that the crack extension is more affected by the contact size and the stress field distribution below the contact.
-Wear quantification is less advanced and formalized. It was nevertheless shown that to identify correctly the wear kinetics under fretting it is essential to consider the elastoplastic response of the material. The shakedown limit is highly dependent on the friction coefficient, and the Archard law, which doesn't integrate this variable, cannot properly quantify the wear kinetics under large friction fluctuations.
An energy approach is introduced allowing better wear quantification and permitting an easier comparison between surface treatments.

This methodology can provide a global description of the influence of surface treatments, and defines coating performance charts which take into account damage evolution as a function of the sliding condition. Future developments will consist in quantifying the effect of plasticity [29], the competitive damage evolutions of cracking and wear, and extending the approach to more complex contact geometry like flat punch with rounded corners [30].

Acknowledgments

This work was supported in part under contracts BRE2-CT92-0224 and BE96-3188 BriteEuram projects. The authors are grateful to Prof. Ky Dang Van, Prof. F. Sidoroff and Prof. Wronski for his helpful comments and suggestions.

References

[1] Waterhouse, R.B., *Fretting Fatigue*, Applied Science publishers, 1981.
[2] Lindley, T.C., Nix, K.J., "The Role of Fretting in the Initiation and Early Growth of Fatigue Cracks in Turbo-Generator Materials", *ASTM STP 853,* 1982, pp. 340.
[3] Hoeppner, D., "Mechanisms of fretting fatigue and their impact on test methods development", *ASTM STP 1159,* 1992, pp. 23-32.
[4] Nowell,D. and Hills, D.A., "Crack Initiation in Fretting Fatigue" , *Wear, 136,* 1990 pp. 329-343.
[5] Blanchard, P., Colombier, C., Pellerin, V., Fayeulle, S., Vincent, L., "Material effect in fretting wear", Metallurgica Transaction, 22A, July 1991, p.1535-1544.
[6] Fouvry, S., Kapsa, Ph., Vincent, L., "Quantification of fretting damage", Wear 200, (1996), p. 186-205.
[7] Johnson, K.L., Contact Mechanics, Cambridge Univ.Press, Cambridge, 1985.
[8] Fouvry, S., "Developments on Fretting Mapping", EUROMAT 2000, ISBN 0-08-042815-0, p.597-602.

[9]   Chateauminois, A., Kharrat, M., Krichen, A., "Analysis of fretting damage in Polymers by Means of Fretting Maps", ASTM STP 1367, 2000, p. 352-366.

[10] Fouvry, S., Kapsa, Ph., Vincent, L., "Developments of fretting sliding criteria to quantify the local friction coefficient evolution under partial slip condition", Eds. D. Dowson et al , Tribology serie, 34 , 1997, p. 161-172.

[11] Fouvry, S., Kapsa, Ph., Sidoroff, F.,Vincent, L., "Identification of the characteristic length scale for fatigue cracking in fretting contacts", J. Phys. IV France 8 (1998), Pr8-159.

[12] Farris, T.N., Szolwinski, M.P., Harish, G., "Fretting in Aerospace Structures and Materials", ASTM STP 1367, 2000, p.523-537.

[13] Neu, R.W., Pape, J.A., Swalla, D.R., "Methodologies for linking Nucleation and Propagation Approaches for Predicting Life Under Fretting", ASTM STP 1367, 2000, p.369-388.

[14] Lykins, C.D., Mall, S., Jain, V., "An evaluation of parameters for predicting fretting fatigue crack initiation" International Journal of Fatigue (2000), p.703-716.

[15] Fouvry, S., Kapsa, Ph., Vincent, L., " Fretting-Wear and Fretting-Fatigue : Relation through Mapping Concept ", ASTM STP 1367, 2000, p. 49-64.

[16] Miller, K.J., "Material science perspective of metal fatigue resistance", Materials Science and Technology, June 1993, Vol. 9, pp. 453.

[17] Archard, J.F., "Contact and rubbing of flat surfaces", J. Appl. Phys., 24 (1953), p. 981-988.

[18] Johnson, K.L., Wear 190, "Contact mechanics and the wear of metals", (1995), 162-170.

[19] Wong, S.K., Kapoor, A., Williams, J.A., "Shakedown limits on coated and engineered surfaces", Wear 203-204, (1997), 162-170.

[20] Hills, D.A., Ashelby, D.W., "The influence of residual stresses on the contact – load-bearing capacity", Wear, 75 (1982), p. 221-240.

[21] Maouche, N., Maitournam, M.H., Dang Van, K., "On a new method of evaluation of the inelastic state due to moving contacts", Wear 203-204, (1997), 139-147.

[22] Fouvry, S., Kapsa, Ph., Vincent, L., "An elastic plastic shakedown analysis of fretting wear", Wear 247, 2001, p. 41-54.

[23] Fouvry, S., Kapsa, Ph., " An Energy Description of Hard Coatings Wear Mechanisms", Surface & Coating Technology, 138 (2001) p.141-148.

[24] Sauger, E., Fouvry, S., Ponsonnet, L., Martin, J.M , Kapsa, Ph, Vincent, L., "Tribologically Transformed Structure In Fretting" Wear 245, 2000, p.39-52.

[25] Kapoor, A., "Wear by plastic ratchetting", Wear 212 (1997), p. 119-130.

[26] Nowell, D, Hills, D.A, and Moobola, R., "Lengh Scale Considerations in Fretting Fatigue" ASTM STP 1367, 2000, p. 141-166.

[27] Dubourg, M.C., Lamacq, V.," Stage II Crack Propagation Direction Determination Under Fretting Fatigue Loading ", ASTM STP 1367, 2000, p. 436-450.

[28] Kondoh, K., Mutoh, Y., "Crack Behavior in the early stage of fretting fatigue fracture", STP 1367, 2000, p. 282-292.

[29] Ambrico, J.M., Begley, M.R.," The role of macroscopic plastic deformation in fretting fatigue life predictions", Int. Journal of Fatigue, 23 (2001), p. 121-128.

[30] Ciavarella, M., Demelio, G.,"A review of analytical aspects of fretting fatigue, with extension to damage parameters, and application to dovetail joints", Int. Journal of Solid and Structures 38 (2001) 1791-1811.

Tomohisa Nishida,[1] Kazunori Kondoh,[2] Jin-Quan Xu,[3] and Yoshiharu Mutoh[3]

Observations and Analysis of Relative Slip in Fretting Fatigue

Reference: Nishida, T., Kondoh, K., Xu, J. Q., and Mutoh, Y., **"Observations and Analysis of Relative Slip in Fretting Fatigue,"** *Fretting Fatigue: Advances of Basic Understanding and Applications, ASTM STP 1425*, Y. Mutoh, S. E. Kinyon and D. W. Hoeppner, Eds., American Society for Testing and Materials International, West Conshohocken, PA, 2003.

Abstract: In-situ observation of the relative slip along the contact surface was conducted on a servo-hydraulic fatigue machine with scanning electron microscope. It was found that the relative slip observed by this method was much smaller than that obtained by a conventional small extensometer. It was also found that the relative slip depended on the rigidity of the contact pad. The higher the rigidity of the pad was, the larger the relative slip became. Numerical analysis with the use of the finite element method (FEM) program ABAQUS was also carried out. The relative slip estimated by the FEM agreed well with the observed one. Therefore, it was proved that the in-situ observation method introduced in this study was effective enough for measuring relative slip at the contact edge.

Keywords: fretting fatigue, relative slip, in-situ observation, FEM analysis

Introduction

Fretting fatigue is well known to occur at contact of combined components, such as joints and bearings. The relative slip along the contact interface between the two components is one of the most important factors to significantly influence the fretting fatigue behavior [1-4]. Fretting fatigue strength is generally degraded with increasing relative slip amplitude. In experiments, the relative slip amplitude has been measured by using specially developed small extensometers [4-7] and a laser extensometer [8]. However, most of the measurements gave only the apparent and macroscopic values. It is also known that partial slip is typical in fretting, which implies that the relative slip amplitude is not uniform along the contact interface. The local relative slip amplitude

---

[1]Numazu College of Technology, Ooka, Numazu-shi 410-8501 Japan.

[2]Topy Industries, Ltd., Ohgami, Ayase-shi 252-1104 Japan.

[3]Nagaoka University of Technology, Kamitomioka, Nagaoka-shi 940-2188 Japan.

especially at contact edge should be important to determine the stick and the micro-sliding regions in the contact area during the fretting fatigue loading.

In the present study, the relative slip amplitude along the contact interface was measured by the in-situ observation using a servo-hydraulic fatigue machine with a scanning electron microscope (SEM). FEM analysis was also carried out to evaluate the relative slip amplitude along the contact interface. From combining the results by experiments and FEM analysis, effect of rigidity of contact pad on relative slip amplitude was discussed in detail. The crack initiation behavior was also discussed based on the stress range distribution along the contact interface.

Experimental procedures

The material used for the fretting fatigue specimens was an aluminum alloy JIS 6063-T6, while both the aluminum alloy JIS 6063-T6 and a Cr-Mo steel JIS SCM420 were used for the contact pads. Chemical composition and mechanical properties of the materials used are shown in Table 1 and 2, respectively. The shape and dimensions of the specimen and the contact pad are shown in Fig. 1. The four different contact pads with pad foot heights L of 0.15, 0.5,1.0 and 2.0 mm with the aligned contact surface were prepared. Fretting was induced by pressing a pair of contact pads onto the specimen gage part using a proving ring. The fretting fatigue tests were performed under a load-controlled tension– compression condition with a stress ratio of R=-1, sinusoidal wave-form and frequencies of 10 to 20 Hz. The tests were carried out at a stress amplitude of 120 MPa with a constant contact pressure of 60 MPa. In-situ observation and measurement of relative slip were conducted using a servo-hydraulic fatigue machine with a scanning electron microscope. The proving ring was tilted against the loading axis, as shown in Fig. 2, so that whole contact region between the specimen and the contact pad can be observed. When measuring the relative slip, loading frequency was reduced to 0.05Hz. The relative slip amplitude was defined as half of the measured relative slip range. After the measurement, the test was stopped, and the fretting scars were observed. It is found that the fretting damages corresponding to four foots are almost the same.

Finite element model

Finite element (FEM) analysis was carried out to estimate the stress distribution near the contact region and the relative slip between the specimen and the contact pad. FEM analysis program ABAQUS [9] was used. As shown in Fig. 3, FEM model with quadrilateral elements was a quarter of the specimen with the contact pad. Boundary conditions were applied along two symmetrical axes, and loading conditions were given as the same as the fretting fatigue tests conducted in the present study. Analysis was carried out assuming a linear elastic body under the

Table 1- *Chemical composition of the material used (wt%).*

	Si	Fe	Cu	Mn	Mg	Cr	Zn	Ti	
6063-T5	0.44	0.20	0.03	0.02	0.53	0.01	0.01	0.02	

	C	Si	Mn	P	S	Cu	Ni	Cr	Mo
SCM420	0.21	0.23	0.70	0.016	0.014	0.13	0.07	1.01	0.15

Table 2- *Mechanical properties of the material used.*

	Tensile strength σ_B(MPa)	Elongation ϕ (%)	Young's modulus E(GPa)	Hardness Hv(MPa)
6063-T5	230	21	65	770
SCM420	940	16	206	2850

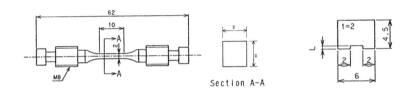

(a)Fretting fatigue specimen   *(b)Contact pad*

Figure 1- *Geometries and dimensions of the fretting fatigue specimen and the contact pad.*

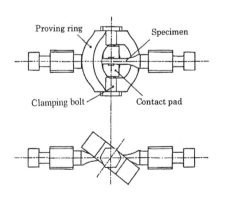

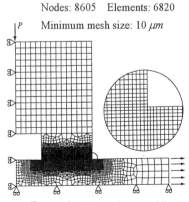

Nodes: 8605   Elements: 6820
Minimum mesh size: 10 μm

Figure 2- *Schematics of fretting fatigue test.*

Figure 3- *FEM analysis model.*

plane strain condition. Based on the FEM analysis results, it is found that only a very small region near the contact edge would be yielded, i.e., it could be considered to be under the small scale yielding condition. Therefore, the linear elastic analysis result is useful for characterizing the fretting fatigue behavior under the present condition. The contact elements located at the sliding interface must fulfil the Coulomb s friction law. Sliding occurs in the contact element when shear stress(τ) on the interface becomes equal to the critical shear stress ($\tau = \mu \, \sigma_n$), where σ_n is the normal stress and μ is the coefficient of friction. Based on the experimental result for the gross sliding case, the coefficient of friction of 0.7 was used in the present analysis.

Results and Discussion

Measurement of relative slip amplitude

Figure 4 shows an example of measurement of relative slip at the contact edge for the aluminum specimen with SCM420 steel contact pad, where the height of pad foot is 0.15 mm. Figure 4(a) and (b) indicate micrographs at the maximum stress and at the minimum stress, respectively. The measurement procedure was as follows: (1) looking for some clear feature marks on both the specimen surface and the contact pad surface, (2) making a base line for measuring relative slip, as shown in Fig. 4(a), and (3) measuring the relative slip as the distance between two base lines on the specimen surface and the contact pad surface, as shown in Fig. 4(b). The relative slip amplitude was given as half of the measured relative slip range.

Three typical points along the contact interface were selected for measuring the relative slip amplitude: at the internal edge, center and external edge. Relationship between relative slip amplitude and number of cycles for the SCM 420 steel contact pad with pad foot height of 0.5 mm is shown in Fig. 5. It has been commonly observed [10, 11] that the relative slip behaves unstable at the initial stage of fretting fatigue. Figure 5 shows that the relative slip amplitude becomes stable after 1×10^2 cycles. The average value after 1×10^2 cycles was therefore defined as the relative slip amplitude for the cases.

The relative slip amplitude measured for the aluminum alloy specimen with the SCM 420 steel contact pad (foot height of 0.5 mm) is shown in Fig. 6. In the horizontal axis, the origin is at the center of contact region and the external edge is in the positive side. In the figure, the relative slip amplitudes estimated by FEM analysis, where the relative slip is determined as the difference of the relative displacements in the horizontal direction of contact points between the pad and the specimen, are shown together with those estimated by assuming the elastic specimen and the rigid contact pad, which is given as [3]

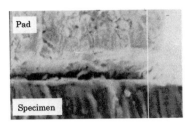

a) At the maximum stress

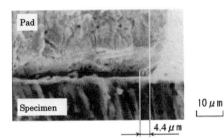

b) At the minimum stress

Figure 4- *Measurement of relative slip at the contact edge for the aluminum alloy specimen with the steel contact pad (L=0.15mm, σ_a=120MPa).*

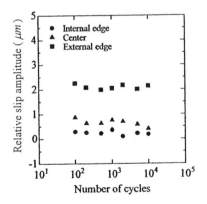

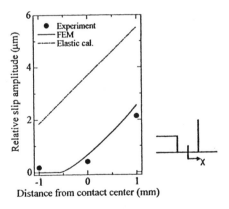

Figure 5- *Relationship between relative slip amplitude and number of cycles for the aluminum alloy specimen with the steel contact pad.*

Figure 6- *Relative slip amplitude distribution for the steel alloy contact pad.*

$$s = \frac{\sigma_a l}{2E} \qquad (1)$$

where s is the relative slip amplitude, σ_a is the amplitude of the cyclic load, l is the span length of contact pad and E is Young s modulus. As can be seen from the figure, the relative slip amplitude at the external edge is the largest, while that at internal edge is the smallest and almost zero. It is obvious that the relative slip amplitude estimated by Eq. (1) with assumption of rigid contact pad is significantly high compared to the measured one. It is also evident that the relative slip amplitude estimated by the FEM analysis agrees with that measured by in-situ SEM observation. The relative slip amplitudes measured by in-situ observation and by FEM analysis for the aluminum alloy specimen with the aluminum alloy contact pad (foot height of 0.5 mm) are shown in Fig.7. The similar behavior to the case of the aluminum alloy specimen with the steel contact pad was observed.

To discuss the difference of relative slip amplitude between the measurements by in-situ SEM observation and the calculation by assuming the rigid contact pad, deformation of the contact pad was estimated by the FEM analysis. The results are shown in Fig. 8. As can be seen from the figure, both the aluminum alloy and the steel contact pads are elastically deformed. It can be easily understood that the difference of the displacements on the top corner and the contact edge of the pad, which is in the order of relative slip amplitude, is not considered in Eq. (1). Because that the relative slip amplitude estimated by assuming the rigid contact pad includes this difference of the displacements, Eq. (1) almost gave the slip amplitude two times of that measured by in-situ SEM observation, which is in reasonable agreement with the result of FEM analysis. The relative slip amplitude has been often measured by using a small extensometer [3,4]. It is obvious that the measured value depends on the position of the measure point at the end surface of the contact pad, the lower the measure point is, the smaller the measured slip value becomes, because the difference of the displacements between the measure point and the contact edge becomes smaller. This also means that the relative slip amplitude measured using a small extensometer will be in between those estimated by assuming the rigid contact pad and measured by the in-situ SEM observation. Therefore, improvement or development of a new method for measuring relative slip amplitude at contact edge in fretting fatigue test is requested.

Effect of contact pad rigidity on relative slip amplitude

The slip amplitude distributions for the aluminum alloy specimen with the steel contact pad with various foot heights are shown in Fig. 9. As can be seen from the figure, the relative slip amplitude at internal edge increases with increasing pad foot height. Relationship between the maximum relative slip amplitude at external edge and pad foot height is shown in Fig. 10. It is found that the maximum relative slip amplitude decreases with increasing pad foot height for both the steel and the aluminum alloy contact pads. It is also obvious that the maximum relative slip amplitude for the aluminum alloy contact pad is smaller than that for the steel contact pad. Therefore, it is concluded that the more rigid contact pad results in the larger relative slip amplitude. The estimated relative slip amplitudes by FEM are also shown in Figs.9 and 10, which indicate that the estimation agrees the experimental results.

The deformation of end surface of the steel contact pad with various pad foot heights is shown in Fig. 11. The contact pad with large foot height is significantly deformed compared to that with small foot height. This behavior is in accord with that the contact pad with larger foot height indicates smaller relative slip amplitude. Therefore, it should be noted that the rigidity of contact pad is an important factor in fretting fatigue.

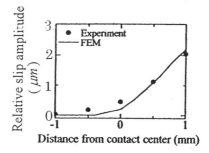

Figure 7- *Relative slip amplitude distribution for aluminum alloy contact pad.*

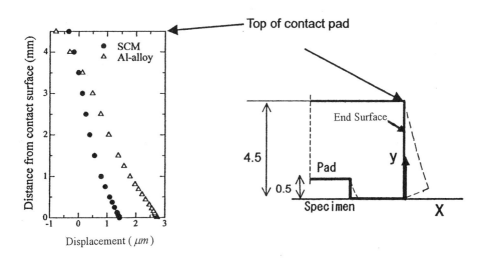

Figure 8-*Deformation along the end surface of contact pad.*

Fretting fatigue crack initiation

From the in-situ SEM observations, fretting fatigue cracks were found to initiate near the external edge of contact region in the very early stage (less than three percent) of fatigue life [12,13]. A fretting fatigue crack observed during a fretting fatigue test of the aluminum alloy specimen with the steel contact pad is shown in Fig. 12. Cracks (with different lengths) were observed for all four external edges. From careful observations, a crack initiated at a very shallow angle to the contact surface and then turned to a higher angle: the former is supposed as the crack initiation stage and the latter is known as the mixed-mode crack propagation stage under the influence of fretting [14].

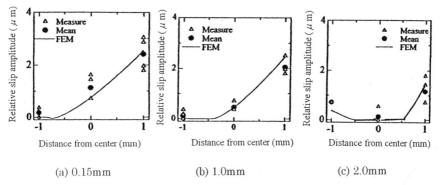

(a) 0.15mm   (b) 1.0mm   (c) 2.0mm

Figure 9- *Effect of pad foot height on relative slip amplitude for the steel contact pad.*

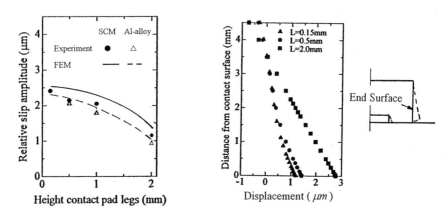

Figure 10- *Relationship between relative slip amplitude and pad foot height.*

Figure 11- *Effect of pad foot height on deformation along the end surface of contact pad for the steel contact pad.*

The FEM analysis mentioned above also can give the stress distribution near the contact interface. Fatigue crack initiation and propagation behavior is known to be controlled by not a maximum stress but a stress range. Figure 13 shows the stress range distributions along the contact interface of the aluminum alloy specimen with the steel contact pad. As can be seen from the figure, the maximum stress range is found at the external edge of contact region for any kind of stresses, where the relative slip amplitude is also the largest.

From the foregoing results, it is concluded that a fretting fatigue crack initiates at the contact edge, where the relative slip amplitude and the stress range are the largest. About the fretting fatigue crack initiation and propagation, we will report in another paper [15].

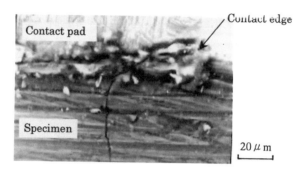

Figure 12· *Observation of a fretting fatigue crack in the specimen with the steel contact pad.*

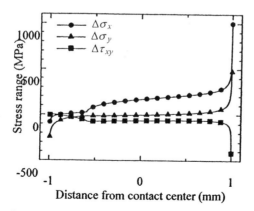

Figure 13· *Stress range distributions along the contact interface of the aluminum alloy specimen with the steel contact pad.*

Conclusions

In-situ measurements and FEM analysis with contact-frictional elements of relative slip amplitude between the specimen and the contact pad were carried out. Effect of rigidity of contact pad was discussed in detail. The main results obtained are summarized as follows:

(1) Relative slip amplitude measured by using an SEM in-situ observation technique was significantly smaller than that measured by using a conventional extensometer.
(2) Relative slip amplitude measured by using an SEM in-situ observation technique agreed well with that estimated by the FEM analysis.
(3) Relative slip at the contact edge increased with increasing rigidity of contact pad.
(4) From the FEM analysis, the relative slip and stress ranges were the largest at the external contact edge, at which fretting fatigue cracks are supposed to initiate. This is in good agreement with the observations of crack initiation in fretting fatigue test.

References

[1] Spink, G. M., Fretting fatigue of a 2.5% NiCrMoV low pressure turbine shaft steel-the effect of different contact pad materials of variable slip amplitude, *Wear,* 136, pp.281-297, 1990.
[2] Duquette, D. J., The role of cyclic wear (fretting) in fatigue crack nucleation in steels, *Strength Metals Alloys*, Vol. 1, pp.213-218, 1979.
[3] Mutoh, Y., Nishida, T., and Sakamoto, I., Effect of relative slip amplitude and contact pressure on fretting fatigue strength, *Zairyo,* 37(417), pp.649-655, 1988.
[4] Fenner, J., and Field, J. E., La fatigue dans les conditions de frottement, *J. E., rev. Met*, 55, p.475, 1958.
[5] Satoh, T., Mutoh, Y., Yada, T., Takano, A., and Tsunoda, E., Effect of contact pressure on high temperature fretting fatigue, *Zairyo*, 42(472), pp.78-84, 1993.
[6] Ochi, Y., Hayashi, H., Tateno, B., Ishii, A., and Urashima, C., Fretting fatigue properties in rail steel and fish plate materials, *JSME*, A63(607), pp.453-458, 1997
[7] Lee, S., Nakazawa, K., Sumita, M., and Maruyama, N., Effect of contact load and contact curvature radius of cylinder pad on fretting fatigue in high strength steel, *ASTM STP1367*, pp.199-212, 1999.
[8] Kondoh, Y., and Bodai, M., Study on fretting fatigue crack initiation mechanism based on local stress at contact edge, *JSME,* A 63(608), pp.669-676, 1997.
[9] ABAQUS/Standard User s Manual, Vol.2, 1999.
[10]Nakazawa, k., Sumita, M., and Maruyama, N., Effect of contact pressure on fretting fatigue of high strength steel and Titanium alloy, *ASTM STP 1159*, pp.115-125, 1992.

[11]Lutynski, C., Simansky, G., and McEvily, A. J., Fretting fatigue of Ti-6Al-4V alloy, *ASTM STP 780*, pp.150–164, 1982.

[12]Zhou, Z.R., Sauger, E., and Vincent, L., Nucleation and early growth of tribologically transformed structure induced by fretting, *Wear*, 212, pp.50-67, 1997.

[13]Hattori, T., Fretting fatigue analysis using fracture mechanics, *JSME Int.*, 31(1988), pp.100–107.

[14]Nix, K.J., and Lindley, T.C., The initiation of fracture and propagation of small defects in fretting fatigue, *Fatigue fracture engineering of material and structure* 8(1985), pp.143–160.

[15]Mutoh, Y., Xu, J.Q., and Kondoh, K., Observations and analysis of fretting fatigue crack initiation and propagation, submitted to *ASTM STP1425*.

Paul N. Clark[1] and David W. Hoeppner[1]

Fretting Fatigue Initial Damage State to Cracking State: Observations and
Analysis

Reference: Clark, P. N. and Hoeppner, D. W., "Fretting Fatigue Initial Damage State
to Cracking State: Observations and Analysis," *Fretting Fatigue: Advances in Basic
Understanding and Applications, STP 1425*, Y. Mutoh, S. E. Kinyon, and, D. W.
Hoeppner, Eds., ASTM International, West Conshohocken, PA, 2003.

Abstract: Interrupted fretting fatigue experiments were performed to
demonstrate the capabilities of a confocal microscope related to characterizing
fretting damage and to correlate that damage with cycles to failure. Fretting
damage was established at stepped down levels of 100% (baseline), 80%, 60%,
40%, 20%, and 10%. The damage levels were calculated as a percentage of the
total cycles to fracture for the baseline fretting fatigue specimen. The baseline
or 100% specimen was subjected to axial fatigue forces and normal forces to
induce a fretting situation for 100% of its life or total cycles to fracture. The
baseline specimen was cycled to fracture without interruption and experienced
fretting due to normal forces throughout the experiment. Other specimens were
subjected to cyclic forces for their respective percentage cycles under fretting
fatigue conditions (i.e. with the applied normal force). After the prescribed
number of cycles was attained each specimen was removed and inspected for
fretting damage and characterized utilizing a confocal microscope and a
scanning electron microscope. Each specimen was then cycled to fracture or
run-out without the applied normal force. A damage threshold was
demonstrated for fretting damage at 60% or less for these experimental
conditions. Fretting was characterized as depth of damage (wear and/or pitting)
and surface cracking. A key factor was the location of the fretting damage, not
necessarily the magnitude of that damage. Fretting damage located near the
edge of the fretting pad contact area was more detrimental than damage near
the center of the fretted area.

Keywords: Fretting fatigue, damage, metrics, confocal microscopy, damage threshold,
fretting damage to crack transition

---

[1] The authors are, respectively, Research Assistant Professor and Professor, Department of Mechanical
Engineering, University of Utah, 50 S. Central Campus Drive, Room 2202, Salt Lake City, UT 84112-
9208.

Introduction:

Fretting is an insidious type of damage that is the result of the interactions between wear and corrosion [1]. This phenomenon is most commonly witnessed between contacting surfaces that experience small amplitude oscillatory loads. These loads can be induced via vibrations or nearly any variety of mechanical force(s). The strain associated with cyclic fatigue forces creates ideal circumstances in which fretting can occur. Fretting fatigue is a particularly vexing issue as the damage caused by fretting is often responsible for the early nucleation of fatigue cracks. This damage is often created through the process of adhesive contact of the asperities on opposing contact surfaces. The contact surfaces then experience the production of debris or the removal/disturbance of oxide layers. This accelerates and compounds the effects of wear and corrosion. Fretting damage varies with particular circumstance. The rate and measure of damage can vary with material, strain or slip amplitude, environment, state of stress, surface finish of contact materials, manufacturing processes and maintenance practices [1-8]. Eden, Rose and Cunningham, who noted a production of oxide between grips and specimens with an accompanied reduction in the fatigue life of their grips, first documented fretting in 1911 [2]. In 1939, Tomlinson published a paper documenting some the first scheduled and controlled experiments on fretting. He noted the small amplitudes of motion could cause surface damage and that corrosion was a compounding effect of fretting [3]. Warlow-Davies was one of the first to suggest that fretting has a deleterious effect on fatigue life. He performed some of the earliest work (circa 1941) examining the effects of fretting on fatigue [4]. Since, then many have contributed to the technical communities knowledge base. The contributors include, among others, Fenner, Wright and Mann, Waterhouse, Nishioka and Hirakawa, Hoeppner and Goss [5-8].

The technical community still lacks the ability to predict the fretting fatigue life of components. A large reason for this is the lack of understanding and characterization of fretting damage. This lack of understanding has made the development of mathematical and mechanical models challenging. A key to understanding fretting fatigue and its implications with regard to reliability and structural integrity is to adequately characterize the fretting damage. This understanding will lead to a more robust understanding of the mechanical phenomenon of fretting, fretting fatigue and damage related to fretting. Fretting can produce, but is not limited to, the following types of damage [9]:

- Pitting corrosion
- Oxides and debris (third body)
- Disturbance of protective oxide layers
- Scratches
- Fretting and/or wear tracks
- Material transfer
- Surface plasticity
- Subsurface cracking and/or voids

- Fretting craters
- Cracks at various angles to the surface(s)

Frequently, this fretting damage will lead to early formation of fatigue cracking and will result in an often times dramatic reduction of fatigue life.

A confocal microscope and scanning electron microscope (SEM) can be utilized to carefully characterize the scope of fretting damage. A Zeiss confocal microscope and Hitachi SEM were used for this study. A SEM provides detailed examination of fretting damage at magnifications as high as 100,000 and a confocal microscope provides detailed images and can characterize damage by measuring and recording depth, cracking and topographical information. This information can lead to significant indications related to the cause of specimen fracture.

Through careful characterization of induced fretting damage and prudent control of experiments, understanding of fretting fatigue damage as a threshold concept or as a threat to structural integrity can begin to be understood. The concept of a fretting fatigue damage threshold has been addressed previously [10-11].

This paper demonstrates fretting damage characterization using a confocal microscope and a scanning electron microscope, followed by tracking the fatigue life of the fretting damaged specimens once the induced fretting damage had been characterized.

Experimental Details

Fretting fatigue tests were performed using a closed loop electro-hydraulic servo-controlled testing system. MTS controllers were used to control specimen mechanical forces. Figure 1 depicts the fretting fatigue experimental apparatus. The actuator visible on the top of the apparatus applied the normal force. Axial forces were applied parallel with the tabletop.

Strain experienced by the fatigue specimen produced a small amplitude oscillatory motion between the fatigue specimen and the fretting pad. This resulted in fretting damage on both the pad and fatigue specimen.

The dog-bone and fretting pad specimens were made from 2024-T3 aluminum 0.063" sheet and 0.250" plate respectively. The dog-bone specimens were manufactured lengthwise in the longitudinal direction. The fretting pad specimens were designed with a flat contact surface. Dog-bone specimens were lap polished to a 0.3μm finish and fretting pads were polished to a 600-grit finish. Specimens were cleaned in acetone for 15 minutes after polishing.

Axial forces were applied in the longitudinal direction while normal load was applied via the fretting pad perpendicular to the dog-bone fatigue specimen. The normal force was applied at a net stress of 3.0ksi. Constant amplitude cyclic forces were controlled at a maximum net stress of 30.0ksi, R=0.1 and f=10Hz. The net stress was calculated based upon the applicable loading divided by the minimum cross sectional area of the dog-bone specimen or maximum contact area between the fretting pad and the faying surface of the dog-bone. Experiments were conducted in laboratory air with no lubrication in the contact area.

Figure 1: Fretting Fatigue Apparatus.

Specimens had fretting damage induced at 100%, 80%, 60%, 40%, 20% and 10% of the maximum fretting fatigue cycles to fracture. 100% was defined as the total cycles to fracture for specimen A. The remaining specimens were subsequently subjected to the applicable number of fretting fatigue cycles that corresponded to the appropriate percentage of approximately 187,020 cycles (e.g: 80% of 187,020 ~ 150,000). The specimens were then microscopically analyzed. This analysis sought damage on the dog-bone specimens in the form of cracking, depth of scarring or pitting, and other significant indicators. The fretting fatigue damaged dog-bone specimens were then loaded to either fracture or run-out under cyclic mechanical forces without fretting (no normal force). Failed specimens were subjected to fractography to identify fatigue cracking and crack origins. This process aided in understanding and characterizing the amount of fretting damage that is sufficient to demonstrate a marked reduction of fatigue life.

Experimental Results

The results for this set of experiments are presented in Table 1. Fretting fatigue cycles and fatigue cycles (post-fretting) were tracked and reported. A graph of the cycles versus percent fretting fatigue life is presented in Figure 2. It is important to note that 4 of the 6 specimens did not fracture. Specimens C, D, E and F were cycled from 1,000,000 to 3,000,000 cycles and did not fracture. This indicates a fretting fatigue damage threshold for these conditions.

Figures 3 and 4 show fretting damage on specimen A (100%FF). Note the detail in the depth record from the confocal microscope. Figure 5 shows the fracture surface and crack origin for specimen A. The left image depicts specimen A at an angle such that the fretting damage on the faying surface and the crack propagation evidence on the fracture surface can be seen concurrently. The right image is a close-up micrograph of the crack origin. The crack origin corresponds with the most severe fretting damage. The damage progression on the fracture surface of specimen A can be seen in Figure 5. The fatigue crack propagated through the specimen. Evidence on the fracture surface indicates the crack originated at the heaviest portion of the fretting damage and propagated to the edge of the specimen. Once the crack grew to the edge of the specimen, it appeared to propagate through the remained of the specimen as a corner crack until a critical crack length was reached and final fracture was eminent.

Figures 6-8 show fretting damage and cracks originating from that damage on specimen B-80%FF. This specimen was analyzed for fretting fatigue damage after 150,000 cycles. There were three fatigue cracks growing from the fretting damage shown in Figure 6. The largest of the three cracks is visible in Figure 6. The largest crack can be seen near the right end of the fretting damage propagating perpendicularly to the applied loading. This crack, measured at approximately 650μm, was the most substantial damage noted during damage analysis and ultimately led to the fracture of specimen B. The other two cracks were measured at 142μm and 34μm from the edge of the fretting damage. The maximum fretting damage on the faying surface was approximately 10μm deep. However, the damage area from which the fatigue cracks originated experienced a maximum damage depth of approximately 4.5μm. Note that the fatigue cracking area was near the edge of the fretting pad contact area, whereas the fretted area that experienced the heavier damage (~10μm) was offset from the contact edge. This is significant due to the complex stresses induced by the fretting contact and the normal force. Cracks will typically nucleate near the contact edge, although this does not occur exclusively.

Figures 9-12 Demonstrate the maximum fretting damage on specimens C-60%FF, D-40%FF, E-20%FF and F-10%FF. Note that the most extensive fretting damage on any of these (C, D, E or F) specimens was on specimen D. Fretting damage as deep as 20μm is shown for specimen D in Figure 10. It is suspected that some form of debris became trapped between the fretting pad and specimen surface creating a plowing situation and creating a more extensive yet localized area of damage on specimen D.

Table 1: Test Results

Specimen	Percent of Induced Fretting Fatigue	Fretting Fatigue Cycles	Fatigue Cycles	Cycles to Fracture	Comments from Post Fretting Analysis
A	100%	187,020	n.a.	187,020	Heavy fretting damage. Deepest damage ~35µm.
B	80%	150,000	74,249	224,249	Moderate fretting damage. Deepest damage ~10µm.
C	60%	112,500	1,000,000	Did not fracture	Moderate fretting. Deepest damage ~8µm.
D	40%	75,000	1,200,000	Did not fracture	Some deep fretting damage near center of pad contact area. Deepest damage ~20µm.
E	20%	37,500	3,000,000	Did not fracture	Light fretting damage. Most damage concentrated at contact edge. Deepest damage ~4µm.
F	10%	18,750	3,000,000	Did not fracture	Moderate fretting damage. Most severe damage on edge of specimen. Deepest damage ~9µm.

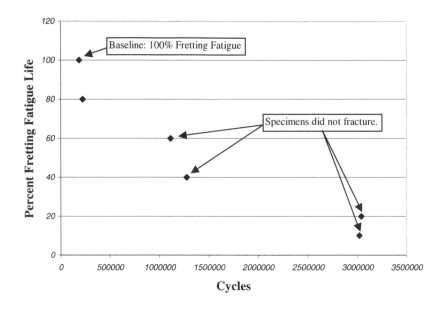

Figure 2: Percent Fretting Fatigue Life vs. Cycle Count.

Figure 3: Fretting Damage to Failed Specimen A-100%FF (100X).
Note: Fracture visible on right side of image.

Figure 4: Fretting Damage to Failed Specimen A-100%FF (200X).
Note: Confocal depth image is a mirror reflection of surface image. (d_{max}~35μm).

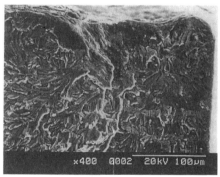

Figure 5: Specimen A-100%FF, Fracture Surface and Crack Origin.

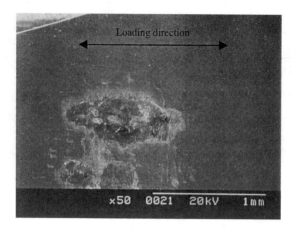

Figure 6: Characterization of Fretting Damage on Specimen B-80%FF (50X).
Note: Crack growing from fretting damage. Crack ~650μm long.

Figure 7: Most Severe Fretting Damage on Specimen B-80%FF (200X magnification).
Note: Confocal depth image is a mirror reflection of surface image. (d_{max}~10μm).

Figure 8: Close up of cracking damage seen in Figure 6, specimen B.
Note: Confocal depth image is a mirror reflection of surface image. (d_{max}~4.5μm).

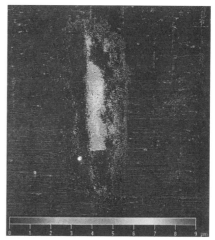

Figure 9: Maximum Fretting Damage on Specimen C-60%FF (200X magnification)
Note: Confocal depth image is a mirror reflection of surface image. (d_{max}~7.5μm).

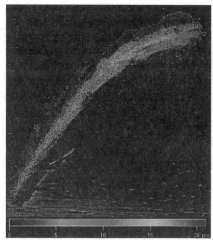

Figure 10: Maximum Fretting Damage on Specimen D-40%FF (100X magnification). Note: Confocal depth image is a mirror reflection of surface image. (d_{max}~20μm).

Figure 11: Maximum Fretting Damage on Specimen E-20%FF (200X magnification). Note: Confocal depth image is a mirror reflection of surface image. (d_{max}~10μm).

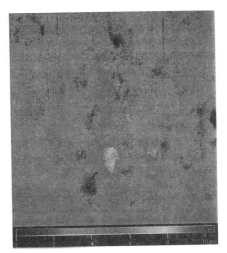

Figure 12: Maximum Fretting Damage on Specimen F-10%FF (400X magnification).
Note: Confocal depth image is a mirror reflection of surface image. (d_{max}~9μm).

Discussion

There exists a fretting fatigue damage threshold for these experimental conditions. This fretting fatigue damage threshold is dependent upon the arrest of further fretting damage. Upon the arrest of further fretting damage, specimens exhibited dramatically increased fatigue life. The baseline test (specimen A-100%FF) and specimen B-80%FF tests were very similar when comparing total cycles to fracture. Characterizing the fretting damage of specimens A and B was dramatically different, however, the location of the fracture origins were similar. The fatigue cracking responsible for fracture originated near the edge of the fretting pad contact area. This is a relatively common phenomenon. Specimens C-F that experienced 60% or less fretting fatigue cycles displayed increased fatigue life upon the arrest of fretting damage. None of the four specimens fractured after at least an additional 1,000,000 cycles. Specimens E and F were cycled for 3,000,000 cycles and did not fracture.

Specimen B, shown in Figures 6-8, was the only specimen where cracking damage was discovered prior to fracture. Note the depth of the fretting damage near the surface cracking was approximately 4.5μm versus the nearly 35μm fretting damage depth on specimen A. This may indicate a few different findings. One, it may be unlikely to be able to correlate the depth of fretting damage with the state of fatigue life of a specimen. Two, the fretting damage on specimen A may be largely due to the increase in displacement (slip amplitude) late in the fatigue life of the specimen due to crack opening displacement. The normal force was applied under load control conditions not displacement control. Therefore, there was no load relief as the fretting pad scarred deeper into the faying surface of the dog-bone specimen.

It is interesting to note that the level of damage for specimen D was relatively extensive. However, the damage was remote from the contact edges and did not nucleate observable fatigue cracking. Although the specimen was cyclically loaded, without fretting, for 1,200,000 cycles, it still showed no evidence of cracking. This is likely due to the lower maximum stress level away from the contact edges. It is believed that some debris became trapped between the fretting pad and faying surface of the dog-bone coupon. Prior to experimentation there was no suggestion of scarring or damage to explain the greater fretting damage experience by specimen D. Although specimen D did experience more pronounced fretting damage than specimens C, E or F, no cracking was discovered within or near the fretting damage region. This result bolsters the suggestion of a fretting fatigue damage threshold.

Specimens C, E and F revealed fretting damage levels from 4-10µm deep and specimen D a maximum fretting damage depth of approximately 20µm, yet none of the specimens revealed fatigue cracking on the surface. Further all four of these specimens were cyclically loaded for more than 1,000,000 cycles, some as many as 3,000,000 cycles, and none fractured. These results indicate that the damage induced by fretting fatigue has a threshold. They suggest the possibility that fretting and fretting fatigue can be managed through prediction, inspection and maintenance techniques. Through understanding the need to arrest fretting and fretting fatigue damage prior to exceeding the damage threshold, components can be maintained and remain in service for prolonged periods versus those where fretting and fretting fatigue damage is disregarded. This is a significant discovery within the scope of these experimental conditions. However, a thorough statistically based experimental program would be necessary to verify these results and develop a model to aid in the prediction and management of fretting, fretting fatigue and related concerns.

It is important to document the threshold that has been demonstrated in this study through adequately designed experiments. This contributes to the technical communities overall understanding of fretting fatigue and the intricate variability of this complex phenomenon.

Conclusions

1. A fretting fatigue damage threshold appears to exist. However, under these experimental conditions the threshold would be dependent upon the arrest of continued fretting damage.
2. At this point no correlation can be made between the depth of fretting damage and the fatigue life of a specimen.
3. Relatively heavy fretting damage that occurs remote from the edge of the contact area has dramatically less impact on the fatigue life of the specimen then damage near the edge of the contact area.

4. Interrupted fretting fatigue experiments are a valuable method for increased understanding of this complex phenomenon.
5. A statistically planned set of experiments is necessary to bolster the results of this study.

Summary

The knowledge that can be gained through interrupted fretting fatigue experiments using confocal and scanning electron microscopy can provide insight into the understanding of this phenomenon. Through interrupted experiments a snap shot in time of damage progression can be achieved. This progression of damage and analysis of that damage can lend significant discovery to aid in the understanding of complex issues. Confocal microscopy and scanning electron microscopy can be utilized to characterize other insidious forms of damage. Studies on pit growth rate, characterizing pitting depth and topography and the transition of pitting to fatigue cracking have been performed at the University of Utah [9,12-14]. These studies have provided knowledge of how corrosion induced damage has affected the fatigue life, including time or cycles to first detectable surface crack and crack growth rates for a variety of aluminum alloys. Interrupted experimentation can lead to an increased knowledge of the transition materials experience from a pristine condition through failure.

Acknowledgement:

The authors would like to thank the University of Utah for the use of their facilities and continued support of intellectual pursuits.

References:

[1] D. W. Hoeppner and G. L. Goss, "The Effect of Fretting Damage on the Fatigue Behavior of Metals", Lockheed Aircraft Corporation, Office of Naval Research Contract #N00014-71-C-0299, 1972.

[2] E. M. Eden, W. N. Rose, and F. L. Cunningham, "The Endurance of Metals," *Proc. Inst. Mech. Eng.*, 875, (1911).

[3] G. A. Tomlinson, P. L. Thorpe, and H. J. Gough, " An Investigation of Fretting Corrosion of Closely Fitting Surfaces," *Proc. Inst. Of Mech. Eng.* 223 (1939).

[4] E. J. Warlow-Davies, "Fretting Corrosion and Fatigue Strength: Brief Results of Preliminary Experiments," *Proc. Inst. Of Mech. Eng.*, 32-38 (1941).

[5] A. J. Fenner, K. H. R. Wright, and J. Y. Mann, "Fretting Corrosion and its Influence on Fatigue Failure," *Proc. Of the Int. Conf. On Fatigue of Metals,* New York, ASME, 386-393 (1956).

[6] R. B. Waterhouse, "Influence of Local Temperature Increases on the Fretting Corrosion of Mild Steel," *J. Iron and Steel Inst.,* 301-305 (1961).

[7] K. Nishioka and K. Hirakawa, "Fundamental Investigation of Fretting Fatigue – Part 3, Some Phenomena and Mechanisms of Surface Cracks," *Bulletin of JSME,* **12**, No. 51, 397-407 (1969).

[8] D. W. Hoeppner and G. L. Goss, "Mechanisms of Fretting Fatigue," Lockheed California Company Report Number LR 24367, (1970).

[9] V. Chandrasekaran, Young In Yoon, and D. W. Hoeppner, "Analysis of Fretting Damage Using Confocal Microscopy", *Fretting Fatigue: Current Technology and Practices, ASTM STP 1367,* D. W. Hoeppner, V. Chandrasekaran and C. B. Elliott, Eds., American Society for Testing and Materials, 1999.

[10] S. Adibnazari and D. W. Hoeppner, "A Fretting Fatigue Normal Pressure Threshold Concept", *Wear, 160* (1993) 33-35.

[11] D. W. Hoeppner and G. L. Goss, "A Fretting-Fatigue Damage Threshold Concept, *Wear, 27* (1974) 61.

[12] V. Chandrasekaran, Y. Yoon, D. W. Hoeppner, "Pit Growth Rates in 2024-T3 Aluminum Alloy", FASIDE International Inc., NCI Information Systems, Subcontract Number, NCI-USAF-9192-005, July, 2001.

[13] P. N. Clark, "The Transition of Corrosion Pitting to Surface Fatigue Cracks in 2024-T3 Aluminum Alloy", University of Utah, Dissertation, 2001.

[14] P. N. Clark and D. W. Hoeppner, "Corrosion Pitting Characterization and Subsequent Transition to Fatigue Cracking", *(forthcoming),* International Committee on Aeronautical Fatigue, 2001.

FRETTING FATIGUE CRACK AND DAMAGE

Yoshiharu Mutoh,[1]   Jin-Quan Xu,[2]   and Kazunori Kondoh[3]

Observations and Analysis of Fretting Fatigue Crack Initiation and Propagation

Reference: Mutoh, Y., Xu, J. Q., and Kondoh, K., **"Observations and Analysis of Fretting Fatigue Crack Initiation and Propagation,"** *Fretting Fatigue: Advances in Basic understanding and Applications, STP 1425*, Y. Mutoh, S. E. Kinyon, and D. W. Hoeppner, Eds., ASTM international, West Conshohocken, PA, 2003.

Abstract: Many factors, such as the amplitude of the cyclic load, the relative rigidity of components and the frictional condition, play important roles in the fretting fatigue process. This study has focused on the local singular stress field at the fatigue crack tip or at the contact edge where the initial crack occurs, based on the numerical analysis results with the use of finite element method (FEM) program ABAQUS. It is proposed that all the effect of test conditions can be contained by the local stress field, which would dominate the fatigue crack initiation and propagation. By comparing with the experimental results, it was found that the initial fatigue crack should occur at the point of maximum shear stress range near the contact edge, and would propagate in the direction of maximum tangential stress range. An analytical simulation of fretting fatigue crack propagation has been carried out. It was found that a fatigue crack was always under the mixed mode conditions, the fracture path could be estimated by the maximum tangential stress theory ($\Delta\sigma_{\theta\max}$), and the propagation curve could be characterized by the stress intensity factor range in this direction ($\Delta K_{\sigma\theta\max}$).

Keywords: fretting fatigue, crack initiation, propagation, simulation, fatigue life

Introduction

From the phenomenological viewpoint, fretting fatigue life depends on factors such as the amplitude of the cyclic load, contact pressure, relative rigidity of the components, frictional condition, the size and geometrical effect, etc. Many studies [1-5] have been reported on the parametric investigations of the effect of these factors on the fretting fatigue life. However, due to the complicated interaction of these factors, it is difficult to grasp the intrinsic features of fretting fatigue by such parametric studies. Therefore, some subsequent parameters [6,7], such as the relative

---

[1] Professor, [2] Associate Professor, [3] Graduate Student, Department of Mechanical Engineering, Nagaoka University of Technology, Kamitomioka 1603-1, Nagaoka, Niigata, Japan, 940-2188.

slip at the contact edge, which partially contains the combined effects of all the phenomenological factors, have also been proposed to characterize the fretting fatigue process. On the other hand, fracture mechanics is expected to assist strongly in the evaluation of fretting fatigue crack initiation and propagation.

The most popular consideration [8-12] for applying fracture mechanics is to introduce a vertical crack to the contact interface, though such a crack can not represent the fretting fatigue fracture path. Giannakopoulos et al [13] considered the kinking of an interface crack on the contact interface, to take the curve path of fretting fatigue into account. Some three dimensional investigations [14,15] on surface cracking due to fretting fatigue are also available. Skeikh et al [16] had analyzed the stress intensity factors of an inclined crack from the contact edge by the finite element method based on the pre-determined surface traction. This study focused on the effect of the local stress field at the contact edge and the fretting fatigue crack tip. Considering that fretting fatigue crack initiation and propagation are dominated by the local stress field, and the interaction effects of the phenomenological factors can be considered as being included in such a local stress field, observations, analyses and evaluations of the fretting fatigue crack have been carried out from the viewpoint of the local stress field. From experimental observations and analysis results, it is found that the direction of the crack initiation agrees well with the maximum shear stress range direction at the contact edge, where the stress range results in singular behavior, and the direction of the crack propagation correlates well with the effective maximum tangential stress range direction. It is also found that the fracture path can be simulated well by the $\Delta\sigma_{\theta\,\text{max}}$ theory, and the fretting fatigue life can be estimated by the effective stress intensity factor range $\Delta K_{\theta\,\text{max}}$ at the fatigue crack tip.

Singular Stress Field around the Contact Edge

The stress range at the contact edge is singular due to the singular stress distribution corresponding to the maximum and minimum loads during one cycle. The singular stress range depends on the deformation type of the contact edge. The deformation of the contact edge (point A) when the specimen subjected to a tension or compression force can be distinguished as following three types, a) gapping, b) sticking and c) slipping, as shown in Fig. 1. Therefore, during one cycle of loading, nine types of deformation exist as shown in Table 1. It is obvious that if the contact edge is under gapping deformation, no singular stress will appear. We denote such a

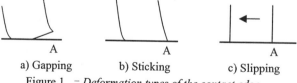

a) Gapping   b) Sticking   c) Slipping

Figure 1   – *Deformation types of the contact edge*

Table 1 – *Basic types of the deformation and stress range field.*

No.	Under maxi. Load	Under mini. Load	Stress range field around the contact edge
1	Gapping	Slipping	$\Delta\sigma_{ij} = \sigma_{ijo}(P_{\max}) - \sigma_{ijs}(r^{1-\lambda_s}, P_{\min})$
2	Gapping	Sticking	$\Delta\sigma_{ij} = \sigma_{ijo}(P_{\max}) - \sigma_{ijb}(r^{1-\lambda_b}, P_{\min})$
3	Gapping	Gapping	$\Delta\sigma_{ij} = \sigma_{ijo}(P_{\max}) - \sigma_{ijo}(P_{\min})$
4	Sticking	Slipping	$\Delta\sigma_{ij} = \sigma_{ijb}(r^{1-\lambda_b}, P_{\max}) - \sigma_{ijs}(r^{1-\lambda_s}, P_{\min})$
5	Sticking	Sticking	$\Delta\sigma_{ij} = \sigma_{ijb}(r^{1-\lambda_b}, P_{\max}) - \sigma_{ijb}(r^{1-\lambda_b}, P_{\min})$
6	Sticking	Gapping	Similar to No. 2
7	Slipping	Slipping	$\Delta\sigma_{ij} = \sigma_{ijs}(r^{1-\lambda_s}, P_{\max}) - \sigma_{ijs}(r^{1-\lambda_s}, P_{\min})$
8	Slipping	Sticking	Similar to No.4
9	Slipping	Gapping	Similar to No.1

non-singular stress field by $\sigma_{ijo}(P)$, here P is the corresponding load. If the contact edge is under sticking deformation, the singular stress field for a V-notch or bonded edge will appear [17-20], as in the form of $\sigma_{ijb}(r^{1-\lambda_b}, P)$, where $1-\lambda_b$ denotes the corresponding singular order. If it is under slipping deformation, the slipping singular stress field $\sigma_{ijs}(r^{1-\lambda_s}, P)$ shall appear, where $1-\lambda_s$ denotes the singular order. The details of the singular stress field will be given next. It can be easily understood that there are six independent basic types of stress range field as shown in the table. Therefore, it is necessary to evaluate the fretting fatigue behavior according to these basic types. Because there is no stress singularity when the contact edge is under gapping deformation, we only need to consider the singular stress field for the cases of sticking and slipping. For the purpose of generalization, we consider the fretting contact model with an arbitrary contact pad as shown in Fig. 2. The material "1" is the contact pad, its shear modulus and Poisson's ratio are denoted by G_1 and v_1. The material "2" is the fretting fatigue specimen whose elastic constants are denoted by G_2 and v_2.

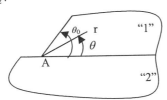

Figure 2 – Analysis model

The Case of Sticking

The theoretical solution of the stress field can be found in literature [17,18] for this case. The general form of the stress field can be expressed as

$$\sigma_{ij} = \frac{K_1 F_{1ij}(\theta)}{r^{1-\lambda_1}} + \frac{K_2 F_{2ij}(\theta)}{r^{1-\lambda_2}} \quad . \tag{1}$$

Here, K_i presents the stress intensity coefficient corresponding to the singular order

λ_i, F is the angular function. The singular orders can be determined from the eigen-equation [17-20], and usually multiple singularities weaker than $1/\sqrt{r}$ appear for such a geometry. If the material of the contact pad is the same of the specimen, the singular order can be determined more simply [17] from

$$\left(-\lambda \sin \theta_0 + \sin \lambda(\pi + \theta_0)\right)\left(\lambda \sin \theta_0 - \sin \lambda(\pi + \theta_0)\right) = 0. \tag{2}$$

The angular function in Eq.(1) can be given as

$$
\begin{aligned}
F_{1\theta} &= (1 + \lambda_1) \cos(1 - \lambda_1)\varphi + \rho_1 \cos(1 + \lambda_1)\varphi \\
F_{2\theta} &= (1 + \lambda_2) \sin(1 - \lambda_2)\varphi + \rho_2 \sin(1 + \lambda_2)\varphi \\
F_{1r\theta} &= (1 - \lambda_1) \sin(1 - \lambda_1)\varphi + \rho_1 \sin(1 + \lambda_1)\varphi \\
F_{2r\theta} &= -(1 - \lambda_2) \cos(1 - \lambda_2)\varphi + \rho_2 \cos(1 + \lambda_2)\varphi
\end{aligned}
\tag{3}
$$

$$\varphi = \frac{\pi - \theta_0}{2} + \theta, \quad \varphi_0 = \frac{\pi - \theta_0}{2}, \quad \rho_1 = -\frac{(1 - \lambda_1) \sin(1 - \lambda_1)\varphi_0}{\sin(1 + \lambda_1)\varphi_0}$$

$$\rho_2 = \frac{(1 + \lambda_2) \sin(1 - \lambda_2)\varphi_0}{\sin(1 + \lambda_2)\varphi_0} \tag{4}$$

The Case of Slipping

The stress field for this case can be expressed as [21]

$$
\begin{aligned}
\sigma_{\theta 2} = K r^{\lambda-1} \Bigg\{ &-\frac{(\lambda+1)(F+\lambda-\cos 2\lambda\pi)}{(F+\lambda-1)\sin 2\lambda\pi}\cos(\lambda-1)\theta + \frac{F(\lambda+\cos 2\lambda\pi)+\lambda^2-1}{(F+\lambda-1)\sin 2\lambda\pi}\cos(\lambda+1)\theta \\
&-\frac{\lambda+1}{F+\lambda-1}\sin(\lambda-1)\theta - \frac{F}{F+\lambda-1}\sin(\lambda+1)\theta \Bigg\}
\end{aligned}
$$

$$
\begin{aligned}
\tau_{r\theta 2} = K r^{\lambda-1} \Bigg\{ &-\frac{(\lambda-1)(F+\lambda-\cos 2\lambda\pi)}{(F+\lambda-1)\sin 2\lambda\pi}\sin(\lambda-1)\theta + \frac{F(\lambda+\cos 2\lambda\pi)+\lambda^2-1}{(F+\lambda-1)\sin 2\lambda\pi}\sin(\lambda+1)\theta \\
&+\frac{\lambda-1}{F+\lambda-1}\cos(\lambda-1)\theta + \frac{F}{F+\lambda-1}\cos(\lambda+1)\theta \Bigg\}
\end{aligned}
\tag{5}
$$

The eigen-equation determining the singular order is

$$\rho\mu(\lambda^2 - 1) - \zeta\mu(\lambda+1) + (\lambda-1+F)\left(\lambda \sin 2\theta_0 + \sin 2\lambda\theta_0\right)$$
$$+ \zeta\mu\left(\lambda \cos 2\theta_0 + \cos 2\lambda\theta_0\right) - \rho\mu(\lambda+1)\left(\lambda \cos 2\theta_0 - \cos 2\lambda\theta_0\right) = 0 \tag{6}$$

where

$$F = -\lambda - 1 - \frac{2 \sin 2\lambda\pi}{\mu(\cos 2\lambda\pi - 1) - \sin 2\lambda\pi}, \quad \rho = \frac{1 + \lambda\alpha + (1-\lambda)\beta + F(\alpha - \beta)}{1-\alpha},$$

$$\zeta = \frac{F[1 - \lambda\alpha + (\lambda-1)\beta] - (\lambda-1)[(\lambda+1)\alpha - (\lambda-1)\beta]}{1-\alpha}. \tag{7}$$

Here μ is the coefficient of friction at the interface. The Dunders' parameters [22] a and β are defined as follows:

$$\alpha = \frac{G_1(\kappa_2 + 1) - G_2(\kappa_1 + 1)}{G_1(\kappa_2 + 1) + G_2(\kappa_1 + 1)}, \beta = \frac{G_1(\kappa_2 - 1) - G_2(\kappa_1 - 1)}{G_1(\kappa_2 + 1) + G_2(\kappa_1 + 1)} \ . \tag{8}$$

If the material of the contact pad is the same as the specimen and the shape angle is $\theta_0 = \pi/2$ (the most popular geometry in fretting fatigue tests), the eigen-equation can then be simplified as

$$2\mu\lambda(\lambda + 1)\sin\lambda\pi - (1 - 2\lambda^2)\cos\lambda\pi + \cos 2\lambda\pi = 0 \ . \tag{9}$$

It can be found that the singular order is dependent on the coefficient of friction. The solution of Eq.(9) is shown in Fig.3. It shall be pointed out that when the contact edge is under slipping deformation, the stress singularity can be stronger than $1/\sqrt{r}$, which corresponds to the energy dissipation due to the friction. Figure 4 shows the dependence of stress singularity on the pad shape for the case that the material of the pad and the specimen are the same. From Figs.3 and 4, it can be found that the higher the frictional coefficient is, the stronger the edge singularity becomes. Therefore, decreasing the coefficient of friction can be expected as an efficient method to improve the fretting fatigue life. It is very interesting that when the frictional coefficient is higher, for example, for the case that $\mu = 1.0$, the pad angle $\theta_0 = \pi/2$ presents the strongest stress singularity, which means that $\theta_0 = \pi/2$ is the severest shape leading to fretting fatigue, even worse than the case of larger angle.

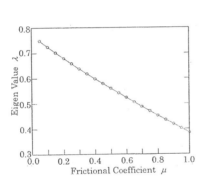

Figure 3 – Stress singularity for various frictional coefficients

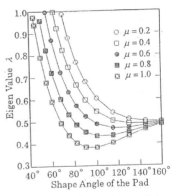

Figure 4 – Dependence of stress singularity on the pad angle

Numerical Analysis

The analysis model and the element division for its one-fourth region are shown in Fig.5. The FEM code ABAQUS was adopted. Frictional contact elements were introduced along the interface. The materials of the pad and the specimen are the same. The material constants were $E = 210\,GPa$ and $\nu = 0.3$. The force P was applied to attain the average contact pressure of $60\,MPa$ on the pad foot. The frictional coefficient is assumed as $\mu = 0.8$. The cyclic load was $\sigma_0 = 180\sin\omega t\,MPa$. Figure 6 shows the deformation of the pad and the specimen under maximum tension and compression. It can be found that the contact edge of the outside deformed as gapping under tension, and as slipping under compression. Therefore, this analysis example corresponds to the type 1 in Table 1. The maximum relative slip range occurring at the edge was about 2.11 μm. Figure 7 shows the stress distribution along the interface. It can be found that there are three regions under the tension load, i.e., the sticking region at the inside, the slip region at the middle part, and the gap region at the outside of the interface. On the contrary, there are only two regions under compression. It shall be pointed out that the size and existence of such regions are dependent on the shapes of the specimen and the pad, the loading condition and the frictional coefficient. It can be found that in the sticking region, the friction shear stress is much smaller than that calculated by the Newton-Columb's friction law. This fact means that the frictional coefficient cannot be calculated directly from the

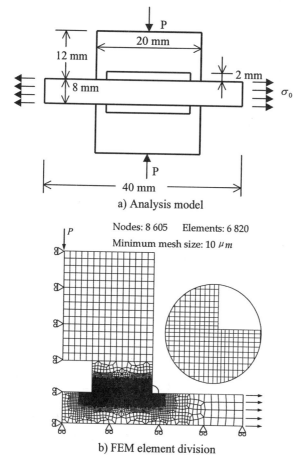

a) Analysis model

b) FEM element division

Figure 5   – *Analysis model and element division*

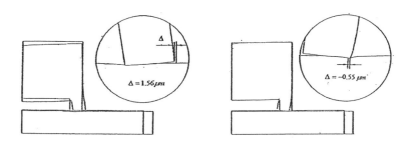

a) Deformation under tension   b) deformation under compression

Figure 6   − *Deformation of the pad and the specimen*

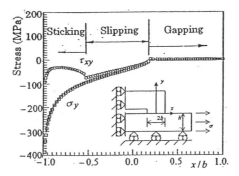

a) Stress on the interface under tension

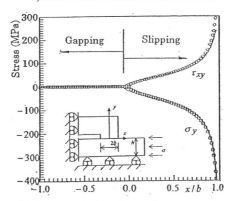

b) Stress on the interface under compression

Figure 7   − *Stress distribution along the contact interface*

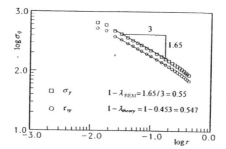

Figure 8   – *Logarithmic distribution and the singular order*

observed shear force in the experiment. Figure 8 shows the logarithmic stress distribution near the edge under the compression. It is found that the singular order obtained numerically agrees with the theoretical result very well.

Fretting Fatigue Test and Its Observations

Fretting fatigue test was carried out for the observation of fretting fatigue crack initiation and propagation. Both the specimen and the contact pad are cut out from a SM430A plate, their geometry sizes are shown in Fig.9. The schematic diagram of the fatigue test is shown in Fig.10. The fatigue test was carried out by an electro-hydraulic servo-controlled fatigue test machine with a capacity of 98 kN under the stress ratio, R, of –1 at frequency of 20 Hz. The average contact pressure is 60 MPa, The coefficient of friction is measured as $\mu = 0.8$ by a completely moving wear test. To observe the fatigue crack initiation and propagation, the fatigue test was stopped at

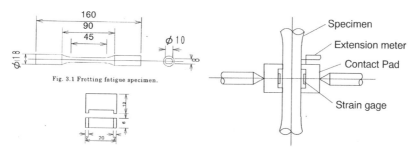

Figure 9   – *Geometry of specimen*   Figure 10   – *Schematic diagram of*
            and pad   *fretting fatigue test*

certain cycles under the cyclic loading of 180MPa. The cross section was observed by SEM. One example of the cross section is shown in Fig.11. It can be seen that the initial crack occurs in a very small angle from the contact interface at the outside contact edge, and then kinks into the direction about 60 degree from the interface at its primary stage, and gradually propagates into the vertical direction to the contact interface. The fracture path was measured by the SEM from multiple specimens which are stopped at different loading cycles, as shown in Fig.12. The crack growth rate was calculated from the well-fitted curve of the fracture path, i.e., the crack growth rate in the tangential direction along the fracture path, not in the vertical direction, is considered in this study. It is noted here that Figs.11 and 12 were drawn under the coordinate system shown in Fig.2.

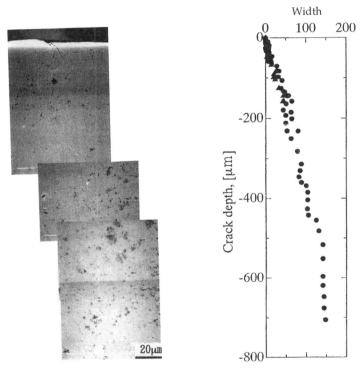

Figure 12 – *Measured fracture path*

Evaluation of the Fretting Fatigue Crack Initiation

In fatigue problems, the important factor is the stress variation range. Due to the edge singularity, the stress variation during one cycle is very complicated. Figure 13 shows the stress range along the interface, obtained by the FEM analysis. It is found that all the range of stress components becomes the maximum at the outside edge. Therefore, it can be estimated that fatigue fracture will occur at the outside edge, which coincides with the experimental observations. Considering that at the outside edge the stress range is singular, the stress range in the polar coordinate is drawn as Fig.14 (note the difference of the coordinate systems in Figs. 2 and 14). It is found that the maximum shear stress direction located in a very small angle near the interface, and the maximum tangential stress occurs at the direction about $\theta = -120°$, i.e., about 60° from the interface. Comparing with the observed results shown in Fig.11, it is possible to propose that the initial fretting fatigue crack may occur at the direction where the shear stress range is maximum, and it will kink into the direction where the tangential stress range is maximum ($\Delta\sigma_{\theta\,max}$ theory) immediately after initiation. This fact means that the fretting fatigue crack initiation can be evaluated by the singular stress range field. However, because the initial crack occurs in the direction very near to the contact interface, due to the severe wearing and plastic deformation of the contact interface, it is difficult to affirm when it occurs and how long it is. It shall be pointed out that the life before crack initiation in

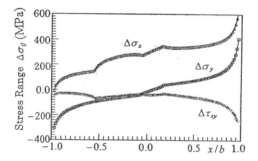

Figure 13 – Stress range along the contact interface

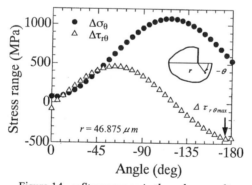

Figure 14 – Stress range in the polar coordinate

fretting fatigue only occupies several percent of the total fatigue life. Therefore, it is reasonable to neglect it in the fatigue life evaluation of fretting fatigue. Because the initial crack orientation has a strong effect on the subsequent crack propagation behavior (such as kinking angle, growth rate, etc.), determining the crack orientation has impact on evaluation of fretting fatigue.

Simulation of the Fretting Fatigue Propagation

By neglecting the life before the crack initiation, we can introduce an initial crack to evaluate the fretting fatigue life as mainly dominated by the crack propagation, in the direction where the shear stress range around the contact edge is the maximum. The initial crack length is considered as the critical smallest crack length [23] for which the linear fracture mechanics can be applied, i.e.,

$$a_0 = \left(\Delta K_{th} / \Delta\sigma_a\right)^2 / \pi \qquad . \tag{10}$$

Here, ΔK_{th} is the threshold stress intensity factor range, and $\Delta\sigma_a$ is the fatigue limit of the material under plain fatigue condition. For SM430A, it can be calculated as $a_0 \approx 80\mu m$. It is easily understood that this initial crack is generally under mixed mode condition, therefore, it should kink continuously until the crack tip loading is mode I dominated. In plain mixed mode fatigue problems, it is well known that the crack shall kink into the maximum tangential stress range direction ($\Delta\sigma_{\theta max}$ theory [23], or in other words, into the direction where $\Delta K_{II} = 0$), and the fatigue properties can be characterized by the stress intensity factor range ($\Delta K_{\theta max}$) in this direction. Based on the $\Delta\sigma_{\theta max}$ theory, the numerical simulation analysis by using FEM program ABAQUS is carried out.

The crack is assumed as propagating $85\mu m$ for every step. The stress range around the crack tip is calculated as the effective stress range, i.e., no overlapping of the crack face is permitted. When the specimen is under compression load, the crack face near the crack tip is dealt with using contact elements. The true contact regions are determined by numerical procedures, assuming the coefficient of friction between two crack faces as zero. Figure 15 shows the simulated crack propagation path. It is obvious that this path coincides well with the fracture path by comparing with Fig.12. Figure 16 shows the fatigue crack growth curve by using the parameter $\Delta K_{\theta max}$.

The plain fatigue test results under mode I are also shown in the figure. It is noted here that $\Delta K_{\theta max} = \Delta K_I$ for single mode I test. It can be found that the fretting fatigue curve agrees well with the plain fatigue curve except for the first step. This means that the fretting fatigue crack propagation can be evaluated as the same as the mixed mode crack propagation (i.e., using $\Delta K_{\theta max}$ instead of ΔK_I) in plain fatigue except for the first step. The first step results correspond to crack initiation and its first kinking step. Many reasons could be considered for its difference from plain fatigue, such as the effects of small crack and the constant term in the stress range field, the error of measured crack length and growth rate, etc. It can be considered that this first step crack propagation has only a minimal effect on the total fatigue life, therefore, we can neglect it in the total fretting fatigue life evaluation.

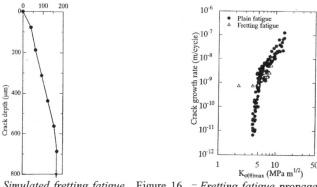

Figure 15 – Simulated fretting fatigue   Figure 16 – Fretting fatigue propagation
              fracture path   curve

Evaluation of the Fatigue Life

Neglecting the first step of the crack propagation, the crack growth curve in Fig.16 (i.e., the plain fatigue curve) can be expressed as

$$\frac{da}{dN} = 1.26 \times 10^{-12} \left(\Delta K_{\theta \max}\right)^{3.65} \quad . \tag{11}$$

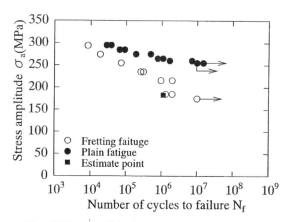

Figure 17 – S-N curve of the fretting fatigue and the predicted life

If we assume that the unstable crack propagation starts at the crack length of $a_f = B/4$ (B is the width of the specimen. Strictly, a_f shall be determined by K_{IC}, here we use this relatively large value just for the convenience), then the fatigue life can be predicted simply as

$$N = \int_{a_0}^{a_f} \frac{da}{1.26 \times 10^{-12} \left(\Delta K_{\theta \max}\right)^{3.65}} = 1.17 \times 10^6 \text{ cycles} \quad . \tag{12}$$

Figure 17 shows the comparison of the experimental results and the predicted result. It can be found that the fretting fatigue life can be well predicted by the $\Delta\sigma_{\theta\,max}^{*}$ theory.

Conclusions

The fretting fatigue crack initiation and propagation are analyzed and observed. The main results can be concluded as follows:

1) There are six basic types of stress variation range with different singularities for different deformations at the contact edge of nine types.

2) The stress variation range becomes the maximum at the outside edge, where a fretting fatigue crack should initiate.

3) A fretting fatigue crack initiates at the direction where the shear stress range is the maximum.

4) Immediately after crack initiation, the crack kinks into the direction where the tangential stress range is the maximum.

5) The fatigue crack is under a mixed mode condition, and should propagate in the direction where the tangential stress range is maximum.

6) The fretting fatigue properties can be well described by using the parameter $\Delta K_{\theta\,max}$, and the fatigue life can be evaluated by substituting ΔK_I with $\Delta K_{\theta\,max}$ in the crack growth rule of the plain mode I fatigue.

References

[1] Waterhouse, R. B. and Lindley, T. C., *Fretting Fatigue*, Mechanical Engineering Publications Limited, 1994.

[2] Hoeppner, D. W., Chandrasekaran, V. and Elliott III, C. B., *Fretting Fatigue,* ASTM STP1367, 2000.

[3] Adibnazari, S. and Hoeppner, D. W., "A Fretting Fatigue Normal Pressure Threshold Concept," *Wear*, 160, 1998, pp.33-35.

[4] Vodopivec, F., Vizintin, J. and Sustarsic, B., "Effect of Fretting Amplitude on Microstructure of 1C-1.5Cr Steel", *Material Science and Technology*, 12, 1996, pp.355-360.

[5] Waterhouse, R. B., *Fretting Wear ASM Handbook*, Vol.18, ASM International, 1992, pp.242-256.

[6] Mutoh, Y., Tanaka, K., and Itoh, S., "Fretting Fatigue Properties of High Strength Stainless Steel," *Journal of the Society of Materials Science, Japan*, (in

Japanese), 37, 1988, pp.643-648.

[7] Berthier, Y., Vincent, L., and Godet, M., "Fretting fatigue and Fretting Wear," *Tribology International*, 22, 1989, pp.235-242.

[8] Nix, K. J., and Lindley, T. C., "The Initiation of Fracture and Propagation of Small Defects in Fretting Fatigue," *Fatigue Fracture Engineering of Material and Structure*, 8, 1985, pp.143-160.

[9] Tanaka, K., Mutoh, Y., Sakoda, S., and Leadbeater, G., Fretting Fatigue in 0.55C Spring Steel and 0.45C Carbon Steel, *Fatigue Fracture Engineering of Material and Structure*, 8(1985), P.129-142

[10] Hattori, T., Nakamura, M., and Watanabe, T., "Fretting Fatigue Analysis by Using Fracture Mechanics," *ASME Paper No.84-WA/DE-10*, 1984.

[11] Li, Y., and Hills, D. A., "Stress Intensity Factor Solutions for Kinked Surface Cracks," *Journal of Strain Analysis*, 25, 1989, pp.21-27.

[12] Edwards, P. R., "Fracture Mechanics Application to Fretting in Joints," *6th International Conference On Fracture*, 1984, pp.3813-3836.

[13] Giannakopoulos, A. E., Lindley, T. C., and Suresh, S., "Aspects of Equivalence Between Contact Mechanics and Fracture Mechanics: Theoretical Connections and A Life-Prediction Methodology for Fretting Fatigue," *Acta. Material* 46-9, 1998, pp.2955-2968.

[14] Dai, D. N., Hills, D. A., and Nowell, D., "Stress Intensity Factors for Three Dimensional Fretting Fatigue Cracks," *Fretting Fatigue*, Ed., Waterhouse R. B., and Lindley T. C., 1994, pp.59-71.

[15] Hirakawa, K., and Toyama, K., "Influence of Residual Stress on the Fatigue Crack Initiation of Press-Fitted Axle Assemblies," *Fretting Fatigue*, Ed., Waterhouse R.B., and Lindley T.C., 1994, pp.461-473.

[16] Sheikh, M. A., Fernando, U. S., Brown, M. W., and Miller, K. J., "Elastic Stress Intensity Factors for Fretting Cracks Using the Finite Element Method," *Fretting Fatigue*, Ed., Waterhouse R. B., and Lindley T. C., 1994, pp.83-101.

[17] Okamura, H., *Analysis of Strength*, (in Japanese), Ohmusya, 1981.

[18] Atkinson, C., et.al., "Stress Analysis in Sharp Angular Notches using Auxiliary

Fields," *Engineering. Fracture Mechanics*, 31, 1988, pp.637-646.

[19] Bogy, D. B., "Two Edge Bonded Elastic Wedges of Different Materials and Wedge Angles under Surface Tractions," *Journal of Applied Mechanics*, 38, 1971, pp.377-386.

[20] Dempsey, J. P., and Sinclair, G. B., "On the Singular Behavior at the Vertex of a Bimaterial Wedge, *Journal of Elasticity*, 11, 1981, pp317-327.

[21] Xu, J. Q., and Mutoh, Y., "Stress Field near the Contact Edge in Fretting Fatigue Tests," *Proc. of APCFS&ATEM'01,* 2001, Sendai, pp.958-963.

[22] Dunders, J., "Effect of Elastic Constants on Stress in a Composite under Plane Deformation," *Journal of Composite Materials*, 1, 1967, pp.310-322.

[23] Kitagawa, H., Yuuki, R., and Tohgo, K., "Fatigue Crack Propagation Behavior under Mixed Mode Condition," *Transaction of JSME*, (in Japanese), A47, 1981, pp.1283-1292.

Toru Kimura[1] and Kenkichi Sato[2]

Stress Intensity Factors K_I and K_{II} of Oblique Through Thickness Cracks in a Semi-Infinite Body Under Fretting Fatigue Conditions

---

Reference: Kimura, T. and Sato, K., "Stress Intensity Factors K_I and K_{II} of Oblique Through Thickness Cracks in a Semi-Infinite Body Under Fretting Fatigue Conditions," *Fretting Fatigue: Advances in Basic Understanding and Applications, STP 1425*, Y. Mutoh, S. E. Kinyon, and D. W. Hoeppner, , Eds. ASTM International, West Conshohocken, PA, 2003.

Abstract: In this paper the formulas for calculating K_I and K_{II} of an oblique crack under fretting fatigue loading conditions have been derived. The scheme of the formulation is based on a method using Green's functions and the principle of superposition, i.e. the Green's functions have been obtained through boundary element analysis for normal and tangential localized forces on oblique crack faces and superimposed their results to calculate two components of mode I and II. The formulas for K_I and K_{II} were expressed as functions of crack angle, crack length and applied loading conditions, such as bulk fatigue stress, contact pressure distribution and friction stress distribution. A user-friendly program based on these formulas, then, has been developed to compute the values of stress intensity factors K_I and K_{II} under arbitrary fretting fatigue conditions. The values of stress intensity factors for both normal and oblique cracks have been illustrated as examples to understand the effects of crack angle and pad load distribution.

Keywords: fretting fatigue, stress intensity factor, boundary element method, Green's function, principle of superposition

Introduction

It has been well known that fatigue strength of machines and structures is markedly decreased due to fretting fatigue. Fretting fatigue is a form of fatigue where micro-slip wear takes place at the contact interface of structural components subjected to contact pressure, friction stress and cyclic fatigue stress. The fretting fatigue crack in the early stage propagates in the direction oblique to the contact surface due to the effects of the above pressure and stresses combined.

In order to estimate fretting fatigue lives, the stress intensity factors of cracks have to be known. Evaluations of the stress intensity factors of fretting fatigue cracks K_I and

---

[1]Graduate Student, Graduate School of Science and Technology, Chiba University, 1-33, Yayoi-cho, Inage-ku, Chiba, 263-8522 Japan.
[2]Associate Professor, Department of Urban Environment Systems, Chiba University, 1-33, Yayoi-cho, Inage-ku, Chiba, 263-8522 Japan.

K_{II}, however, are rather difficult because of the combined stress fields and crack features. Since it is a conventional but complicated way to determine the stress intensity factors using finite element method or boundary element method, the formulas proposed by Rooke and Jones [1] have been used assuming that fretting fatigue crack is normal to the contact surface. However, because of simplification and complexity of determining mode II stress intensity factor K_{II} the mode I stress intensity factor K_I has been only adopted in the conventional analysis for fretting fatigue problems. In order to evaluate the oblique cracks in fretting fatigue, it is important to consider the effect of mode II stress intensity factor as well.

In this paper, therefore, the authors have expanded the method of Rooke and Jones to determine stress intensity factors K_I and K_{II} for oblique fretting fatigue cracks. The Green's functions for oblique cracks [2] to adopt the principle of superposition were obtained through boundary element analysis. The general equations for K_I and K_{II} have been expressed as multi-term equations including parameters of crack angle, crack length, plate width, contact pressure, friction stress and fatigue stress.

Formulation of stress intensity factors

Modeling of fretting fatigue

In this paper, a rectangular plate subjected to contact pressure, friction stress and fatigue stress was used for a fretting fatigue model as shown in Figure 1. A single straight crack at the contact edge was modeled according to the analysis model of Edwards [3]. In order to evaluate the stress intensity factor, the principle of superposition was used assuming the stress–free crack, as shown in Figure 2. In Figure 2, case (a) refers to a normal load acting on the edge and is equivalent to the sum of case (b), with the same load acting on the edge of the plate without a crack, and case (c), with the distributed stresses induced by case (b) and acting on the line corresponding to the crack site of case (a). Because the value of the stress intensity factor of case (b) is zero, cases (a) and (c) have the same value as the stress intensity factor. By using this principle, case (a), the stress intensity factor due to external load, can be replaced by case (c), the stress intensity factor due to the internal stress distribution along the crack surface.

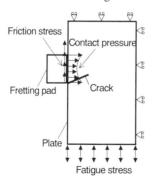

Figure 1 - Fretting fatigue analysis model.

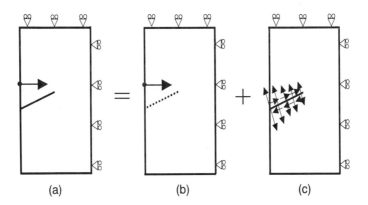

Figure 2 – *principle of superposition.*

Boundary element analysis

The magnitude of the normal force per unit thickness, p, which acts on the crack surface, on the stress intensity factor was analyzed by the boundary element model shown in Figure 3. The magnitude of the tangential force per unit thickness, q, was also analyzed by the model. In the model, w and h are width and height of the plate respectively, b is the loading point on the crack surface, a and θ are length and angle of the edge crack respectively. To evaluate the effect of unit force, the values of p and q were taken to be 1 N per unit thickness. In this paper the stress intensity factors due to unit forces p and q are called the differential stress intensity factors dk_I and dk_{II}, for modes I and II respectively. They were analyzed for various ratios of a/w and b/a under several values of crack angle θ.

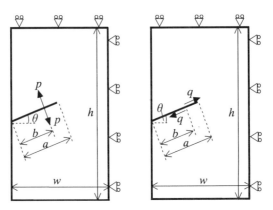

Figure 3 - *Boundary element analysis for the oblique crack model subjected to localized forces p and q acting on the crack surface.*

Derivation of Green's Functions

The values of the differential stress intensity factor obtained from the boundary element analysis were summarized as Equations (1) and (2) by interpolation:

$$dk_I = \frac{p}{\sqrt{\pi a}} \cdot F_{Ip} \left(\frac{a}{w}, \frac{b}{a}, \theta\right) + \frac{q}{\sqrt{\pi a}} \cdot F_{Iq} \left(\frac{a}{w}, \frac{b}{a}, \theta\right) \tag{1}$$

$$dk_{II} = \frac{p}{\sqrt{\pi a}} \cdot F_{IIp}\left(\frac{a}{w}, \frac{b}{a}, \theta\right) + \frac{q}{\sqrt{\pi a}} \cdot F_{IIq} \left(\frac{a}{w}, \frac{b}{a}, \theta\right) \ , \tag{2}$$

where F_{Ip}, F_{Iq}, F_{IIp} and F_{IIq} are functions formulated as multi-terms.

Stress intensity factors due to localized forces on the edge of the plate

Equations (1) and (2) can be used as Green's functions to obtain the stress intensity factors k_I and k_{II} due to normal and tangential stresses distributions, $\sigma(b)$ and $\tau(b)$ respectively, acting along the oblique crack surface. Since the differential normal and tangential forces per unit thickness, acting on the differential region db at the location b are expressed as $\sigma(b)db$ and $\tau(b)db$ respectively, the stress intensity factors due to the whole distribution of $\sigma(b)$ and $\tau(b)$ are given by integration of Equations (1) and (2) with respect to b:

$$k_I = \frac{1}{\sqrt{\pi a}} \int_0^a \sigma(b) \cdot F_{Ip} \, db + \frac{1}{\sqrt{\pi a}} \int_0^a \tau(b) \cdot F_{Iq} \, db \tag{3}$$

$$k_{II} = \frac{1}{\sqrt{\pi a}} \int_0^a \sigma(b) \cdot F_{IIp} \, db + \frac{1}{\sqrt{\pi a}} \int_0^a \tau(b) \cdot F_{IIq} \, db \ , \tag{4}$$

where F_{Ip}, F_{Iq}, F_{IIp} and F_{IIq} are functions of a/w, b/a and θ.

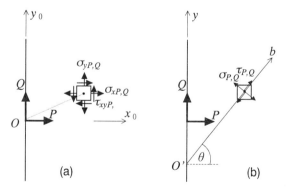

Figure 4 - *The stresses in an uncracked semi-infinite plate subjected to localized forces P and Q acting on the boundary.*

The values of $\sigma(b)$ and $\tau(b)$ are calculated according to Timoshenko and Goodier [4] in the following manner. The internal normal stresses σ_{xP}, σ_{yP} and tangential stress τ_{xyP} due to external normal force per unit thickness, P, in Figure 4(a), are calculated by Equations (5) to (7) given below, where the origin of the coordinate is set at the applied point of the force P:

$$\sigma_{xP} = -\frac{2P}{\pi} \frac{x_0^3}{(x_0^2 + y_0^2)^2} \tag{5}$$

$$\sigma_{yP} = -\frac{2P}{\pi} \frac{x_0 y_0^2}{(x_0^2 + y_0^2)^2} \tag{6}$$

$$\tau_{xyP} = -\frac{2P}{\pi} \frac{x_0^2 y_0}{(x_0^2 + y_0^2)^2} \ . \tag{7}$$

The internal normal stress σ_P and tangential stress τ_P are on the assumed crack line, inclined at the angle θ to the normal to the boundary. The coordinate b is set along the assumed crack line from the origin O', as shown in Figure 4(b). By substituting the following relations,

$$x_0 = b \cos \theta \tag{8}$$
$$y_0 = b \sin \theta - y \ , \tag{9}$$

into Equations (5) to (7), and using Equations (10) and (11),

$$\sigma_P = \sigma_{xP} \sin^2 \theta + \sigma_{yP} \cos^2 \theta - 2\tau_{xyP} \sin \theta \cos \theta \tag{10}$$
$$\tau_P = (\sigma_{yP} - \sigma_{xP}) \sin \theta \cos \theta + \tau_{xyP} (\cos^2 \theta - \sin^2 \theta) \ , \tag{11}$$

then stresses σ_P and τ_P are given by the following functions of b, θ and y:

$$\sigma_P = P \cdot f_{P\sigma}(b, \theta, y) = \frac{-2Pby^2 \cos^3 \theta}{\pi(b^2 + y^2 - 2by \sin \theta)^2} \tag{12}$$

$$\tau_P = P \cdot f_{P\tau}(b, \theta, y) = \frac{2Pby \cos^2 \theta (b - y \sin \theta)}{\pi(b^2 + y^2 - 2by \sin \theta)^2} \ . \tag{13}$$

In a similar way, the internal stresses due to external tangential force per unit thickness, Q, are given by

$$\sigma_{xQ} = -\frac{2Q}{\pi} \frac{x_0^2 y_0}{(x_0^2 + y_0^2)^2} \tag{14}$$

$$\sigma_{yQ} = -\frac{2Q}{\pi} \frac{y_0^3}{(x_0^2 + y_0^2)^2} \tag{15}$$

$$\tau_{xyQ} = -\frac{2Q}{\pi}\frac{x_0 y_0^2}{(x_0^2 + y_0^2)^2} \quad . \tag{16}$$

Finally the stresses σ_Q and τ_Q are given as follows:

$$\sigma_Q = Q \cdot f_{Q\sigma}(b,\theta,y) = \frac{2Qy^2\cos^2\theta(y - b\sin\theta)}{\pi(b^2 + y^2 - 2by\sin\theta)^2} \tag{17}$$

$$\tau_Q = Q \cdot f_{Q\tau}(b,\theta,y) = \frac{Qy\cos\theta\{by(-3 + \cos 2\theta) + 2\sin\theta(b^2 + y^2)\}}{\pi(b^2 + y^2 - 2by\sin\theta)^2} \quad . \tag{18}$$

Stresses $\sigma(b)$ and $\tau(b)$ in Equations (3) and (4) are replaced by σ_P and τ_P of Equations (12) and (13), then the differential stress intensity factors due to normal localized force P acting on the boundary of the plate, dK_{IP} and dK_{IIP} are given by

$$dK_{IP} = \frac{P}{\sqrt{\pi a}}\int_0^a (f_{P\sigma} \cdot F_{1p} + f_{P\tau} \cdot F_{1q})db \tag{19}$$

$$dK_{IIP} = \frac{P}{\sqrt{\pi a}}\int_0^a (f_{P\sigma} \cdot F_{IIp} + f_{P\tau} \cdot F_{IIq})db \quad . \tag{20}$$

The differential stress intensity factors due to tangential localized force Q acting on the edge of the plate, dK_{IQ} and dK_{IIQ} are also given using stresses σ_Q and τ_Q of Equations (17) and (18) as stresses $\sigma(b)$ and $\tau(b)$ in Equations (3) and (4) respectively,

$$dK_{IQ} = \frac{Q}{\sqrt{\pi a}}\int_0^a (f_{Q\sigma} \cdot F_{1p} + f_{Q\tau} \cdot F_{1q})db \tag{21}$$

$$dK_{IIQ} = \frac{Q}{\sqrt{\pi a}}\int_0^a (f_{Q\sigma} \cdot F_{IIp} + f_{Q\tau} \cdot F_{IIq})db \quad . \tag{22}$$

Stress intensity factors due to contact pressure and friction stress on the boundary of the plate

Therefore, the stress intensity factors due to distributed contact pressure and friction stress in the contact surface can be obtained using Equations (19) through (22). By integrating Equations (19) to (22) with respect to y assuming that contact pressure between y_1 and y_2 is represented as $\sigma(y)$ in Figure 5, the stress intensity factors subjected to contact pressure, $K_{I\sigma}$ and $K_{II\sigma}$, are obtained:

$$K_{I\sigma} = \int dK_{IP} = \frac{1}{\sqrt{\pi a}}\int_{y_1}^{y_2}\int_0^a \sigma(y) \cdot (f_{P\sigma} \cdot F_{1p} + f_{P\tau} \cdot F_{1q})db\,dy \tag{23}$$

$$K_{II\sigma} = \int dK_{IIP} = \frac{1}{\sqrt{\pi a}}\int_{y_1}^{y_2}\int_0^a \sigma(y) \cdot (f_{P\sigma} \cdot F_{IIp} + f_{P\tau} \cdot F_{IIq})db\,dy \quad . \tag{24}$$

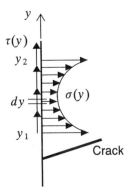

Figure 5 - *Stress distribution in the contact region.*

Similarly, by adopting friction stress $\tau(y)$ in Figure 5, stress intensity factors due to friction stress, K_{Ir} and K_{IIr}, are given by the following expressions:

$$K_{Ir} = \int dK_{IQ} = \frac{1}{\sqrt{\pi a}} \int_{y_1}^{y_2}\int_0^a \tau(y)\cdot(f_{Q\sigma}\cdot F_{Ip} + f_{Q\tau}\cdot F_{Iq})dbdy \tag{25}$$

$$K_{IIr} = \int dK_{IIQ} = \frac{1}{\sqrt{\pi a}} \int_{y_1}^{y_2}\int_0^a \tau(y)\cdot(f_{Q\sigma}\cdot F_{IIp} + f_{Q\tau}\cdot F_{IIq})dbdy \quad . \tag{26}$$

Stress intensity factors due to fatigue stress

The stress intensity factors, K_{If} and K_{IIf}, due to fatigue stress applied at the end of the plate, are obtained from boundary element analysis separately, and they are formulated as the following relations:

$$K_{If} = \sigma_f \sqrt{\pi a}\cdot F_{If}\left(\frac{a}{w},\theta\right) \tag{27}$$

$$K_{IIf} = \sigma_f \sqrt{\pi a}\cdot F_{IIf}\left(\frac{a}{w},\theta\right) \quad . \tag{28}$$

The functions F_{If} and F_{IIf} were formulated as multi-term expressions.

Total stress intensity factors due to fretting fatigue loading

The total stress intensity factors, K_I and K_{II}, are obtained by the sum of three components for each mode. They are given by

$$K_I = K_{I\sigma} + K_{Ir} + K_{If} \tag{29}$$

$$K_{II} = K_{II\sigma} + K_{IIr} + K_{IIf} \quad . \tag{30}$$

Detailed expressions

Results of boundary element analysis

The boundary element analysis was carried out to obtain Equations (1) and (2) for the following conditions: at the constant of $h = 4w$, the ratio a/w of 0.005, 0.010, 0.016, 0.025, 0.040, 0.0625, 0.10, 0.16 and 0.25, crack angle θ of 0, 15, 30 and 45 degrees, and the twenty sets of b/a ranging from 0.025 to 0.975 were assumed. For determining the exact values of stress intensity factors by boundary element method, the proportional stress method proposed by Yuki et al. [5] was adopted. Figure 6 shows an example of the relations between the functions F_{Ip}, F_{Iq}, F_{IIp} and F_{IIq} in Equations (1) and (2) and loading point on crack surface, b/a, changing with θ at the constant $a/w = 0.005$.

Expression of Green's functions by interpolation

The differential stress intensity factors, dk_I and dk_{II}, obtained from the boundary element analysis were summarized by interpolation using a numerical software *Mathematica*. The multi-terms interpolation with a/w, b/a and θ for four factors of F_{Ip}, F_{IIq}, F_{Iq}, and F_{IIp} in Equations (1) and (2) was determined by the least-squares method. The expressions are given by the following power series functions:

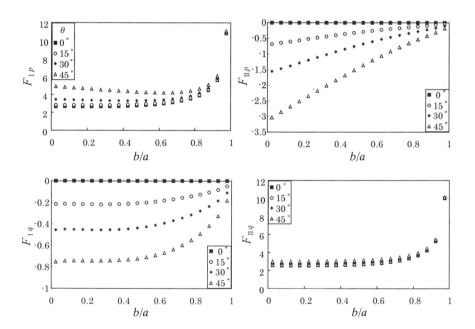

Figure 6 - *Functions of differential stress intensity factors obtained from boundary element analysis.*

$$F_{\mathrm{Ip,\ IIq}}\left(\frac{a}{w},\frac{b}{a},\theta\right) = \frac{1}{\sqrt{1-\left(\frac{b}{a}\right)^2}} \sum_{i,j,k} c_{\mathrm{Ip,IIq}\ ijk}\left(\frac{a}{w}\right)^i\left(\frac{b}{a}\right)^j \theta^k \qquad (31)$$

$$F_{\mathrm{Iq,\ IIp}}\left(\frac{a}{w},\frac{b}{a},\theta\right) = \sum_{i,j,k} c_{\mathrm{Iq,IIp}\ ijk}\left(\frac{a}{w}\right)^i\left(\frac{b}{a}\right)^j \theta^k \quad , \qquad (32)$$

where the indices i, j and k are integer values of 0 to 3, 0 to 8, and 0 to 3, respectively, and c_{Ipijk}, c_{IIqijk}, c_{Iqijk} and c_{IIpijk} are constants.

Finally, the Green's functions to obtain stress intensity factors subjected to contact pressure and friction stress are given by the following expressions with the term of $y/(y+a)$:

$$dK_{\mathrm{IP,\ IIP}} = \frac{P}{\sqrt{\pi a}} \cdot \sum_{i,j,k} C_{\mathrm{IP,IIP}\ ijk}\left(\frac{a}{w}\right)^i \theta^j \left(\frac{y}{y+a}\right)^k \qquad (33)$$

$$dK_{\mathrm{IQ,\ IIQ}} = \frac{Q}{\sqrt{\pi a}} \cdot \sum_{i,j,k} C_{\mathrm{IQ,IIQ}\ ijk}\left(\frac{a}{w}\right)^i \theta^j \left(\frac{y}{y+a}\right)^k \quad , \qquad (34)$$

where i, j and k are integer values of 0 to 4, 0 to 3, and 0 to 10, respectively. Figure 7 shows curves for the case of a/w=0.005, for Equations (33) and (34).

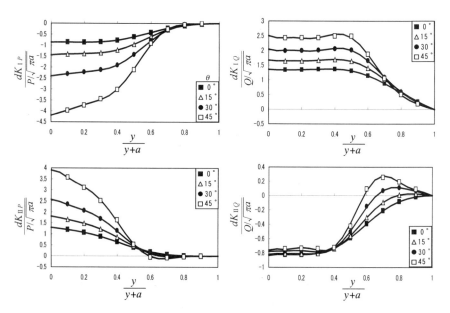

Figure 7 - *Stress intensity factors due to localized forces.*

The functions F_{If} and F_{IIf} for the fatigue stress in Equations (27) and (28) are given as

$$F_{If,IIf}\left(\frac{u}{w},\theta\right) = \sum_{i,j} d_{I,II\ ij}\left(\frac{a}{w}\right)^i \theta^j \quad , \tag{35}$$

where i and j are integers assuming values 0 to 4, and d_{Iij} and d_{IIij} are constants.

Introducing contact pressure and friction stress

In the present paper, the distribution of contact pressure was assumed by the following fourth order polynomial:

$$\sigma(y) = p_0 + p_1 y + p_2 y^2 + p_3 y^3 + p_4 y^4 \tag{36}$$

where p_0, p_1, p_2, p_3 and p_4 are constants. The friction stress is simply expressed using coefficient of friction μ, that is,

$$\tau(y) = \mu \cdot \sigma(y) \quad . \tag{37}$$

By introducing Equations (36) and (37) into Equations (23) to (26), adopting Equations (33) and (34) as dK_{IP}, dK_{IIP}, dK_{IQ} and dK_{IIQ}, and then executing the integration between y_1 and y_2, fretting pad area, and adding the components of fatigue stress, given by Equation (35), the total values of stress intensity factors, K_I and K_{II}, by Equations (29) and (30) can be obtained.

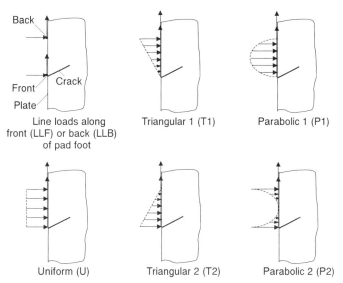

Figure 8 - *Assumed distribution of frictional and normal pad loads.*

Calculation of stress intensity factors in fretting fatigue

In order to compute the stress intensity factors from the above general expressions a computer program has been developed by using Visual Fortran (ver. 6.5). The stress intensity factors K_I and K_{II} obtained from this program have been compared with the results in reference [3]. Figure 8 shows the assumed pad load distributions in reference [3], and Figure 9 shows the relations between mode I stress intensity factors, $K_{I\sigma}$ and K_{Ir}, and crack length a under assumed distributions of normal and frictional pad load respectively. The values of these stress intensity factors were obtained using the parameters of plate width of 25 mm, pad width of 1.27 mm and thickness of 8mm, assuming in each case a load of 1 N, and the results agreed with those of reference [3]. Figures 10 and 11 show the cases of oblique crack under the same loading condition. Figure 10 shows the mode I stress intensity factors for crack angle of 30 degrees, and a comparison with Figure 9 shows that the absolute values of stress intensity factors for 30 degrees crack are higher than those for normal crack. Figure 11 is the case of mode II, and the different curves for each distribution are represented.

Figures 12 (a) and (b) show the total stress intensity factor of mode I for crack angles of 0 degrees and 30 degrees under normal pad load of 1250 N, frictional pad load of 700 N, and bulk fatigue stress of 35 MN/m^2. Each of the curves represents the stress

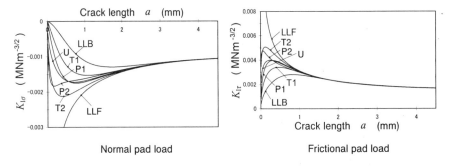

Figure 9 - *Mode I stress intensity factors of normal crack for different distributions.*

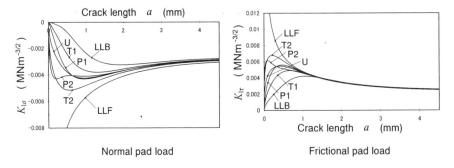

Figure 10 - *Mode I stress intensity factors of oblique crack for different distributions.*

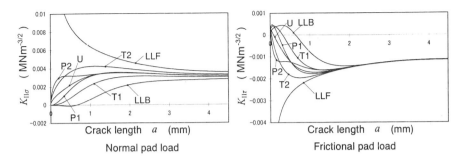

Figure 11 - *Mode II stress intensity factors of oblique crack for different distributions.*

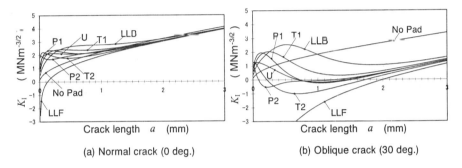

Figure 12 - *Total stress intensity factors K_I.*

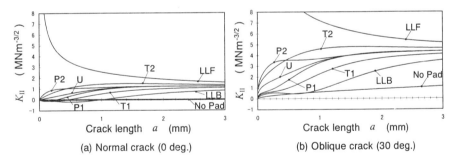

Figure 13 - *Total stress intensity factors K_{II}.*

intensity factor for assumed pad load distributions and also for plain fatigue without fretting. It was found that the stress intensity factor K_I in fretting fatigue, except for assumed line load at the front of the pad foot (LLF), has greater value in the short crack region than that without fretting, but in the long crack region their relation is reversed, and the reversing points vary with crack inclination angle. In Figure 12 (b), the values of

K_I initially increase then decrease markedly with increasing crack length, due to the greater negative values of stress intensity factor $K_{I\sigma}$ in normal pad load in Figure 10. This may change the crack propagation direction or form the non-propagating crack. Figures 13 (a) and (b) show the total stress intensity factor of mode II, K_{II}. As can be seen, the value of K_{II} changes markedly depending on the assumed pad load distributions and crack angles. As the results, the combination of K_I and K_{II} affects the crack initiation and propagation morphologies in fretting fatigue.

The above calculations show that the values of stress intensity factors are significantly affected by crack angle and also pad load distribution. That is, in Figures 12 and 13, the two extreme pad load distributions, LLF and LLB, at 30 degrees give the maximum and the minimum values of stress intensity factors of mode I and mode II, and other pad load distributions give the values in the range. It is noted that the studies on fretting fatigue crack growth must be carried out by using the actual pad load distributions. The program developed in this paper can very easily obtain the values of stress intensity factors for normal and oblique cracks by inputting the pad load distribution as a function of a fourth order polynomial. Unfortunately, a finite or boundary element stress analysis for determining pad load distribution is required before using this program separately. Therefore, the authors are now trying to develop the advanced program which can determine the actual pad load distributions automatically in a program execution.

Conclusions

In order to obtain stress intensity factors, K_I and K_{II}, in fretting fatigue conditions easily, the authors have derived the general expressions formulated through determining Green's functions and boundary element analysis, and executing their integrals. By using the general expressions, stress intensity factors can be calculated by inputting the values of crack length, crack angle, magnitude and distribution of contact pressure and friction stress, coefficient of friction, fatigue stress and plate size. The general expressions are applicable to the study of fretting fatigue crack propagation mechanism.

References

[1]   Rooke, D. P. and Jones, D. A., "Stress Intensity Factors in Fretting Fatigue," *Journal of Strain Analysis*, Vol. 14, No. 1, 1979, pp. 1-6.
[2]   Kimura, T. and Sato, K., "Stress intensity factors for oblique cracks in fretting fatigue," *Computational Methods in Contact Mechanics IV*, L. Gaul and C. A. Brebbia, Ed., WIT Press, Southampton, 1999, pp. 303-312.
[3]   Edwards, P. R., "The Application of Fracture Mechanics to Predicting Fretting Fatigue," *Fretting Fatigue*, R. B. Waterhouse, Ed., Applied Science Publishers, London, 1981, pp. 67-97.
[4]   Timoshenko, S. P. and Goodier, J. N., *Theory of Elasticity third edition*, McGraw-Hill, 1970, pp. 97-100.
[5]   Kisu, H., Yuki, R. and Kitagawa, H., "Boundary element analysis for stress intensity factors of surface crack," *Transactions*, Japan Society of Mechanical Engineers (A), Vol. 51, 1985, pp. 660-669.

D. R. Swalla[1] and R.W. Neu[1]

Characterization of Fretting Fatigue Process Volume Using Finite Element Analysis

REFERENCE: Swalla, D. R. and Neu, R. W., **"Characterization of Fretting Fatigue Process Volume Using Finite Element Analysis,"** *Fretting Fatigue: Advances in Basic Understanding and Applications, STP 1425*, Y. Mutoh, S. E. Kinyon, and D. W. Hoeppner, Eds., ASTM International, West Conshohocken, PA, 2003.

ABSTRACT: Fretting fatigue damage is characterized by the nucleation of cracks very early in the life of a component along a critical plane at an oblique angle to the contact surface. The number of cycles to fretting fatigue crack nucleation as well as the orientation of the initial crack critical plane is strongly influenced by the large stress and strain gradients that occur within a small volume near the fretting contact. Therefore, it is necessary to average the stresses and strains along a critical plane within a fretting fatigue process volume (FFPV) to obtain more reliable crack nucleation predictions. An approach for computing damage in a FFPV using finite element analysis is presented. The life and crack angle prediction is consistent with experimental observations of fretting fatigue tests on PH 13-8 Mo stainless steel. This averaging method allows for the introduction of a length scale for crack nucleation prediction.

KEYWORDS: fretting fatigue, crack nucleation, finite element analysis, critical plane approaches

Introduction

Many different methods have been employed to predict damage and initial fretting fatigue crack location and/or orientation. Empirical approaches based on continuum field variables correlate the combination of relevant macroscopic variables such as stress, strain, and slip range with observed crack nucleation sites. An example is the fretting fatigue damage parameter (FFDP) developed by Ruiz et al. [1] in the mid 1980s in which the site of crack nucleation was determined based on a combination of the tangential stress component, σ_T, the shear stress, τ, and the relative slip amplitude, δ, in the form FFDP = $\sigma_T \tau \delta$. The FFDP reliably predicts the location of the crack nucleus along the fretting interface; however, no relevance is given to the orientation of the nucleated crack.

Cracks have been found to form very early in the life of a component due to fretting fatigue loading. These cracks tend to nucleate and grow along particular planes dependent upon the cyclic state of stress and the material's microstructure [2,3]. A

---

[1] Graduate research assistant and associate professor, respectively. The George W. Woodruff School of Mechanical Engineering, Georgia Institute of Technology, Atlanta, GA, 30332-0405, USA.

micrograph of a PH 13-8 Mo stainless steel fretting fatigue specimen interrupted during testing at 200 cycles is shown in Figure 1 [4]. The experimental conditions represented a nominally cylindrical-on-flat contact with normal force, P = 343.4 N, remote fatigue stress amplitude, σ_a = 217 MPa, and stress ratio, R = 0.1. In Figure 1, the fretting fatigue crack extends to a depth of approximately 10 μm into the specimen, with a total length equal to about 30 μm at an angle of about 65° ± 5° as measured counter-clockwise from a line perpendicular to the surface.

Figure 1- *Subsurface cracking observed after 200 cycles.*

A fretting fatigue crack tends to change direction as it grows further into the specimen [3,5], as observed in Figure 2 [8]. In general, the fretting fatigue process can be divided into three regimes. The nucleation regime is located within a depth of approximately 50 μm from the surface. Once the crack has grown to a depth of about 50 μm, the crack angle changes from about 65° ± 5° (see Figure 1) to about 55° ± 5°. The next regime is an intermediate region, which is still strongly influenced by the frictional force. It is located between 50 μm and 200 μm from the surface and the crack angle is near 25° ± 5°. The third regime is a crack growth region located greater than 200 μm from the surface, with diminishing frictional force influence. The crack gradually reduces to 0°, which is the direction normal to the fatigue loading. Once the crack is beyond about 50 μm in this high strength steel, the growth rate can be characterized by fracture mechanics [6]. Additional experimental tests (not shown here) show that the observed changes in crack angle with increasing depth are nearly identical for different stress ratios when the normal pad force and contact geometry are the same [7,8]. These observations point to the existence of a critical volume of material near the contact that is highly influenced by the fretting loads, as evidenced by the change in crack angle as it grows further into a fretting fatigue specimen. This critical volume is referred to as the Fretting Fatigue Process Volume (FFPV).

Figure 2 - *Micrograph of primary fretting fatigue cracks in PH 13-8 Mo stainless steel.*

Critical plane approaches for describing multiaxial fatigue have been used to predict fretting fatigue damage, initial crack location and orientation [8,9,10]. A wide variety of multiaxial fatigue models exist in the literature, differing mainly in the selection of stress and/or strain component combinations based on whether the fatigued material fails in a shear or tensile manner [11,12,13,14]. Usually some fitting factor is applied in order to correlate these stresses and/or strains with the number of cycles necessary to nucleate a fatigue crack based on stress-strain life curves from either cyclic uniaxial or torsional experimental tests. It can be quite difficult to choose an appropriate prediction model because of the range of required assumptions related to issues that most directly affect the integrity of fretting fatigue analyses. Some of these issues involve (a) the properties of the interface between the two components (i.e., the coefficient of friction), (b) the material model (i.e., elastic or elastic-plastic and homogeneous vs. heterogeneous), and (c) the appropriate scale of the fretting fatigue process volume and crack length.

In recognition of these problems, the analyses presented in this paper emphasize comparison of individual stress/strain components to more easily elucidate how the combination of these components making up some common multiaxial fatigue models differ with changing FFPV depth. Two multiaxial fatigue models studied in this research are the Fatemi-Socie-Kurath (FSK) [11,12] and Smith-Watson-Topper (SWT) [15] models. The FSK model is often used for materials that tend to fail predominantly by shear-cracking. In this model, fatigue life is dependent on the maximum cyclic shear strain amplitude ($\Delta\gamma/2$) on a particular plane modified by the maximum normal stress ($\sigma_{n,max}$) to this plane.

$$\frac{\Delta\gamma}{2}\left(1+k\frac{\sigma_{n,max}}{\sigma_y}\right) = \frac{\tau_f{'}}{G}(2N_i)^{b_o} + \gamma_f{'}(2N_i)^{c_o} \qquad (1)$$

The loading parameters defined on the plane experiencing the largest range of cyclic shear strain are represented on the left-hand side of the equation. The $\sigma_{n,max}$ is normalized by σ_y, the cyclic yield strength, to preserve the dimensionless features of strain, and k is a material dependent parameter. The ratio σ_y/k is often replaced by the fatigue strength coefficient ($\sigma_f{'}$), which has been found to be nearly equal. The strain-life curve generated from fully-reversed torsional fatigue tests is represented by the right-hand side, where $\tau_f{'}$ is the shear fatigue strength coefficient, $\gamma_f{'}$ is the shear fatigue ductility coefficient, b_o is the shear fatigue strength exponent, c_o is the fatigue ductility exponent, G is the shear modulus, and $2N_i$ is the number of reversals to the formation of a surface crack of a certain length.

Alternatively, the SWT model is used for materials that fail predominantly by tensile cracking and is defined as follows:

$$\frac{\Delta\varepsilon}{2}\sigma_{n,max} = \frac{\sigma_f{'}^2}{E}(2N_i)^{2b} + \sigma_f{'}\varepsilon_f{'}(2N_i)^{b+c} \qquad (2)$$

where $\Delta\varepsilon/2$ is the maximum principal strain amplitude on a plane and $\sigma_{n,max}$ the maximum normal stress during the cycle on this plane. The right-hand side of the equation is similar in form to the Fatemi-Socie-Kurath model (Eq. 1), but is correlated with strain-life data generated from uniaxial testing.

Both models were developed for fatigue prediction of cracks about 1 mm in length. In this case, one could argue that an arbitrary length scale has already been applied (e.g.

crack length equal to 1 mm). However, fretting fatigue crack nucleation occurs at a much smaller scale approaching the grain size for a given material. Microstructural approaches that are currently being investigated hold much promise to predict damage at this scale, but are not sufficiently well-developed at this time [16].

In the past, most fretting fatigue damage prediction methods relied on calculation of stress and strain results obtained at a point along the fretting fatigue contact surface. However, the mechanisms that drive crack nucleation under fretting fatigue conditions are well understood to be dependent upon a volume of material near the contact, not just the stresses and strains at a point [17,18,19,20]. An engineering level analysis using macroscopic material properties is still just an approximation of fretting fatigue mechanisms occurring at a microstructural scale. For that reason, a number of averaging methods, such as "line" averaging, critical distance, surface averaging, and volume averaging, have been developed in an attempt to "smear" the effect of high stress/strain gradients over some finite damage area/volume [21,22]. These methods allow for the introduction of a length scale to engineering level analysis, albeit in a simplistic way. The use of "line" averaging of stresses and strains in a 2-D analysis or plane averaging in 3-D analyses seems to be more relevant when evaluating multiaxial fatigue parameters to predict maximum fretting fatigue damage, initial crack location, and crack orientation along a particular critical plane.

Emphasis is placed on prediction of the fretting fatigue crack orientation because it is desirable to be able to predict the fretting fatigue crack orientation from both a model validation standpoint as well as from an actual physical standpoint. First, when conducting analysis of any kind it is necessary to correlate as many results as possible with experimental test observations in order to validate the model. The ability to predict both the location of crack nucleation and orientation gives one faith that the model captures the "right" mechanisms. Second, depending on the geometry in actual components, the direction of crack growth may mean the difference between cracking that results in catastrophic failure or alternatively, a crack that may eventually stop propagating.

In this paper, the stress and strain results of an elastic finite element analysis (FEA) are averaged along a potential critical plane for a range of depths into the specimen, representing a FFPV. This volume averaging approach is used to determine the relative influence of individual stress/strain components on crack angle prediction at various depths and will be discussed in detail. Specifically, critical values of maximum average shear strain amplitude ($\overline{\Delta\gamma_c}/2$), maximum average normal strain amplitude ($\overline{\Delta\varepsilon_N}/2$), and average maximum normal stress ($\overline{\sigma}_{N,\max}$), as well as the FSK and SWT multiaxial fatigue parameters are calculated. These results are then compared to interrupted fretting fatigue experimental results to show how the individual stress/strain components that make up each parameter affect the crack angle and subsequent life prediction values within a certain volume. Finally, the stress amplitude, stress ratio, and material model (elastic vs. elastic/plastic) are varied in the FEA to determine their influence on life and crack angle prediction within a FFPV using a volume averaging approach. It will be shown that the initial crack angle prediction is dependent upon the size of the FFPV under consideration and the dominant stress or strain component calculated within this volume.

Finite element model and experimental configuration

Finite element modeling was used to simulate a fatigue specimen clamped between two bridge-type fretting pads and loaded using a calibrated proving ring. The finite element model shown in Figure 3(a) represents a symmetric section of this fretting fatigue assembly. Experimental test details and specimen dimensions are listed elsewhere [7,23]. The fretting pad is cylindrical with a radius of curvature equal to 15 mm. A mapped mesh with element width of 10 μm was used near the area of contact in a region 1 mm (length) x 0.5 mm (depth), and a free mesh was used elsewhere. A close-up of the mesh near the contact is shown in Figure 3(b) along with the angle convention used in the analyses. The positive angle direction is counter-clockwise. The 0° reference is perpendicular to the bulk fatigue loading (σ) direction.

The ABAQUS users manual states that a body initially in contact at one point, which is the case for a cylindrical contact, must be grounded with soft springs or dashpots to eliminate rigid body motion in the early steps of the analysis [24]. A springy layer of reduced stiffness elements has been added above the pad to perform this function, as well as to model the compliance of the proving ring assembly in the experimental configuration [25]. The elements along the top of the springy layer are fixed in the x- and y-directions. The pad and specimen are fixed in the x- and y-directions along lines of symmetry. ABAQUS CPE8R elements, which are 8-noded biquadratic quadrilateral 2-D solid plane strain reduced integration elements, were used. Conditions of plane strain were considered appropriate because the area of contact is much smaller than the width of the specimen. Contact pair elements with small sliding assumption are used for the contact elements between the fretting pad and the specimen. The specimen is defined as the master surface and the fretting pad is defined as the slave surface. A Lagrange multiplier formulation is used to enforce the sticking constraints along the interface between the two contacting surfaces.

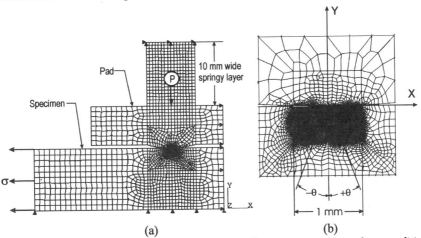

(a)   (b)

Figure 3 - *Finite element model showing (a) overall geometry and boundary conditions, and (b) mesh near contact area.*

The material properties of the pad and specimen are representative of PH 13-8 Mo stainless steel. All material constants (fatigue, kinematic hardening, and elastic constants) were obtained through experimental tests on the same heat of material used in the fretting fatigue experimental tests [23]. The elastic constants are E = 192 GPa and ν = 0.3. The modulus of the springy layer was set to 0.001 * E, which was determined by trial-and-error to match the ideal elastic Hertzian contact solution before addition of the bulk fatigue loading. The theoretical Hertzian contact width (2a) is equal to 0.26 mm for these loading conditions (2a/R = 0.017). There are approximately 30 eight-noded elements encompassing the theoretical Hertzian contact width. A mesh density of this order has been proven to capture the qualitative behavior of plastic strain evolution [26].

The elastic solution before addition of the bulk fatigue loading was verified via Hertzian contact theory [27,28]. The error in the stresses along the interface was less than 1.0%. For our exact experimental configuration and boundary conditions, there are no analytical solutions that are presently available to perform further direct validation. The classic Mindlin solution used by others performing similar analyses [25,29,30] does not exactly apply because of the presence of the bulk fatigue loading, which cannot be uncoupled from the shearing (or frictional) force in our model. Furthermore, when the bulk stress is included in the analytical analysis, it is assumed to be uniform in the fatigue specimen [30]. In bridge-type apparatus the bulk stress varies along the length of the contact. We chose to more closely represent our experimental conditions in order to make closer links to experimental and model response; therefore, the mesh refinement method was used to assure convergence for the elastic and elastic-plastic solution after addition of the bulk loading.

The plastic behavior for selected analyses is described using a nonlinear kinematic hardening model available in ABAQUS [24] with the evolution equation having the form:

$$\dot{a} = C\frac{1}{\sigma^0}(\sigma - \alpha)\dot{\bar{\varepsilon}}_{pl} - \gamma\alpha\dot{\bar{\varepsilon}}_{pl} \qquad (3)$$

The constants are obtained using the cyclic stress-strain curve. For PH13-8 Mo stainless steel, C = 157.3 GPa, γ = 381.8 MPa, and σ^0 = 909 MPa. The 0.2% offset monotonic yield strength is 1286 MPa [23].

Stress and strain averaging methodology within a FFPV

The averaging approach is illustrated in Figure 4. Mathematically, this line averaging method is similar to that used by Lamacq *et al.* [19] except that the stress/strain field in the specimen is obtained via FEA in this case.

A potential critical plane is represented by a line with a particular length (L), starting at a point on the surface. The number of interpolation points (N) is then selected, making sure that the minimum resolution is at least equal to the FEA mesh density. The stress and strain values are extracted from the FEA at the centroid of the elements for every increment in a cycle, and then extrapolated to equally distributed points on the line. Increasing the number of these points to a resolution that is much denser than the mesh size in the FEA does not result in greater accuracy because the stress/strain data can only be obtained at discrete points (such as element centroids) in the FEA. However, steps were taken to insure that changing the number of points would not change the averaging results. For example, doubling the number of points beyond the minimum resolution resulted in a percent difference of no more than 0.6% for line lengths up to 400 μm and up to 100 points.

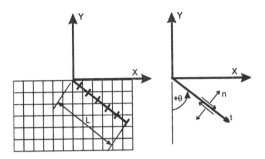

Figure 4 - *Line averaging method*

The stress/strain values at the points are then transformed at a particular angle and averaged along the "plane" for every increment in the cycle. This sequence is repeated for angles corresponding to -90° to 90° in increments of 2° to encompass a FFPV of depth equal to radial length L. Finally, the predicted *critical* plane orientation is the angle at which the average stress/strain, or combination thereof (depending on the chosen multiaxial fatigue parameter), is a *maximum* along a plane of length L. Results shown in subsequent sections are presented in terms of the FFPV radial length L, which corresponds directly with the FFPV depth into the specimen. By varying the radial length in the analyses, the influence of individual stress/strain components on the accuracy of different multiaxial fatigue parameters used in fretting fatigue damage prediction for a given crack length scale can be evaluated.

Both the FSK and SWT models were used to determine the critical plane using the line averaging method just presented. In this case, the stress and strain components in either model are replaced by their respective averaged values.

Exercises and Discussion

Critical plane and life prediction

For the results in this section, the stress/strain response was obtained using an FEA with an elastic material model and prescribed loading conditions matching the experimental fretting fatigue conditions described earlier (see Figs. 1 and 2). The normal force (P) equal to 343.4 N, was first imposed and then three full cycles were modeled by incrementing ten load steps during the analysis. Three complete load cycles are considered sufficient to capture steady-state response [26,31]. The remote fatigue stress amplitude was 217 MPa with stress ratio, R = 0.1, which corresponds to maximum and minimum fatigue loads of 34.9 kN and 3.49 kN, respectively. The prescribed coefficient of friction is 1.5; previous analyses have shown that this value gives results that correlate well with the observed experimental deformation field [32].

As a baseline, the stress/strain response at discrete points along the interface is evaluated in order to subsequently illustrate the importance of considering a volume of material near the contact. The values of the maximum shear strain amplitude ($\Delta\gamma/2_{max}$), FSK, and SWT parameter, along different planes at the point of maximum damage along the interface are shown in Figure 4. The angle convention is shown in Figure 3(b). Previous research [8] showed that the $\Delta\gamma/2_{max}$, FSK, and SWT parameters predict the point of maximum damage at x/a = -0.72 ± 0.10.

The orientation of the predicted critical plane is located by determining where the maximum damage value for a given parameter (y-axis) coincides with a particular angle (x-axis). For instance, the ($\Delta\gamma/2_{max}$) predicts two planes having damage that are nearly equal when calculated at a point. The orientation of these critical planes are 65° and -30°. The SWT parameter predicts a critical angle of 15°, which is much steeper than what is indicated by experiments (see Figs. 1 and 2). The FSK parameter predicts maximum damage at an angle of -25°, which also is not observed experimentally for this case.

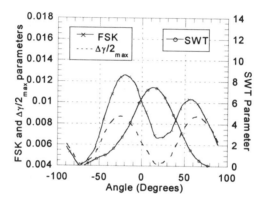

Figure 4 - *Values of $\Delta\gamma/2_{max}$, FSK, and SWT parameters on different planes at the point of maximum damage on the interface (x/a = -0.72 ±0.10).*

The full-field $\Delta\gamma/2_{max}$ calculated at the centroid of the FEM elements is shown in Figure 5. Near the fretting interface, the gradient in strain amplitude is quite large. The calculation of stress/strain at a point on the surface does not capture the physical significance of the effect that this steep gradient has on fretting fatigue damage within a critical volume near the contact. The results shown in Figures 4 and 5 illustrate why using cyclic stresses and strains at a point do not lead to accurate initial angle predictions and also brings into question whether using stress/strain response at a point can be successfully used for life prediction as well. The remainder of this section is devoted to showing how critical values of maximum average shear strain amplitude ($\overline{\Delta\gamma_c}/2$),

maximum average normal strain amplitude ($\overline{\Delta\varepsilon_N}/2$), average maximum normal stress

($\overline{\sigma}_{N,max}$), as well as the FSK and SWT multiaxial fatigue parameters influence initial crack angle and life prediction using the volume averaging approach outlined in the previous section.

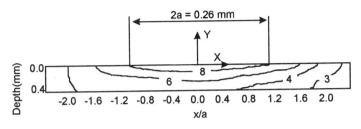

Figure 5 - *Full-field $\Delta\gamma/2_{max}$ (x 10^{-3}) near fretting fatigue contact.*

The $\overline{\Delta\gamma_c}/2$ and $\overline{\Delta\varepsilon_N}/2$, calculated for a FFPV radii equal to L = 20, 40, 100, and 400 µm, are shown in Figure 6, along with the critical angle predictions. Note that the lines describing the $\overline{\Delta\gamma_c}/2$ and $\overline{\Delta\varepsilon_N}/2$ magnitudes have similar slope and drop significantly with depth into the specimen. At any length, the magnitude of $\overline{\Delta\gamma_c}/2$ is nearly double that of $\overline{\Delta\varepsilon_N}/2$. Furthermore, the values of either $\overline{\Delta\gamma_c}/2$ or $\overline{\Delta\varepsilon_N}/2$ at L = 20 µm are double their respective values when L = 400 µm. The critical angle is the angle at which the peak damage is predicted (similar to Figure 4). The number of points averaged on a given plane are 5, 10, 15, and 50 points, respectively, for each of these line lengths. The starting point on the interface is located at x/a = -0.72, which was earlier calculated as the point of maximum damage. The angle prediction at L ≤ 40 µm is 70° using $\overline{\Delta\gamma_c}/2$, which corresponds well to experimental observations of fretting cracks of this length (see Figs. 1 and 2). Recall that two critical planes are predicted, as shown earlier in Figure 4, when $\Delta\gamma/2_{max}$ is calculated at a point. However, when using $\overline{\Delta\gamma_c}/2$ at any given depth, only one critical value is predicted.

As the radius is increased beyond 100 µm, the predicted critical angle approaches 90° for $\overline{\Delta\gamma_c}/2$, which indicates that the severity of the shear strain gradient *into* the specimen has diminished significantly. For $\overline{\Delta\varepsilon_N}/2$, the predicted critical angle stays relatively constant (between 20° and 30°) for any L ≤ 400 µm and correctly predicts the crack angle once it has grown to L > 100 µm. It is not predictive when L < 100 µm.

The $\overline{\sigma}_{N,max}$ along with the critical angle prediction for FFPV radii ranging in length from 20 µm ≤ L ≤ 400 µm is shown in Figure 7. The value of $\overline{\sigma}_{N,max}$ at L = 20 µm is nearly double the value at L = 400 µm, which is similar to the trend in shear and normal strains (see Figure 6). However, much of the reduction (about 60%) in $\overline{\sigma}_{N,max}$ occurs within L < 100 µm, unlike the strains, which exhibit a more gradual change in slope.

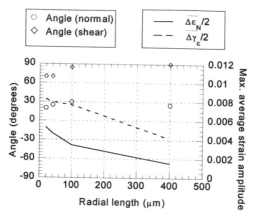

Figure 6 - *The influence of the size of the process volume on the values of $\overline{\Delta\gamma_c}/2$ and $\overline{\Delta\varepsilon_N}/2$ and the associated angle of the critical plane.*

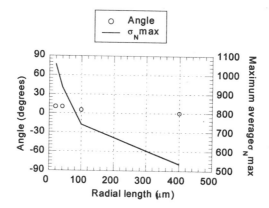

Figure 7 - *The influence of the size of the process volume on the value of $\overline{\sigma}_{N,max}$ and the angle of the critical plane.*

The predicted critical plane angle is equal to $10°$ for $L \le 40$ µm, changing to $5°$ when $L = 100$ µm, and finally to $0°$ at $L = 400$ µm. The critical angle calculated using $\overline{\sigma}_{N,max}$ is not predictive for crack lengths less than 400 µm, but does correspond to the $0°$ angle observed in experiments once the crack has grown to a length that is well outside the FFPV.

The critical plane angle predictions obtained using the FSK and SWT models are shown in Figure 8 (a) and (b) for $L = 20$ µm to $L = 400$ µm. For comparison, the critical plane predictions using $\overline{\Delta\gamma_c}/2$, $\overline{\Delta\varepsilon_N}/2$, and $\overline{\sigma}_{N,max}$ are also shown. The schematics Figure 8 (b)) illustrate how the predicted critical angle changes with radius length. Note that the predicted critical angle is negative for L ≤ 40 μm using the FSK parameter, but switches to 65° at 50 μm. When L=100 μm and greater, the FSK and $\overline{\Delta\gamma_c}/2$ critical angle predictions tend to converge. Although the FSK parameter is primarily a shear-strain based parameter, it is also influenced by the $\overline{\sigma}_{N,max}$.

The critical angle predicted using the SWT parameter is 15° when L ≤ 40 μm, jumps up to 25° when L = 50 μm, then back down to 10° or 15° for 100 μm ≤ L ≤ 400 μm. From this plot, it is not clear whether the $\overline{\Delta\varepsilon_N}/2$ or $\overline{\sigma}_{N,max}$ has a stronger influence on the SWT parameter critical angle prediction. The crack angle prediction using either parameter corresponds with the experimentally observed crack angle when 100 μm ≤ L ≤ 400 μm.

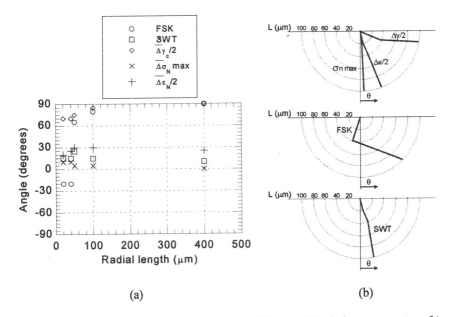

(a)   (b)

Figure 8 - (a) Predicted critical planes using different critical plane parameters (b) Schematic illustrating the angle of the critical plane with radius length.

The trend in predicted life for different parameters is shown in Figure 9. The fatigue constants were obtained from uniaxial fatigue tests on smooth specimens conducted on the same heat of material at Georgia Institute of Technology (constants given in Table 1).

Table 1- *Fatigue constants for PH 13-8 Mo stainless steel.*

σ_f'	1997 MPa
ε_f'	0.525
b	-0.074
c	-0.737

To approximate the shear-based fatigue constants from the uniaxial constants given in Table 1, the von Mises assumption is used. The relationships between the shear and normal components are:

$$\tau_f' = \frac{\sigma_f'}{\sqrt{3}} \qquad (3)$$

$$\gamma_f' = \sqrt{3}\varepsilon_f' \qquad (4)$$

where σ_f' is the fatigue strength coefficient, ε_f' is the fatigue ductility coefficient, $b_o = b$, and $c_o = c$.

The damage and crack lengths studied here are much smaller than the crack scale in standard low cycle fatigue (LCF) testing. For that reason, it is more relevant to discuss the relative differences between the various models rather than focus exclusively on the life values on individual data points. As an example, interrupted fretting fatigue experiments indicate that cracks nucleate and grow to a length of about 30 μm in as few as 200 cycles. Cracks considered long enough to indicate crack "initiation" in LCF tests are often on the order of 1 mm in length.

There is a strong correlation between the $\overline{\Delta\gamma_c}/2$ and the FSK parameter and the predicted life using $\overline{\Delta\varepsilon_N}/2$ or SWT parameters are nearly identical when 20 μm ≤ L ≤ 400 μm. The latter suggests that $\overline{\Delta\varepsilon_N}/2$ is the predominate component influencing life prediction when using the SWT parameter.

Both $\overline{\Delta\gamma_c}/2$ and FSK are shear stress/strain dominated parameters, and both $\overline{\Delta\varepsilon_N}/2$ and SWT are normal stress/strain dominated parameters. When L < 40 μm and at L = 400 μm, all of the parameters predict similar lives. However, there is a certain volume (around 100 μm in depth) in which the shear stress/strain and normal stress/strain parameters diverge.

Given these results, it is clear that the most physically realistic damage parameter is dependent on the size of the FFPV considered. First, it is clear that the $\overline{\Delta\gamma_c}/2$ is the primary component influencing the initial crack angle prediction and therefore it seems likely that it would be a logical parameter for correlating cycles to nucleation for this material when L< 40 μm. When L = 40 to 100 μm, a transitional region exists in which the normal stresses/strains, such as $\overline{\sigma}_{N,max}$ and $\overline{\Delta\varepsilon_N}/2$, become increasingly dominant, as evidenced by the crack angle predictions. This is also the region in which the trend in life prediction using a shear stress/strain or normal stress/strain parameter diverges significantly. At this scale, use of a multiaxial fatigue model that incorporates both the shear stress/strain and normal stress/strain, such as the FSK parameter, seems desirable. When L > 100 μm, the normal stresses resulting from the bulk fatigue load are clearly dominant. In this region, a tensile cracking multiaxial fatigue model similar to the SWT model seems to be more appropriate. At this length a much greater proportion of the crack length is oriented in a normal direction. Therefore, the optimal multiaxial fatigue model depends on the length scale of the initial fretting fatigue crack in question, or in other words, the size of the FFPV.

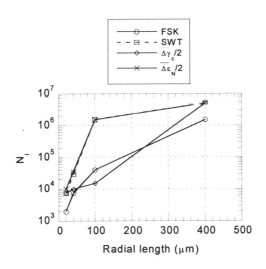

Figure 9 - *Predicted life using different critical plane parameters.*

Parametric Study

The influence of stress amplitude, stress ratio, and material model (elastic vs. elastic/plastic) on life and crack angle prediction within a FFPV using a volume averaging approach is now considered. Unless explicitly stated otherwise, all stress/strain results were obtained using elastic material properties.

Using a constant stress amplitude (σ_a) of 100 MPa, the influence of stress ratio (-1.0 ≤ R ≤ 0.5) was considered with results shown in Figure 10. Both the critical angle prediction and magnitude are rather insensitive (± 5°) to the stress ratio. On the other hand, when the stress amplitude is varied for a constant R= 0.1 (Figure 11), the difference in magnitude of $\overline{\Delta\gamma_c}/2$ is quite noticeable, with increasing stress amplitude leading to larger values of $\overline{\Delta\gamma_c}/2$. However, the difference in critical angle prediction is not significant when L > 20 μm. Even when L = 20 μm, the difference between the angle prediction when 100 MPa ≤ σ_a ≤ 250 MPa is only about 10°.

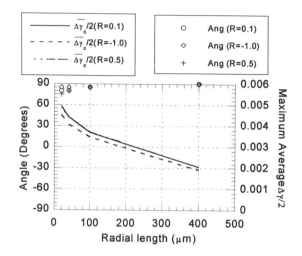

Figure 10 - *Influence of stress ratio (R) with σ_a=100 MPa on critical plane angle.*

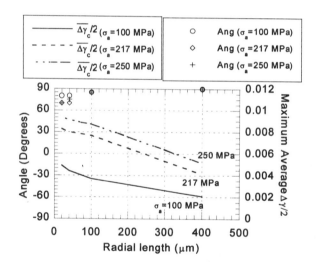

Figure 11 - *Influence of stress amplitude (σ_a) with R=0.1 on critical plane angle.*

The influence of using an elastic-plastic material model versus an elastic model on $\overline{\Delta\gamma_c}/2$ and corresponding predicted crack angle is shown in Figure 12(a). The life prediction using constants given in Table 1 is shown in Figure 12(b). The difference in $\overline{\Delta\gamma_c}/2$ between the two material models correlates with a difference in life ranging from about 5000 cycles when L = 20 μm to about 10000 cycles when L = 100 μm. Similarly, the influence of the material model on $\overline{\Delta\varepsilon_N}/2$ and the corresponding predicted angle is shown in Figure 13. The $\overline{\Delta\varepsilon_N}/2$ is less sensitive than $\overline{\Delta\gamma_c}/2$ to the choice of material modelThe FSK parameter follows similar trends as $\overline{\Delta\gamma_c}/2$ and the SWT follows the $\overline{\Delta\varepsilon_N}/2$ trends, suggesting that FSK parameter is more sensitive to an elastic-plastic material model. As shown here and in earlier figures, the use of an elastic material model may result in non-conservative life prediction. However, even though the difference in strain magnitude is somewhat significant in this case, the angle prediction is relatively insensitive to the choice of material model using any parameter.

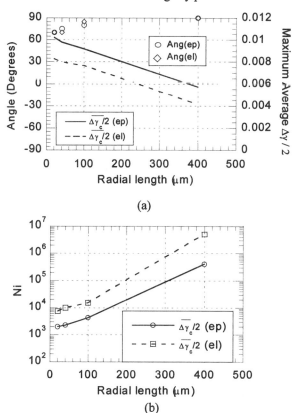

Figure 12 - *Influence of material model (elastic vs. elastic-plastic) using $\overline{\Delta\gamma_c}/2$ on (a) critical plane angle and (b) life prediction.*

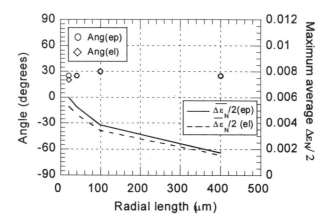

Figure 13 - *Influence of material model (elastic vs. elastic-plastic) on critical plane angle using* $\overline{\Delta \varepsilon_N}/2$.

Conclusions

Determination of the extent of the FFPV by averaging the stresses/strains along a critical plane and examining their influence on predicted critical angle orientation and life prediction for a given radial length allows for the introduction of a length scale in engineering level fretting fatigue analyses. Use of this approach results in an initial crack angle orientation prediction that correlates well with experimental observations of PH 13-8 Mo stainless steel within a certain FFPV, provided the appropriate damage parameter is used, unlike calculations using stresses/strains at a point. The angle predictions are relatively insensitive to stress ratio and stress amplitude using a homogeneous elastic or elastic/plastic material model.

The $\overline{\Delta \gamma_c}/2$ is the primary component affecting the initial crack angle prediction and cycles to nucleation for the stainless steel studied when L< 40 μm. At this scale, either the $\overline{\Delta \gamma_c}/2$ or FSK parameters do a better job of predicting the crack orientation than the $\overline{\Delta \varepsilon_N}/2$ or SWT parameters. When the FFPV encompasses a depth between L = 40 to 100 μm, a transitional region exists in which the normal stresses/strains become increasingly dominant, as evidenced by the crack angle predictions. At this scale, the use of a multiaxial fatigue model that incorporates both the shear stress/strain and normal stress/strain seems desirable. When L > 100 μm, the normal stresses resulting from the bulk fatigue load are dominant. In this region, a tensile cracking multiaxial fatigue model similar to the SWT model appears to be more appropriate. Based on these results, the most appropriate multiaxial fatigue models for fretting fatigue damage prediction from a physical point of view is dependent on the size of the FFPV considered in the analysis.

Acknowledgements

This work was part of a multi-university research initiative on Integrated Diagnostics funded by the Office of Naval Research through grants N00014-95-1-0539 and N00014-98-1-0532 with Dr. Peter Schmidt serving as program manager. The assistance of John Pape is greatly appreciated.

References

[1]   Ruiz, C., Boddington, P.H.B., and Chen, K.C., "An Investigation of Fatigue and Fretting in a Dovetail Joint", *Experimental Mechanics,* Vol. 24, No. 3, 1984, pp. 208-217.

[2]   Dubourg, M.-C., and Lamacq, V. "Stage II Crack Propagation Direction Determination Under Fretting Fatigue Loading. A New Approach in Accordance with Experimental Observations," *Fretting Fatigue: Current Technology and Practices, ASTM STP 1367,* D.W. Hoeppner, V. Chandrasekaran, and C.B. Elliott, Eds., American Society for Testing and Materials, 1999.

[3]   Endo, K., and Goto, H. "Initiation and Propagation of Fretting Fatigue Cracks," *Wear,* Vol. 38, 1976, pp. 311-324.

[4]   Pape, J.A., and Neu, R.W., "Fretting fatigue damage accumulation in PH13-8 Mo stainless steel", Submitted to International Journal of Fatigue, October 2000.

[5]   Adibnazari, S. and Hoeppner, D.W. "Study of fretting fatigue crack nucleation in 7075-T6 aluminum alloy", *Wear*, Vol. 159, 1992, pp. 257-264.

[6]   Hills D.A. and Nowell, D., *Mechanics of Fretting Fatigue.* Dordrecht, The Netherlands: Kluwer Academic Publishers, 1994.

[7]   Pape, J.A., "Design and implementation of an apparatus to investigate the fretting fatigue of Ph13-8 Mo Stainless Steel", Masters Thesis, Georgia Institute of Technology, 1997.

[8]   Neu, R.W., Pape, J.A., and Swalla, D.R., "Methodologies for Linking Nucleation and Propagation Approaches for Predicting Life under Fretting Fatigue," *Fretting Fatigue: Current Technology and Practices, ASTM STP 1367.* D.W. Hoeppner, V. Chandrasekaran, and C.B. Elliott, Eds., American Society for Testing and Materials, 2000.

[9]   Glinka G, Shen G, Plumtree A, A multiaxial fatigue strain energy density parameter related to the critical fracture plane, *Fatigue Fract. Engng. Mater. Struct.*, Vol.18, No. 1, pp. 37-46, 1995.

[10] Szolwinski, M, Farris, T., "Mechanics of fretting fatigue crack formation", *Wear*, Vol. 198, 1996, pp. 193-107.

[11] Fatemi,A, Kurath,P.Multiaxial fatigue life predictions under the influence of mean-stresses.Transactions of the ASME., Vol. 110, 1988, pp.380-388.

[12] Fatemi, A. and Socie, D., "A Critical Plane Approach to Multiaxial Fatigue Damage Including Out-of-Phase Loading, " *Fatigue and Fracture of Engineering Materials and Structures*, Vol. 11, No. 3, 1988, pp. 145-165.

[13] Socie, D., "Multiaxial Fatigue Damage Models," *J.of Eng. Mat. and Tech.* ASME, Vol. 109, 1987, pp. 293-298.

[14] Papadopoulos, I.V., "Critical Plane Approaches in High-cycle Fatigue: on the Definition of the Amplitude and Mean Value of the Shear Stress Acting on the Critical Plane," ", *Fatigue & Fracture of Engineering Materials & Structures*, Vol. 21, 1998, pp. 269-285.

[15] Smith, R.N., Watson, P., and Topper, T.H., "A Stress Strain Function for the Fatigue of Metals," *J. of Materials JMLSA*, Vol. 5, No. 4, 1970, pp. 767-778.

[16] Goh, C,-H., Wallace, J.M., Neu, R.W., and McDowell, D.L., "Polycrystal Plasticity Simulations of Fretting Fatigue," International Conference on Fatigue Damage of Structural Materials III, September 18-22, 2000, Hyannis, MA; papers to be published in International Journal of Fatigue, submitted for review, September 2000.

[17] Dang Van, K., (1993) "Macro-Micro Approach in High-Cycle Multiaxial Fatigue", *Advances in Multiaxial Fatigue, ASTM STP 1191,* D.L. McDowell and R. Ellis, Eds.

[18] Fouvry, S., Kapsa, P., and Vincent, L., "A multiaxial fatigue analysis of fretting contact taking into account the size effect", *Fretting Fatigue: Current Technology and Practices, ASTM STP 1367,* D.W. Hoeppner, V. Chandrasekaran, and C.B. Elliott, Eds., American Society for Testing and Materials, 1999.

[19] Lamacq, V., Dubourg, M., Vincent, L., "A theoretical model for the prediction of initial growth angles and sites of fretting fatigue cracks", *Tribology International*, Vol. 30, No. 6, 1997, pp. 391-400.

[20] Mura, T., and Nakasone, Y., "A theory of fatigue crack initiation in solids", J. *Appl. Mech.*, 1990, Vol. 57, pp. 1-6.

[21] Slavik, D.C., Dunyak, T., Griffiths, J., Kurath, P., "Crack Initiation Modeling in Ti-6Al-4V for Smooth and Notched Geometries", proc. 5th National Turbine Engine High Cycle Fatigue Conference (HCF '00), Chandler, Arizona, 7-9 March 2000.

[22] Krgo, A., Kallmeyer, A.R., Kurath, P., "Evaluation of HCF Multiaxial Fatigue Life Prediction Methodologies for Ti-6Al-4V. proc. 5th National Turbine Engine High Cycle Fatigue Conference (HCF '00), Chandler, Arizona, 7-9 March 2000.

[23] Pape, J.A., and Neu, R.W., "Influence of contact configuration in fretting fatigue testing", *Wear*, (225-229), 1999, pp. 1205-1214.

[24] Hibbitt, Karlsson & Sorensen Inc. ABAQUS User's Manual, Version 5.8, 1998, West Lafayette, IN.

[25] McVeigh, P.A. and Farris, T.N., "Finite Element Analysis of Fretting Stresses", *J.Tribology*, Vol. 119, 1997, pp. 797-801.

[26] Ambrico, J.M., Begley, M.R., "Plasticity in fretting contact", J. *Mech. Phys. Solids*, Vol. 48, 2000, pp.2391-2417.

[27] Johnson, K.L., *Contact Mechanics*, Cambridge University Press, Cambridge, 1985.

[28] Swalla DR. Fretting Fatigue Damage Prediction using Multiaxial Fatigue Criteria. Masters Thesis, Georgia Institute of Technology, 1999.

[29] Nowell, D. and Dai, D.N., "Analysis of Surface Tractions in Complex Fretting Fatigue Cycles Using Quadratic Programming", *J. Tribology*, Vol. 120, 1998, pp. 744-49.

[30] Nowell, D. and Hills, D.A., "Mechanics of Fretting Fatigue Tests", *Int. J. Mech. Sci.*, Vol. 29, No. 5, 1987, pp. 355-65.

[31] Petiot, C., Vincent, L., Dang Van, K., Maouche, N., Foulquier, J., Journet, B., "An analysis of fretting fatigue failure combined with numerical calculations to predict crack nucleation", *Wear*, Vol. 181,1995, pp.101-11.

[32] Swalla, D.R. and Neu, R.W., "Influence of Coefficient of Friction on Fretting Fatigue Crack Nucleation Prediction", submitted to Tribology International, Sept. 2000.

Michele Ciavarella,[1] Daniele Dini,[2] Giuseppe P. Demelio[3]

A Critical Assessment of Damage Parameters for Fretting Fatigue

REFERENCE: Ciavarella, M., Dini, D., and Demelio, G. P., "A Critical Assessment of Damage Parameters for Fretting Fatigue," *Fretting Fatigue:Advances in Basic Understanding and Applications, ASTM STP 1425*, Y. Mutoh, S. E. Kinyon, and D. W. Hoeppner, Eds., ASTM International, West Conshohocken, PA, 2003.

ABSTRACT: Fretting Fatigue (FF) has been considered as fatigue in a region of stress concentration due to the contact accelerated by mechanical erosion (possibly enhanced by metal transformations and/or chemical reactions in an aggressive environment). Whether the effect of erosion is significant or not is not clear. However, recently a more precise quantification of the effect of stress concentration has been attempted (some authors have used the terminology "crack analogue" and "notch analogue"). Most practical cases are concerned with finite stress concentration, so that a "notch analogue" criterion seems more appropriate, like for example in the Hertzian FF set of experiments like those on Aluminium alloy (Al/4%Cu, HE15-TF) by Nowell in the late '80s, and (Al2024) by Farris in late 90s. Application of the "notch analogue" with a simple stress concentration criterion is usually overconservative for a notch (and indeed it is for the experiments considered, although it may not in general because of mechanical erosion effect in the FF case), so that "averaging" methodologies have been proposed where unfortunately the best-fitting averaging constant turns to be not just a material constant. In fact, in the present paper the "best fitting distance" is shown to be much smaller than what expected from existing Kf criteria.

KEYWORDS: Fretting fatigue, fatigue, gas turbine engines, HCF, LCF, surface damage, stress concentration, SCF

Introduction

As generally defined, Fretting Fatigue (FF) is a phenomenon caused by the synergy between mechanical erosive surface damage and initiation of cracks due to the stress field (although this synergetic effect has never been precisely quantified). Attention has been for long time focused on minute surface tangential motion arising between components pressed together by normal forces and subjected to oscillatory loads or vibratory motion (one of the sites of greatest concern with FF is in the compressor stages of gas turbine engines, and particularly blade-disk dovetail joints).

FF is responsible to many in service "mishaps" in gas turbine engines, for which HCF (High Cycle Fatigue) is itself the largest single cause of failure. Despite the remarkable progress, FF is still one of the most "inexact" areas of fatigue, where effort is now being

[1] Associate Professor, DIMeG-Politecnico di Bari, V. Gentile 182, 70126, Bari, Italy.
[2] D.Phil Student, Oxford University, Department of Engineering Science, Parks Road, OX1 3PJ, Oxford, UK.
[3] Full Professor, DIMeG-Politecnico di Bari, V. Gentile 182, 70126 Bari, Italy.

108

focused. In particular, it is not clear, to date, if FF is (i) a surface damage phenomenon, or (ii) it is simply fatigue from a stress raiser feature. Even though surface damage parameters can be computed analytically or quasi-analytically in a wide range of contact FF conditions [1–6], attempts to use method (i) have so far remained very empirical and, as it will be shown in the present paper, only a vague qualitative correlation with fatigue life is obtained; moreover, a precise and quantitative correlation is hardly possible in the future as connection to the standard fatigue law understanding is not foreseeable. Attempts to use method (ii) have surprisingly been very limited until recently (crack and notch "analogues" have been proposed by the MIT group in [7,8]), but the use of notch analogues seems to be the most promising. We discuss aspects of both approaches in the following paragraphs.

Surface Damage Parameters

One of the very first parameters used to identify FF has been slip amplitude δ_{max}: it appeared that FF was most detrimental for δ_{max} around 15 μm (Nishioka et al [9,10]). A second choice was then the frictional energy dissipation parameter, $D=F_1=(\tau\delta)_{max}$, which being an energy dissipation may well be correlated to the damage mechanisms on the fretted surface, and indeed has the merit of being significant also in the gross slip regime as Archard's wear law parameter. Finally, it has been also shown to give a measure of the cumulated shear strain according to the Bower-Johnson's ratchetting mechanism (see Nowell & Hills [11]). Further, the Oxford group (Ruiz et al. 1984, 1986 [12,13]) suggested a "Ruiz's composite surface damage parameter," $R=F_2=(\sigma\tau\delta)_{max}$, which takes empirically into account the evidence that cracks are more likely to develop in a region of tension rather than compression, therefore accounting also for the early stages of propagation. Nowell & Hills [11] obtained a value of $R \bullet 4N^2/mm^3$ as the lower bound for initiation and failure within 10^7 cycles for the particular material examined (the 4% Cu-Al alloy known as HE15-TF) and for a Hertzian 2D plane fretting contact.

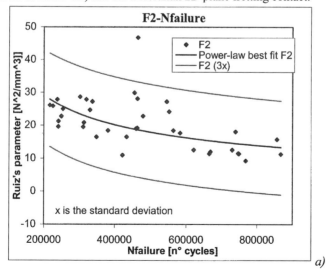

a)

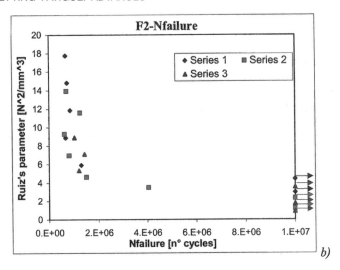

FIG. 1— *F2 against fretting fatigue life of specimens: a) experimental data [14]; b) experimental data [15].*

In the present paper, we computed these parameters for the Hertzian experiments in (Farris & Szolwinski, 1998, [14]) and (Nowell & Hills, 1988, [15][4]). The original stress analysis in [14, 15] has also been improved to consider finite thickness of the specimen,[5] with a minor correction for [14], but quite large for [15] in terms of increase of the surface tensile stresses. Also, in [15] there was no surface damage parameter calculation. Results are synthetically shown in Fig. 1 for the Ruiz' parameter, F2, showing a power-law correlation similar to fatigue S-N curves (the scales are different and [14] have much lower fatigue life and concentrated in the range 200-900 x 10^3 cycles, whereas [15] span a wider range with run-out in several cases (smaller contact areas) $>10^7$ cycles.

Fretting as a Fatigue Phenomenon

Pape et al. and Neu et al. (1999) [17,18] review different fatigue criteria, Szolwinski et al. (1998) [14] propose a K_t (Stress Concentration Factor) initiation criterion based on the multiaxial Smith-Watson-Topper equation (being limited on surface uniaxial, the SWT is used in a rather simplified form).

---

[4] Farris' specimens material is Al2024-T351 aluminium alloy and Nowell's one is HE15-TF aluminium alloy.

[5] This was made by applying the Kelly's theory (based on Fourier transform functions) for elastic layer resting on a rigid substrate [16] that allows to obtain the corrected solution simply by separating the semi-plane hertzian solution and the corrective one from the total internal stress field.

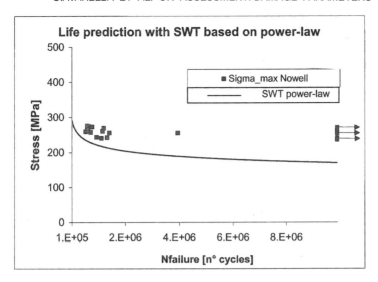

FIG. 2— *Life prediction with SWT based power-law.*

Despite a quite large scatter, the data seem to fit the simple model proposed in [14]. However, this is surprising if we consider that a K_t criterion is usually largely over-conservative when the gradient of the stress field is high as in the contact region. Using this simple stress criterion for the full set of the Nowell and Hills' data [15] would be reasonable for the large contact area sizes, but extremely conservative for the set of smaller contact areas where much higher fatigue life is found. A simple SWT approach is then inadequate to predict the size effect fretting fatigue life. This is clearly pointed out in Fig. 2 where, by comparing the maximum tensile stress with the SWT_{stress} (see Eq. (3) written below), the difference between the real maximum stresses and SWT_{stress} is larger for the "run-out" experiments (small contact areas) than for the finite life ones (large contact area) failing at small number of cycles. This effect can be qualitatively explained with the stress gradient increment of the internal tensile stress associated with the decrement of the contact area shown in Fig. 3.

One possibility to consider a less *over*-conservative prediction and take size effect into account is based on an empirical averaging of the stress field [19,20]. However, Refs. [19,20] havefound that there is no unique "material constant" distance over which the results fit the data in Nowell and Hills [15] reasonably well.

Various criteria exist for the fatigue knock down factor K_f, which is by definition $1 < K_f < K_t$, where K_t is the stress concentration factor. An alternative to K_f is often defined via a notch sensitivity, $0 < q < 1$:

$$q = \frac{(K_f - 1)}{(K_t - 1)} \tag{1}$$

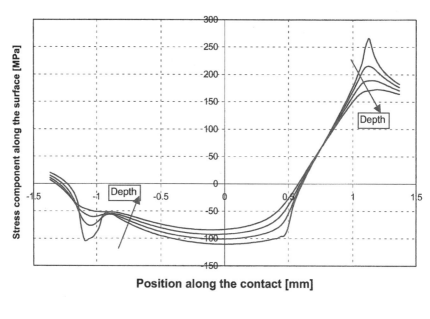

a)

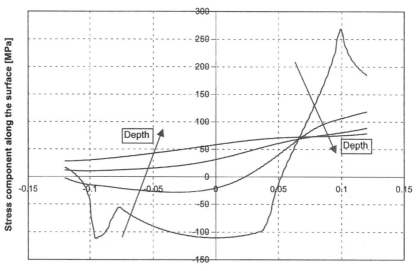

b)

FIG. 3— *Internal stress trend at different distances from the surface (the arrows show lines of increasing depth: 0,25,50,75 μm): a) contact semi-width: a = 1.14 [mm]. b) contact semi-width: a = 0.1 [mm] (Nowell's data [15], Series 1).*

Recent work by Taylor [21,22] has shown that Neuber and Peterson's notch sensitivity factors belong to the averaging techniques. In particular, in order to make them consistent with fracture mechanics, the averaging distance needs to be connected to the *intrinsic material defect* constant (also knows as "Topper" constant), depending on fatigue limit $\Delta\sigma_l$ and threshold ΔK_{th}, as:

$$a_0 = \frac{1}{\pi}\left(\frac{\Delta K_{th}}{\Delta\sigma_l}\right)^2 \qquad (2)$$

Specifically, the simplest method is to use the stress at distance $a_0/2$ from the trailing edge of the contact, which is sometimes called the "hot spot" stress ($\sigma_{hot\text{-}spot}$). Either ways, the stress has to be compared with the predicted SWT_{stress} [14]:

$$SWT_{stress} = \sqrt{\left[\frac{\left(\sigma_f{}'\right)^2}{E}\left(2N_f\right)^{2b} + \sigma_f{}'\varepsilon_f{}'\left(2N_f\right)^{b+c}\right]\frac{E}{1-\nu^2}} \qquad (3)$$

where N_f is the failure life and E, ν, $\sigma_f{}'$, $\varepsilon_f{}'$, a, and b are material constants that could be extracted from a uniaxial fatigue experiment and indeed are known for Al2024-T351 and HE15-TF to present values shown in Table1. The results are plotted in Fig. 4.

Table1— *Al2024-T351 and HE15-TF aluminium alloys material constants used in [14,15].*

Material constant	Constant's description	Constant's value	
		Al2024-T351	HE15-TF
$\sigma_f{}'$	Fatigue strength coefficient	741 [MPa]	1015 [MPa]
b	Fatigue strength exponent	-0.078	-0.11
$\varepsilon_f{}'$	Fatigue ductility coefficient	0.166	0.21
c	Fatigue ductility exponent	-0.538	-0.52
E	Young's modulus	74.1 [GPa]	68.9 [GPa]
ν	Poisson's ratio	0.33	0.33

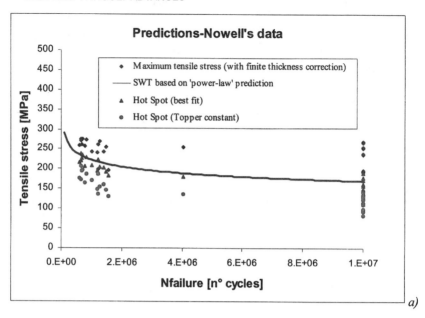

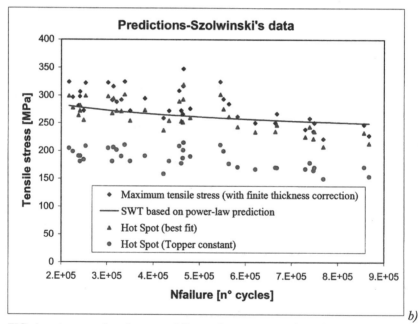

FIG. 4—*A comparison between different fatigue approaches: a) Nowell's experimental data; b) Szolwinski' experimental data ("Best fit" corresponds to $a_0/2 = 15$ μm).*

The hot spot results obtained using the "Topper" constant (we used $a_0 = 91$ μm for Nowell's material, see [20] and $a_0 = 129$ μm for Szolwinski's data, see [22]) are still very *under*-conservative, as the best fit would only occur for much smaller hot spot constants (around 15 μm).

This result does not necessarily imply that a prediction based purely on fatigue is incorrect: it is possible, for example, that this averaging constant is not just a material constant, but depends on geometrical features. Therefore, there is not enough evidence to suggest that the difference is due to fretting *per se*, returning to the well-known concept that FF is a combination of both surface damage and fatigue.

Similar conclusion is drawn from Fig. 5 where only the best-fit hot spot constants (*15 μm*) is used to fit correctly the size effect experimental results.

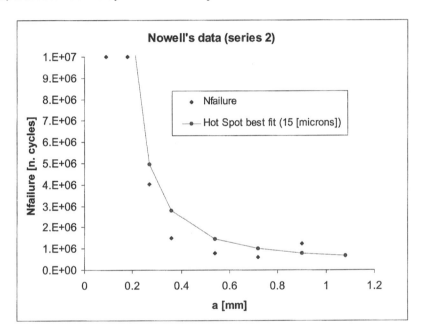

FIG. 5—*Best fitting prediction of fatigue life size effect.*

Conclusions

Surface damage parameters have been found to be qualitatively correlated to FF life for both sets of experiments [14,15], particularly F1 and F2. Obviously, these parameters are of empirical nature and suffer from the drawback of being unrelated to any other fatigue parameter. Further, they would depend on surface conditions, particularly friction coefficient of the pair of contacting materials, as typical in tribology. On the other hand, "pure fatigue" lifing methodology also requires some caution. In [15] it was apparently found that a simple Kt approach would make conservative results. However, we have shown that this may be just a coincidence of fretting effect nearly cancelling the notch sensitivity effect, which we expect because of high stress gradient. Therefore, there is no reason to believe that the "additional" fretting effect on fatigue life would be always so simply taken into account. In other words, a Kt approach neglecting the additional fretting damage contribution may result in underconservative prediction under appropriate circumstances: in particular, when the notch sensitivity of the material is not so high and the fretting damage is large.

A specific FF-related notch sensitivity, or alternative additional parameters, depending again on surface conditions, friction and materials in contact, may be required.

Fatigue criteria are more likely to be successful in the future, but if a methodology is to be expected to be conservative, fundamentally solid and amenable to standardisation, a lot better understanding of the "notch sensitivity" effect is needed. Use of stress concentration alone, as suggested in some existing "notch analogues" criteria (Farris, [15], Giannakopoulos et al., 2000 [8]), is not necessarily a conservative approach, and could be made arbitrarily over-conservative when the Kt factor becomes large. Introducing a Kt factor design limit would probably render inadequate the majority or existing designs! On the other hand, using existing "best-fitting" averaging methodologies (Nowell & Araújo, 2000 [19], Araújo, 2000 [20]) has proved to be questionable and empirical. In fact, the "best-fitting" distance turns out to be very much dependent not just on material constants but also on loading conditions, and it is to be expected that other dependencies (particularly on geometry), not emerged in existing experimental validation, could be also found.

References

[1]   Ciavarella, M. and Demelio, G., "A review of analytical aspects of fretting fatigue, with extension to damage parameters, and application to dovetail joints," *International Journal of Solids and Structures*, 2001, Vol. 38 (10–13), pp. 1791–1811.

[2]   Ciavarella, M., "The Generalized Cattaneo Partial Slip Plane Contact Problem. I-Theory", *International Journal of Solids and Structures*, 1998, Vol. 35/18, pp. 2349-2362. II-Examples, pp. 2363–2378.

[3]   Ciavarella, M., Demelio, G., and Hills, D. A., "The use of almost complete Contacts for Fretting Fatigue Tests", *Fatigue and Fracture Mechanics*: twenty-ninth Volume, ASTM STP 1332, T.L.Panontin and S.D.Sheppard, Eds., ASTM International, West Conshohocken, PA, 1999a, pp. 696–709.

[4]   Ciavarella, M. and Hills, D. A., "Brief note: some observation on oscillating tangential forces and wear in general plane contacts," *European Journal of Mechanics*, Part A: Solids, 1999, 18, pp. 491–497.

[5]   Ciavarella, M., Demelio, G., and Hills, D.A, "An Analysis of Rotating Bending Fretting Fatigue Tests using "Bridge" Specimens." *Fretting Fatigue: Current Technologies and Practices, ASTM STP 1367*, D.W. Hoeppner, V. Chandrasekaran and C. B. Elliot, Eds., ASTM International, West Conshohocken, PA, 1999b.

[6]   Ciavarella, M., Hills, D. A., and Monno, G, "The influence of rounded edges on indentation by a flat punch." *IMechE part C – Journal of Mechanical Engineering Science*, 1998, Vol. 212, No. 4, pp. 319–328.

[7]   Giannakopoulos, A. E., Lindley, T. C., and Suresh, S., "Aspect of equivalence between contact mechanics and fracture mechanics: theoretical connections and a life-prediction methodology for fretting-fatigue," *Acta Metallurgica Inc.,* Elsevier

Science Ltd., *Acta Materialia*, 1997, Vol. 46, No 9, pp. 2955-2968.

[8]   Giannakopoulos, A. E., Lindley, T. C., Suresh, S. and Chenut, C., "Similarities of stress concentrations in contact at round punches and fatigue at notches: implication to fretting fatigue crack initiation," *Fatigue and Fracture of Engineering Material and Structures,* 2000, Vol. 23, pp. 561–571.

[9]   Nishioka, K., Nishimura, S., and Hirakawa, K, "Fundamental investigation of fretting fatigue part 1. On the relative slip amplitude of press-fitted axle assemblies," *Bullettin of JSME*, 11, 1968, pp. 437–445.

[10] Nishioka, K., and Hirakawa, K, "Fundamental investigation of fretting fatigue - part 3. Some phenomena and mechanisms of surface cracks - part 4. The effect of mean stress - part 5. The effect of relative slip amplitude," *Bullettin of JSME*, 12, 1969, pp. 397–407, 408–414, 692–697.

[11] Nowell, D., and Hills, D. A, "Crack initiation criteria in fretting fatigue," *Wear 136*, 1990, pp. 329-343.

[12] Ruiz, C., Buddington, P. H. B., and Chen, K.C., "An investigation of fatigue and fretting in a dovetail joint," *Experimental Mechanics*, Vol. 24, No. 3, 1984, pp. 208–217.

[13] Ruiz, C., and Chen, K. C., "Life assessment of dovetail joints between blades and discs in aero-engines." *Proc. International Conference of Fatigue*, Sheffield, ImechE, London, 1986.

[14] Szolwinski, M. P., and Farris, T. N., "Observation, analysis and prediction of fretting fatigue in 2024-T351 Aluminium Alloy." *Wear 221*, 1998, pp. 24–36.

[15] Hills, D. A., Nowell, D., and O'Connor, J. J., "On the mechanics of Fretting Fatigue," *Wear 125*, 1988, pp. 129–156.

[16] Kelly, P. A., "An elastic layer resting on a rigid substrate," OUEL Report 2039/94, August 1996.

[17] Pape, J. A., and Neu, R.W., "Influence of contact configuration in fretting fatigue testing," *WEAR 229*, 1999, No. 2, pp.1205–1214.

[18] Neu, R.W., Pape, J. A., Swalla, D. R., "Methodologies for Linking Nucleation and Propagation Approaches for Predicting Life under Fretting Fatigue", *Fretting Fatigue: Current Technologies and Practices, ASTM STP 1367*, D.W.Hoeppner, V. Chandrasekaran and C. B. Elliot, Eds., ASTM International, West Conshohocken, PA, 1999.

[19] Nowell, D., and Araújo, J. A., "Prediction of fretting fatigue life using volume averaged multiaxial initiation parameters," pp. 223–234, *Proc. 4th International Conference on Modern Practice in Stress and Vibration Analysis*, Nottingham, 5–7 Sept 2000, A.A. Becker, Ed., EMAS, West Midlands, 2000.

[20] Araújo, J. A., "Prediction of fretting fatigue crack initiation using multiaxial fatigue criteria," Chapter 4, *DPhil thesis*, Oxford University, 2000.

[21] Taylor, D., "Geometrical effects in fatigue: a unifying theoretical model," *International Journal of Fatigue*, 21, 1999, pp. 413–420.

[22] Taylor, D., Wang, G., "The validation of some methods of notch fatigue analysis," *Fatigue and Fracture of Engineering Material and Structures,* May 2000, Vol. 23, No.5, pp. 387–394.

LIFE PREDICTION

Carlos Navarro,[1] Mercedes García,[1] and Jaime Domínguez[2]

An Estimation of Life in Fretting Fatigue Using an Initiation-Propagation Model

Reference: Navarro, C., García, M., and Domínguez, J., **"An Estimation of Life in Fretting Fatigue Using an Initiation-Propagation Model,"** *Fretting Fatigue: Advances in Basic Understanding and Applications, STP 1425,* Y. Mutoh, S. E. Kinyon, and D. W. Hoeppner, Eds., ASTM International, West Conshohocken, PA, 2003.

Abstract: This paper proposes a method for estimating the total fatigue life in fretting fatigue. It compares the fatigue life of some specimens obtained in various series of fretting fatigue tests with the estimates of total fatigue life using a model that separately takes into account the initiation and propagation of the crack. The experimental tests used a spherical contact and were conducted on an aluminum alloy (Al 7075-T6). The initiation of the crack is considered using the multiaxial fatigue criterion of McDiarmid together with the theoretical distribution of stresses for the given loads applied in each test. The propagation of the crack is considered using two different laws of propagation, Paris law and a curve given by Lankford that would represent the upper limit for the crack propagation velocity (lower limit for fatigue life) when the crack is short. These two estimates are an upper and lower bound respectively of the fatigue life of the specimen; the actual life lies in between.

Keywords: fretting fatigue, life estimation.

Introduction

Fretting fatigue is a phenomenon that can be observed everyday in service in many mechanical systems elements such as riveted and bolted joints, shrink-fitted couplings, metal ropes and cables, etc. [*1*]. It appears in systems where service loads cause relative displacement of very small amplitude between contact surfaces. This displacement generates friction forces associated to the load normal to the contact surface. These local stresses add to the bulk load of the system producing a similar effect to the stress concentrators. When these forces vary cyclically, cracks initiate by fatigue mechanisms. This initiation process is faster compared to the case when there is no contact between the surfaces. These cracks may stop or eventually propagate until final fracture of the component.

An important issue in the design of mechanical systems that suffer fretting fatigue is

---

[1]Graduate student, Departamento de Ingeniería Mecánica, ESI, Universidad de Sevilla, Camino de los Descubrimientos s/n, 41092 Sevilla, Spain.
[2]Professor, Departamento de Ingeniería Mecánica, ESI, Universidad de Sevilla, Camino de los Descubrimientos s/n, 41092 Sevilla, Spain.

the prediction of the endurance under the design loads and many efforts and resources are dedicated to it. This is a difficult problem and there are different approaches that are aimed at solving it [2,3].

There is a general agreement that the fretting fatigue process has two different phases: the fatigue crack initiation and crack propagation, which are due to different mechanisms. Some methods to estimate the fretting fatigue life are based on the initiation mechanism, and use the data obtained in tests with smooth specimens under completely reversed loading, combined with various multiaxial fatigue criteria. These methods are more frequently used when the initiation phase dominates over propagation. Another approach consists in the use of linear elastic fracture mechanics. In this case the initiation life is assumed to be small compared to propagation [4]. Another method is to combine initiation with propagation using a fixed crack initiation length [3]. This length tries to determine the separation of the process of initiation and propagation. The total life would be the sum of the number of cycles required to initiate a crack and the number of cycles to propagate a crack of this length until final fracture [5]. A step forward in this study is, as in the latter case, to combine initiation and propagation, but instead of previously choosing a fixed initiation length, it will come out as a result of the calculations. There are different ways to make these calculations developed by several authors [6,7,8] for notched specimens, although they have not been applied to fretting fatigue.

This paper presents a similar method to the one proposed by Socie [8], which combines initiation and propagation defining a non-arbitrary crack initiation length, to estimate the life obtained in a series of fretting fatigue tests with spherical contact.

Initiation-Propagation Model

The method used in this paper is a modification of Socie, Morrow and Chen [8] for fatigue life analysis of notched members. This modification takes into account the multiaxial stress field developed in fretting fatigue with spherical contact. The authors calculated two curves as a function of crack depth, i.e., along the hypothetical crack path: (a) a crack growth rate caused by crack initiation mechanisms from low cycle fatigue concepts and (b) the damage rate during the crack propagation phase from linear elastic fracture mechanics (LEFM) concepts. The first one is obtained estimating, for every point along that path, the number of cycles needed to initiate a crack and then calculating the derivative. Hence, something that could be considered as a crack propagation rate due to initiation is obtained. This function decreases with depth, and for points close to the surface it is higher than the classical crack propagation rate due to fracture mechanics, which increases with depth. Therefore, a hypothetical crack close to the surface would grow faster by initiation than by propagation. At a certain depth, the two curves cross, and from then on the crack grows faster by propagation. According to their proposal, this point sets what they call the crack initiation length. The crack initiation life, N_i, would be the one corresponding to this point and the crack propagation life, N_p, would be the number of cycles needed to propagate a crack from that point until final fracture.

Instead of calculating the growth rates as explained, a similar but numerically equivalent method is used [9]. Along the hypothetical crack path, the fatigue life at each point based on low cycle fatigue is calculated, N_i. This would be the number of cycles needed to initiate a crack at that point. Also, for each point, the number of cycles needed to grow a crack from that point to the final crack length, N_p, are calculated using LEFM.

The sum of N_i and N_p gives the total life associated to each point, considered as the point where the growth law changes from initiation to crack propagation. Therefore, the life of the specimen would be the minimum of that curve, (Figure 1). The point at which the minimum is produced also defines the crack initiation length, which coincides exactly with the one calculated using the procedure explained in the former paragraph [8].

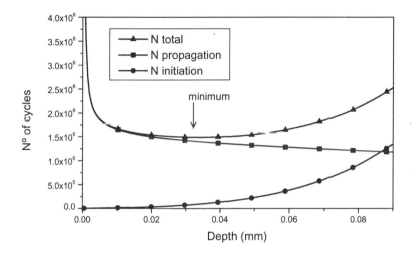

Figure 1 - *Initiation, propagation and total life estimates.*

Application to Fretting Fatigue with Spherical Contact

The modification of the method formerly explained consists in applying it to fretting fatigue with spherical contact, considering a three-dimensional stress field. The stresses produced by spherical contact are calculated analytically using the explicit expressions obtained by Hamilton [10]. Additionally, the eccentricity of the stick zone due to the bulk stress is taken into account as shown in [11].

The first step in order to apply this method is to localize the site of the initiation of the crack. Experimentally, it has been observed that cracks initiate at points located very close to the edge of the contact zone [5]. Figure 2 shows the scar produced in a specimen in a fretting fatigue test with spherical contact. It is also shown the crack path on the surface where it can be seen that it appears close to the limit of the contact zone. Different multiaxial stress criteria (Von Mises, McDiarmid, Smith-Topper-Watson) applied at the surface and along the axis of symmetry of the contact zone show that the critical point is the edge of the contact zone. Applying these criteria at a small depth, the most critical point is not the limit of the contact zone but close to it.

The cracks start to grow at a small angle to the surface, but close to the surface they rotate to an angle of 70°-80°, (Figure 3). This angle is not very different from 90°, hence, for simplicity, in the application of this method the crack will be supposed to grow perpendicular to the surface. The application of this method to cracks growing perpendicular to the surface and from points close to the edge of the contact zone give initiation curves that are similar but give higher values of the initiation life when compared to the crack located exactly on the edge. This implies that, supposing the crack grows perpendicular to the surface, the most critical crack seems to initiate at the edge of the contact zone, therefore this location will be chosen for all the calculations.

Figure 2 - *Fretting scar.*

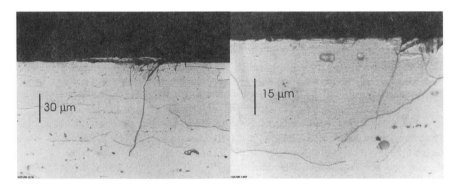

Figure 3 - *Cross section of cracks.*

In order to calculate the initiation curve, McDiarmid multiaxial fatigue criterion will be applied [*12*]

$$\frac{\Delta\tau_{max}}{2} + \frac{t}{2\sigma_{TS}}\sigma_{max} = \sigma_{eq} \tag{1}$$

where $\Delta\tau_{max}$ is the maximum increment of tangential stresses, σ_{max} is the maximum normal stress in the direction perpendicular to the plane where $\Delta\tau$ is maximum ($\Delta\tau_{max}$), t is the fatigue limit in torsion and σ_{TS} is the ultimate stress. This criterion is used together with the strain-life relationship

$$\frac{\Delta\varepsilon}{2} = \frac{\sigma'_f}{E}(2N_i)^b + \varepsilon'_f(2N_i)^c \tag{2}$$

where $\Delta\varepsilon$ is the normal strain obtained in tests with smooth specimens under completely reversed loading, E is the Young modulus, N_i is the number of cycles to the initiation of a crack and the remaining parameters are constants that depend on the material and are determined experimentally. The values for the constants are taken from [9] for the material Al 7075-T6, $\sigma'_f = 1917$ MPa, $b = -0.176$, $\varepsilon'_f = 0.8$, $c = -0.839$. In order to combine both equations, (1) and (2), McDiarmid criterion is applied to a fatigue test of a smooth specimen under completely reversed loading, $\pm\sigma$. The plane where the maximum shear stress can be found is at 45° from the loading direction. In this plane, the amplitude of shear stresses is $\sigma/2$ and the maximum normal stress in the direction perpendicular to such plane is $\sigma/2$. The value of the equivalent stress is

$$\sigma_{eq} = \frac{\sigma}{2} + \frac{t}{2\sigma_{TS}}\frac{\sigma}{2} = \sigma \cdot f \tag{3}$$

where

$$f = \frac{1}{2}\left(1 + \frac{t}{2\sigma_{TS}}\right) \tag{4}$$

Writing (2) in terms of stress instead of deformation and multiplying both terms by (4) yields

$$\frac{\sigma_{eq}}{E} = f \cdot \left(\frac{\sigma'_f}{E}(2N_i)^b + \varepsilon'_f(2N_i)^c\right) \tag{5}$$

The latter equation gives the number of cycles to initiation given the equivalent stress calculated with eqn. (1).

As it was said earlier, the crack initiates at a small angle and rapidly rotates to an angle of 70°-80°. In order to calculate the stress intensity factor (SIF), the crack is presumed to be perpendicular to the surface below that point. The fact that the first part of the kinked crack, a (Figure 4), is very different from 90° will not affect the calculation of the SIF in the second part, b (Figure 4), even for high values of a/b (10 ~ a/b) [13].

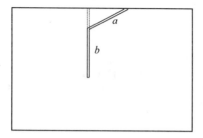

Figure 4 - *Kinked crack.*

There is mixed mode loading at the crack tip, but only K_I is going to be used because the crack is nearly perpendicular to the surface, and in this case mode II stress intensity factors are small compared to mode I. Therefore the crack growth will not be significantly changed [4]. In order to calculate the SIF in mode I for a crack perpendicular to the surface, the following weight function calculated for an edge through crack in a two-dimensional problem proposed by Bueckner [14] has been used.

$$w(t) = \frac{1}{\sqrt{t}} (1 + m_1 \cdot \frac{t}{a} + m_2 \cdot \left(\frac{t}{a}\right)^2) \tag{6}$$

where a, t and W are shown in Figure 5 and m_1 and m_2 are functions which depend on the ratio a/W. Using this weight function, the stress intensity factor can be obtained from the expression

$$K_I = \frac{\sqrt{2}}{\pi} \int_0^a w(t) \cdot \sigma_x(t) \, dt \tag{7}$$

where σ_x is the normal stress in the direction perpendicular to the crack plane. This stress intensity factor calculated has to be corrected to take into account the shape of the crack and that it is a three-dimensional problem. The aspect ratio of the crack is approximately 0.5 and this gives a correction factor of 0.78, [15].

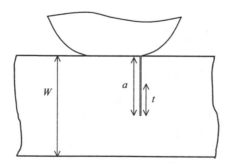

Figure 5 - *Cross section of the specimen*

At this point different propagation laws can be used. The data for crack growth in NASA/FLAGRO program [16] for Al 7075-T6 was chosen. This data coincides with other sources for long cracks [17, 18] but not for short cracks where there is crack growth below the long crack threshold. The function used in the NASA/FLAGRO program, eqn. (8), includes the effect of the stress ratio, R, the critical stress intensity factor, K_c, and the threshold stress intensity factor range for long cracks, ΔK_{th}.

$$\frac{da}{dN} = \frac{5.3465(1-R)^{-2.84} \Delta K^{2.836} (\Delta K - (1-R)\Delta K_{th})^{0.5}}{((1-R)K_c - \Delta K)^{0.5}} \tag{8}$$

This threshold can be corrected to include the small crack effect multiplying by a factor which depends on the crack length eqns. (9) and (10), [19].

$$\Delta K_{th} = \Delta K_{th\infty} \cdot \sqrt{\frac{a}{a+d}} \tag{9}$$

$$d = \frac{1}{\pi}\left(\frac{\Delta K_{th\infty}}{\Delta \sigma_f}\right)^2 \tag{10}$$

where a is the crack length, $\Delta K_{th\infty}$ is the threshold stress intensity factor range for long cracks and a stress ratio of $R = 0$ and $\Delta \sigma_f$ is the fatigue limit. But this correction can be still insufficient as will be seen in the results. Figure 6 shows the propagation rate versus the stress intensity factor range for four different models. The first one is directly the eqn. 8 with the data from the program NASA/FLAGRO (long crack), and the second is the same one but modified for short cracks using eqns. (9) and (10), (modified long crack). The third is simply the Paris law for this material for $R = -1$ and the data from [16],

$$\frac{da}{dN} = 4.2151 \cdot 10^{-12} \, \Delta K^{3.517} \tag{11}$$

The fourth is taken from Lankford [20] where different curves for Al 7075-T6 from different sources for short cracks are presented. From these curves, different points lying in the limit for short cracks with high stress levels were chosen. The curve shown in Figure 6 is the result of doing a best fit of the points extracted from [20]. The mathematical expression of this curve is

$$\frac{da}{dN} = 1.8416 \cdot 10^{-7} + \frac{3.2706 \cdot 10^{-9} - 1.8416 \cdot 10^{-7}}{1 + e^{\frac{\Delta K - 15.267}{2.03103}}} \tag{12}$$

With lower stresses this curve would present a local minimum in the short crack region and the depth of it will depend, among other factors, on the stress level. Therefore, using this data to calculate the crack growth would give shorter lives. The four curves coincide for $\Delta K > 8$ MPa m$^{1/2}$, that is long cracks, but below that value they are very different.

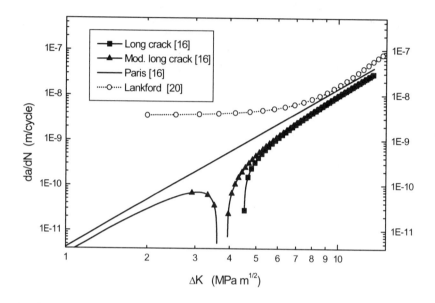

Figure 6 - *Different crack growth laws*

Experimental Data

The experimental data were obtained from Wittkowsky et al [21]. The material used in the tests was Al 7075-T6 and its principal mechanical properties are listed below, (Table 1). The spherical surfaces and plane specimens were machined from extruded bars with a diameter of 25.4 and 12.7 mm, respectively. Specimens were flat-sided plates with a rectangular gage section and the radius of the spherical pads was 25.4 mm. The grain size perpendicular to the surface was approximately 35 μm, measured according to ASTM Standard Test Methods for Determining Average Grain Size (E 112-88).

Table 1 - *Mechanical properties of the Al 7075-T6 extruded bars*

UTS (MPa)	$\sigma_{y0.2}$ (MPa)	E (GPa)	ΔK_{th} (R= 0) MPa m$^{-1/2}$
572	503	72	2.2

The loads applied in each test are listed in Table 2 and a schematic of the test setup is shown in Figure 7, where the load Q and P, which gives rise to the bulk stress σ, are cyclic loads that vary in phase, while N is a constant load. This test setup for fretting fatigue is further explained in [22].

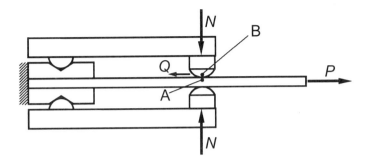

Figure 7 - *Fretting test loads*

Table 2 - *Loads in the fretting tests*

Test N°	N (N)	Q (N)	σ (Mpa)	N_f (cycles)
T1	30	±15	±85	480000
T2	20.8	±15	±83	449500
T3	15.6	±15	±85	395000
T4	12.5	±15	±83	361000
T5	18.5	±13.6	±77	551000
T6	16	±11.7	±83	530000
T7	13.9	±10	±83	803000
T8	10.3	±7.5	±83	2940000
T9	8.33	±8	±83	616500
T10	20	±15	±83	549000
T11	20	±15	±70	516000

Results

The lives of these tests have been calculated using the Paris law, eqn. (11), and the curve obtained from Lankford, eqn. (12). The other two laws give an infinite life and the reason is that these laws are useful for long cracks, eqn. (8), and physically small cracks, eqn. (8) modified with eqns. (9) and (10), but they do not describe the behavior of microstructurally short cracks. The results obtained with the Paris law and with the curve obtained from Lankford are summarized in Table 3. This table shows the crack initiation length, a_i, the initiation life, N_i, the propagation life, N_p, and the total estimated life using the Paris law, N_{TP}. The next column shows the experimental life, N_f and last one shows the total life using the law obtained from Lankford, N_{TL}. Just one value of life is given because in this case the initiation life is zero.

Table 3 - *Estimated lives and crack initiation lengths using Paris.*

Test	a_i (μm)	N_i	N_p	N_{TP}	N_f	N_{TL}
T1	26.6	133649	1806984	1940633	480000	323795
T2	27.9	98269	1634839	1733109	449500	319309
T3	33.1	88536	1319871	1408407	395000	307052
T4	38	86386	1209737	1296123	361000	304908
T5	26.8	123927	2088149	2212076	551000	347254
T6	25.5	125118	1970793	2095911	530000	328957
T7	26.2	177453	2205340	2382793	803000	332302
T8	23.7	237248	2711965	2949213	2940000	354296
T9	26.9	194659	2270958	2465617	616500	321603
T10	29.6	105917	1593345	1699262	549000	321584
T11	27.5	128692	2353014	2481705	516000	371971

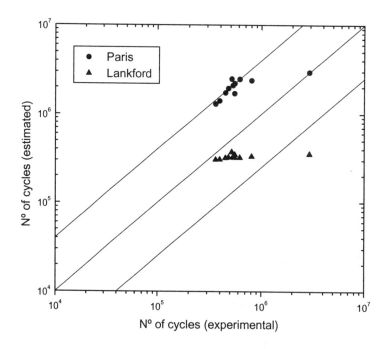

Figure 8 - *Estimated lives.*

McDiarmid multiaxial fatigue criterion was used in this paper, but any other could be used. As can be seen in Table 3, the life spent in initiation is considerably smaller than in propagation, so at this stage of the model and with these stress levels, other multiaxial fatigue criteria would have given similar results in the total estimated life. This means that for those stress levels more attention has to be paid to the propagation process. The crack initiation length calculated, 25-38 µm long, is longer than the depth at which the crack turns (about 20 µm), this means that stage I of the crack, where it grows at small angles, is included in the initiation process. When the curve obtained from Lankford (Figure 5) is used to estimate the fatigue lives of the specimens the results show that the initiation length is null and everything is propagation. This law underestimates the fatigue life of the specimens. The total estimated lives using these two laws are shown in Figure 8. Real lives lie in the middle and this is because the crack growth rates shown in Figure 5 do not represent the behavior for short cracks correctly. The Paris law does not show the behavior of short cracks and give too low crack growth rates, on the other side, the curve obtained from Lankford [20] shows what seems to be the limit for short cracks and give too high rates. The latter would be closer to reality for higher stresses (shorter lives), as can be seen in Figure 8.

Conclusions

An estimated life for a specimen subjected to fretting fatigue with spherical contact has been obtained considering the crack initiation and propagation, together with a non-arbitrary crack initiation length. It has been shown that for the cases analyzed the initiation of the crack takes place very soon compared to the total life and that it includes the early stage of the crack where it grows at small angles.

The comparison of these estimates with the results of a series of tests shows that life is overestimated by a factor around 4 in the range of stresses studied when Paris crack growth law is considered. This law gives better results for long lives. The law obtained from the results of Lankford is an upper bound for crack growth rate (lower bound for life estimate), therefore the lives obtained are underestimated and it gives better results for shorter lives (higher stresses).

More attention has to be paid to the growth of short cracks in order to obtain better results for the fatigue life. In future analysis the model will be expanded including mixed mode crack growth and the inclination of the crack in stage II and the exact point of initiation.

Acknowledgments

The authors wish to thank the Ministerio de Educación y Cultura for their financial support through the investigation project with reference PB97-0696-C02-01.

References

[1] Waterhouse, R. B. and Lindley, T. C., Eds., *Fretting Fatigue, ESIS 18*, Mech. Eng. Publ., 1994.
[2] Harish G. and Farris T.N., "Shell Modeling of Fretting in Riveted Lap Joints," *AIAA Journal*, Vol. 36, 1998, pp. 1087–1093.

[3]   Ruíz C., Boddington P.H.B. and Chen K.C., "An investigation of the fatigue and fretting in a Dovetail Joint," *Experimental Mechanics*, Vol. 24, 1984, pp. 208–217.
[4]   Faanes, S. and Fernando, U.S., "Life Prediction in Fretting Fatigue using Fracture Mechanics," *Fretting Fatigue, ESIS* 18, Mech. Eng. Publ.,1994, pp. 149–159.
[5]   Szolwinski, M.P. and Farris, T.N., "Observation, analysis and prediction of fretting fatigue in 2024-T351 aluminum alloy," *Wear*, Vol. 221, 1998, pp. 24–36.
[6]   Dowling N.E., "Notched member fatigue life predictions combining crack initiation and propagation," *Fatigue of Engineering Materials and Structures*, Vol. 2, 1979, pp. 129–138.
[7]   Smith, R.A. and Miller, K.J., "Prediction of fatigue regimes in notched components," *International Journal of Mechanical Sciences*, Vol. 20, 1978, pp. 201–206.
[8]   Socie, D.F., Morrow, J. and Chen, W.C., "A procedure for estimating the total fatigue life of notched and cracked members," *Journal of Engineering Fracture Mechanics*, Vol. 11, 1979, pp. 851–859.
[9]   Chen, W.C., PhD Thesis, University of Illinois at Urbana-Champaign, U.S.A., 1979.
[10] Hamilton, G.M., "Explicit equations for the stresses beneath a sliding spherical contact," *Proceedings of the Institution of Mechanical Engineering,* 197C, 1983, pp. 53–59.
[11] Navarro, C. and Domínguez, J., "Contact conditions and stresses induced during fretting fatigue," *Computational Methods in Contact Mechanics IV*, WIT Press, 1999, pp. 453–462.
[12] McDiarmid, D.L., "A shear stress based critical-plane criterion of multiaxial fatigue failure for design and life prediction," *Fatigue and Fracture of Engineering Materials and Structures*, Vol. 17, 1994, pp. 1475–1484.
[13] Hills D.A., Kelly, P.A., Dai D.N. and Korsunsky A.M., *Solution of Crack Problems: The Distributed Dislocation Technique*, Kluwer Academic Publishers, 1996.
[14] Bueckner, H.J., "Weight functions and stress-intensity factors," *Methods of analysis and solutions of crack problems (Chapter 5, Appendix II)*, Sih, G.C., Ed., Noordhoff International Publishing, 1973, pp. 306–313.
[15] Suresh, S., *Fatigue of Materials*, Cambridge University Press, 1998.
[16] Fatigue Crack Growth Computer Program, NASA/FLAGRO, L.B. Jonhson Space Center, JSC-22267, 1986.
[17] *Fatigue and Fracture*, ASM Handbook, Vol. 19, 1996.
[18] Taylor D. and Jianchun L., *Sourcebook on Fatigue Crack Propagation: thresholds and crack closure,* EMAS, 1993.
[19] El Haddad, M. H., Topper, T. H. and Smith, K. N., "Prediction of non propagating cracks," *Engineering Fracture Mechanics,* Vol. 11, 1979, pp. 573–584.
[20] Lankford, J., "The growth of small fatigue cracks in 7075-T6 aluminum," *Fatigue of Engineering Materials and Structures*, Vol. 5, 1982, pp. 233–248.
[21] Wittkowsky, B.U., Birch, P.R., Domínguez, J. and Suresh, S., "An Experimental Investigation of Fretting Fatigue with Spherical Contact in 7075-T6 Aluminum Alloy," *Fretting Fatigue: Current Technology and Practices, ASTM STP 1367*, 2000, pp. 213–227.
[22] Wittkowsky, B.U., Birch, P.R., Domínguez, J. and Suresh, S., "An apparatus for quantitative fretting-fatigue testing," *Fatigue and Fracture of Engineering Materials and Structures,* Vol. 22, 1999, pp. 307–320.

David Nowell[1] and J. Alex Araújo[1]

Application of Multiaxial Fatigue Parameters to Fretting Contacts with High Stress Gradients

Reference: Nowell, D. and Araújo, J. A., **"Application of Multiaxial Fatigue Parameters to Fretting Contacts with High Stress Gradients,"** *Fretting Fatigue: Advances in Basic Understanding and Applications, ASTM STP 1425*, Y. Mutoh, S. E. Kinyon, and D. W. Hoeppner Eds., ASTM International, West Conshohocken, PA, 2003.

Abstract: Many practical fretting contacts give rise to high stress concentrations or singularities that cause high stress gradients. Straightforward application of initiation-based life prediction methods can result in conservative life predictions as the high stress levels are not sustained over a critical volume. This paper uses experimental results that show a size effect with the Hertzian contact geometry to investigate the application of Smith, Watson, Topper, and Fatemi-Socie multiaxial initiation parameters to fretting fatigue. It is concluded that, for cases where a high stress gradient exists, an averaging procedure is required in order to produce an acceptable prediction of initiation life. This approach is compared to an alternative procedure based on short crack arrest and the similarities and differences are discussed.

Keywords: fretting fatigue, initiation, multiaxial parameters

Introduction

The prediction of fretting fatigue life is particularly important in safety-critical applications such as occur in the aerospace or nuclear industries. The designer needs to know how well the interface will perform under the imposed loading conditions and, in particular, whether either of the contacting components is likely to fail during the operating lifetime or inspection interval. This is a complex problem, since there are a number of coupled phenomena present. The imposed load may, itself, depend on the interface response, in particular the level of frictional damping present. Friction coefficients may vary with position and with time, and wear may lead to changes in geometry and contact tractions as well as removing initiated cracks. The range of situations in which fretting occurs is quite wide [1] and it is generally accepted that fretting can play a part in accelerating crack initiation, short crack growth, and long crack growth. The most significant effect may well depend on the relative magnitudes of a number of length scales [2]. Since the mid 1970s a significant amount of effort has

---

[1] Department of Engineering Science, University of Oxford, Parks Road, Oxford, OX1 3PJ, UK.

gone into the analysis of the long crack problem using fracture mechanics based techniques e.g. [3, 4]. It is clear that this aspect of the problem may be solved with the available tools, viz. Fracture mechanics and the Paris Law [5], although it should be recognized that the loading is mixed-mode, non-proportional and there may be a significant amount of crack closure.

Unfortunately, solution of the long crack problem does not seem to have brought us much closer to a practical life prediction method for the majority of fretting situations and component failures are still a practical concern [6]. The reason for this is, of course, that in many cases most of the life is consumed in crack initiation and propagation. Attempts have been made to extend a propagation-based analysis into this regime [7] by using an "effective initial flaw size" methodology. Such methods are in common use for critical parts (e.g. gas turbine disks) under plain fatigue, but have met with limited success in the fretting situation. The reason for this is not entirely clear, but is likely to be associated with the treatment of initiation as Paris Law propagation from an effective initial flaw. In the case of high stress gradients this assumption will give different weight to the initiation phase than in the case of more modest gradients. In recent years, attention has therefore turned to the initiation and short crack phases of crack development. Several attempts [8, 9] have been made to predict fretting fatigue thresholds using short crack methods based on the Kitagawa-Takahashi diagram [10]. These show some promise, but can be lengthy and time-consuming to implement. Others [11, 12, 13] have concentrated on the initiation approach and have attempted to apply widely accepted initiation parameters to the fretting problem. This approach seems to work reasonably well for large contacts of Hertzian-type geometry, where stress gradients are not excessively high. However, these are not necessarily typical of actual component geometries, where complete or nearly-complete contacts are frequently used and stress gradients can be very high in the neighbourhood of stress concentrations or singularities.

Even with the Hertzian geometry (where there is no stress singularity), the high stress gradients present in small contacts can lead to anomalous results. Fouvry et al [13] found that, applying the Dang Van criterion [14] to Hertzian contact of a sphere on a flat produced an under-estimate of specimen life unless the Dang Van parameter was averaged over a characteristic volume. The need for this averaging process was ascribed to the high stress gradients present. The difficulty is illustrated particularly well experimentally in experiments first carried out by Bramhall [15], and repeated by Nowell [16, 17]. Here a Hertzian geometry of a cylinder on flat was used and the contact geometry and loading were varied in such a way as to produce a constant maximum stress across a range of contacts, but to vary the size of contact and therefore the size of subsurface region affected. It was found that large contacts produced low lives, whereas for small contacts the life appeared very long (greater than 10^7 cycles). The experiments in [15-17] were carried out with an aluminium alloy, but similar results have been obtained with other materials.

In practical design situations, the designer is often interested in the fatigue threshold under fretting. He or she must design the contact so that fretting fatigue is unlikely to be a life-limiting phenomenon for the components. We believe that experiments of the type described above provide an important input for this type of safe-life design. Transition to the long-life regime in the Hertzian experiments might enable us to determine the critical volume which must be subjected to high stresses in order to initiate cracks in the material. Further, they provide a demanding, but well-controlled, test of the applicability of various life prediction methods in fretting fatigue. In this paper we intend to describe the application of multiaxial initiation parameters (together with Paris Law propagation) to the prediction of total fretting fatigue life for these type of experiments. The methods developed should provide a means of analysing other contacts where high stress gradients are present. There are parallels, of course with our earlier work on short crack growth, and we shall return to this point in the discussion.

Experiments

The main features of the experiment have been described elsewhere (e.g. [15-17]), but a brief overview will be given here for readers unfamiliar with the configuration. A schematic diagram of the experimental geometry is shown in Figure 1. Cylindrical fretting pads are brought into contact with a plane specimen under the influence of a constant normal force, P, and a cyclic tensile stress, $\sigma_0\sin(\omega t)$, is then applied to the specimen. This causes extension of the specimen, so that the contact point moves. The pads are restrained by springs and these cause the development of a tangential fretting force, Q, as the contact displaces. Hertzian contact takes place between the pads and the specimen and it can be shown [19] that the peak contact pressure, p_0, is proportional to P and to the pad radius, R, as follows

$$p_0 \propto \sqrt{\frac{P}{R}} \tag{1}$$

Similarly, the semi-width of contact, a, also varies with P and R, but in a different combination:

$$a \propto \sqrt{PR} \tag{2}$$

Hence by varying P and R, so that their ratio remains constant it is possible to carry out a series of experiments where the peak contact pressure remains constant, but the size of the contact varies. This was done for a number of different values of the salient parameters (p_0, σ_0 and Q/P). In each case small contacts produced a life greater than 10^7 cycles,[2] whereas larger contacts produced lives in the region of 10^5 to 10^6 cycles. Typical results are shown in Figure 2 for HE15-TF, and Al-4%Cu alloy. The transition between long and short life regimes may readily be seen. It should be noted that all tests were carried out in partial slip at a constant frequency of 20Hz.

Analysis

In developing an analysis method, the first decision to be made is which initiation parameter or parameters are to be used. A wide variety of empirically based parameters have been used in practice. Some [20] have been specifically developed for fretting

---

[2] Tests were stopped at 10^7 cycles if no failure had occurred.

situations, but more recent approaches have attempted to use parameters which have been more widely adopted in other fatigue situations. A full discussion of suitable

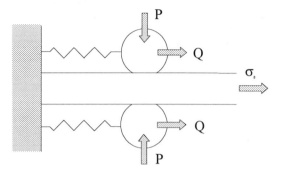

Figure 1 – *Schematic of experimental fretting fatigue configuration*

parameters is given in [*18*] and will not be repeated here. However we should note that, in principle, cracks may initiate at any point beneath the contact and a multiaxial stress state may therefore exist. Further, critical plane parameters are attractive, since they provide information regarding the location of crack initiation and the initial direction of growth. Several of these have been used in fretting situations by others. In particular the Smith Watson Topper (SWT) parameter [*21*] has been used by Farris [*11*] and Neu [*12*]. This parameter is appropriate for tensile crack initation. The value of the parameter, Γ_{SWT} on a particular plane may be related to the strain range, $\Delta\varepsilon$, experienced normal to the plane and the maximum tensile stress, $\Delta\sigma_{max}$ across the plane by

$$\Gamma_{SWT} = \sigma_{max}\left(\frac{\Delta\varepsilon}{2}\right) \qquad (3)$$

The value of this parameter may be related empirically to the fatigue life, N_f, by a relationship of the form

$$\Gamma_{SWT} = \frac{\sigma_f^2}{E}(2N_f)^{2b} + \sigma_f\varepsilon_f(2N_f)^{b+c} \qquad (4)$$

where E is Young's Modulus and σ_f, ε_f, b and c are material constants, which may be derived from the S-N curve in plain fatigue. The above parameter is not, however, appropriate for materials which exhibit shear mode initiation and a similar parameter, suggested by Fatemi and Socie [*22*] is an appropriate alternative in this case. The Fatemi-Socie parameter Γ_{FS} is related to the range of shear strain, $\Delta\gamma$, as well as $\Delta\sigma_{max}$ and is given by

$$\Gamma_{FS} = \frac{\Delta\gamma}{2}\left(1 + \alpha\frac{\sigma_{max}}{\sigma_y}\right) \qquad (5)$$

where σ_y is the tensile yield stress of the material and α a constant which approaches unity for long lives. Again the parameter may be related empirically to fatigue life using the empirical equation

$$\Gamma_{FS} = \frac{\tau_f}{G}(2N_f)^d + \gamma_f(2N_f)^e \qquad (6)$$

where G is the shear modulus and τ_f, γ_f, d, and e are material parameters, again derived from the plain fatigue curve. This parameter has been used by Neu et al. [12] to predict fretting fatigue life.

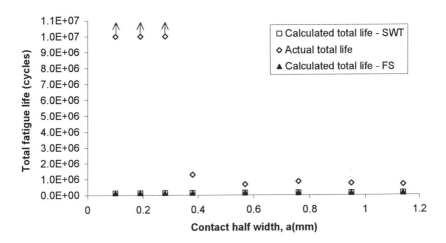

Figure 2 – *Comparison of experimental life and predicted total life as a function of contact size – SWT and FS parameters*

In a given fretting situation one does not necessarily know a priori whether initiation will take place in tensile or shear mode, so an appropriate design philosophy might be to predict the initiation lives by both methods and take the lowest prediction. In order to analyse our experimental configuration it is necessary to determine the stresses and strains in the neighbourhood of the contact. The contact pressure is essentially Hertzian and in partial slip the shear tractions are similar to those described by Mindlin [23], with a modification (essentially a shift of the stick zone) for the presence of tension in the specimen [24]. By treating the experiment as a half plane, it is possible to use Muskhelishvili potentials to determine the stress components [25]. Some caution is needed in the use of half-plane methods and a modification to the stress component parallel with the surface is required in some cases [26, 27]. Once the stresses have been obtained, values of the initiation parameters may be determined by evaluating the stress history for a complete cycle at a grid of points underneath the contact. It should be noted that a full range of candidate critical planes needs to be examined at each point so that the number of calculations required is quite large. Figure 3 shows contours of the two initiation parameters beneath the contact for Series 1 experiments (p_0 = 157 MPa, σ_0 = 93 MPa, Q/P = 0.45). It will be seen that in both cases there is a concentration of the parameter towards the trailing edge of contact where experimental failures are observed.

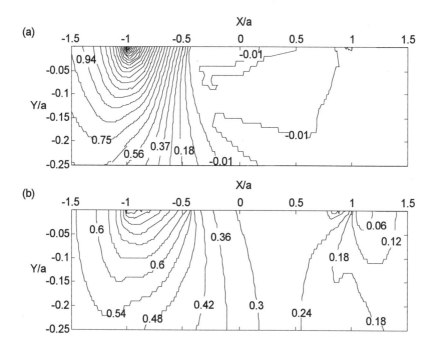

Figure 3 - *Contours of (a) $\Gamma_{SWT}E/p_0^2$ and (b) $\Gamma_{FS}G/p_0$ for Al 4%Cu Series 1 experiments (p_0 = 157 MPa, σ_0 = 93 MPa, Q/P = 0.45)*

The experiments were run to 10^7 cycles or to failure (whichever was earlier) so that information is obtained on total life. In order to compare the experimental total life with predicted values, materials data for equations (4) and (6) was obtained from plain fatigue tests reported in the literature.[3] This gave initiation life to a crack size of 1mm. In order to obtain total life, the propagation life from a 1mm crack to failure was estimated using

---

[3] For details, please see reference [18].

the Paris Law [5] and added to the predicted initiation life.[4]   In general it was found that the predicted initiation life was much less than the propagation life. Figure 2 shows the predictions of total life according to the SWT and FS parameters together with the experimental results. It can be seen that the predicted lives are similar for both parameters, and that both approaches give rise to reasonable predictions of life for large contacts. The transition to long lives at small contacts is, however, not predicted by either approach. The reason for this is clear; the magnitude of the stresses has remained unchanged between the different experiments, whereas the extent of the stress field has changed. Since the parameters and the predictions of life are based only on the most severe stress state, there will be no difference in predicted life. It is clear that, in order to predict the size effect properly, it will be necessary to develop a method which introduces a length scale into the problem.

Averaging procedures

From a physical basis it might be argued that a high value of an initiation parameter at a single point is insufficient to cause crack initiation. Cracks will not propagate by LEFM until they have reached a sufficient size for the stress intensity factor range to equal the long crack threshold. Thus, one might argue that for a crack to initiate and propagate to this 'onset size', high values of stresses must be sustained over a sufficient distance for the crack to become established. In terms of the multiaxial initiation parameters discussed here, we may propose that the average value of the parameter over a critical length or volume is a better measure of the likelihood of crack initiation. Accordingly two averaging methods were investigated:

Method 1: averaging along the critical plane - Here we first found the most highly stressed point (i.e. the maximum of the parameter) and established the critical plane. It was argued that the crack would initiate on this plane and values of the parameter were therefore averaged over a distance d_c along the direction of the critical plane. Several values of d_c were chosen in order to see which would provide the best fit to the experimental results.

Method 2: averaging over a volume - In this case we chose to define a critical volume ($V_c = d_c^3$).[5]   Averaging of the stress components was carried out over the volume before calculation of the initiation parameters. Different volumes were chosen around the most heavily loaded point in order to establish the one which gave the largest value of averaged parameter. This was used to predict the initiation life. Once again different values of d_c were used.

Figures 4 and 5 show the results obtained for each method for different values of d_c using the SWT parameter. Once again total life was calculated as the sum of initiation life (to 1mm) and propagation life. Results were also calculated for the F-S parameter, but were rather similar. A full presentation of the results, including those for different series of experiments and for another material (Ti-6/4) is given in [28]. It can be seen

---

[4] Stress intensity factors were calculated for a two-dimensional geometry (i.e. assuming a through-thickness crack) under contact loading using the dislocation density method (see [16, 28]).

[5] In practice the stress analysis used here is two-dimensional, so the actual averaging was carried out over an area in this case.

from the figure that there is little difference between the results from the two averaging procedures. Choice of d_c in the region of 50 µm seems to give a reasonable fit to the experimental results over the full range of contact sizes. The value must, of course, be chosen by reference to a series of tests like these before it can be used in a design situation and it remains to be seen whether it may be regarded as a material constant. However it is worth noting that the value of the parameter is of the same order as the grain size of the material (about 100 µm in the alloy used, which has a highly elongated grain structure in the extrusion direction (along the specimen axis)). A similar effect was noted with later experiments using Ti6/4 [29], where the best fit value for d_c was about 10 µm, compared with a grain size of 5 µm.

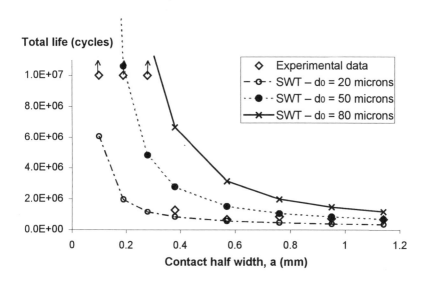

Figure 4 – *Experimental and predicted total life for series 1 experiments using averaging method 1*

Discussion and Conclusions

The work carried out here has shown that simple application of multiaxial initiation parameters can lead to under-prediction of fatigue life in cases where there are high stress gradients. Averaging the parameters over a volume gives a much better estimate of the fatigue life and can be used to explain the size effect noted in the Hertzian contact geometry. A number of different techniques for averaging are available, but the results do not seem to be very sensitive to the method used. The size of the averaging volume appears to be of the same order as the grain size for the two materials which have been investigated so far, but more work needs to be done to establish whether this is the case for other materials. Other workers have, however, found similar effects; Fouvry et al [13] have carried out fretting fatigue tests on a low alloy stainless steel using the Dang

Van initiation parameter [*14*]. The found that averaging of the parameter over a cube of side 5μm was necessary in order to predict the fatigue life correctly. This dimension was of the same order as the grain size. As part of the current program of work [*28*] we have also employed the Dang Van parameter, but as noted by Fouvry, the results are rather similar to the Fatemi-Socie predictions, but the computations are more lengthy. We have not, therefore, carried out the averaging procedure for the Dang Van parameter

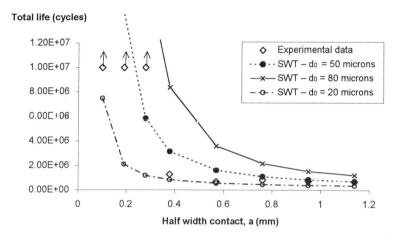

Figure 5 – *Experimental and predicted total life for series 1 experiments using averaging method 2*

There are a number of implications for life prediction of fretted components. First, some form of averaging procedure is likely to be necessary in cases of high stress gradients. In particular, applications such as dovetail blade roots and spline joints in gas turbine engines give rise to very localised stress concentrations and life prediction on the basis of peak stress values will almost certainly produce underestimates of fatigue lives. At present it is not clear how to choose the averaging dimension, but Hertzian tests in which the normal load and contact radius are varied so that the P/R ratio remains constant provide a potentially useful means of establishing the appropriate value. Secondly, although multiaxial parameters were used in the current study, it should be noted that failures actually took place on the surface close to the edge of contact, where the shear component of loading is small. Direct stress amplitude in a direction parallel to the surface may therefore provide a simpler procedure yielding satisfactory results. In order for the proposed method to gain widespread acceptance it will be necessary to validate the procedure using more practical contact geometries. We are currently undertaking an experimental program of work on the dovetail and spline geometries in order to provide the necessary information on fatigue life.

It should be noted that averaged initiation parameters are not the only means of accounting for the size effect discussed here. Our earlier work has proposed that short crack arrest methods based on the Kitagawa-Takahashi diagram could be used to similar effect [*8*]. In a general sense there is not much fundamental difference between the two approaches in that they both require a high level of stress to be sustained over a sufficient distance so that crack initiation and early growth can take place. The method described here approaches the problem from an initiation standpoint, whereas the short crack work uses a growth threshold. Neither approach is likely to be completely correct, since initiation and short crack growth are *both* required in order for failure to take place. Further, the mechanics of these processes are not well understood and the use of empirical initiation parameters, stress range, or ΔK to characterise the 'crack driving force' is questionable. More fundamental research into the mechanics of initiation and short crack work is clearly required. Nevertheless, designers still require improved methods of life prediction in fretting fatigue, and the method described above provides a possible solution, at least in the short to medium term.

Acknowledgments

Alex Araújo would like to acknowledge the support of the Brazilian government (CAPES).

References

[1] Waterhouse, R. B., "Fretting Fatigue," *International Materials Reviews*, 37, 2, 1992, pp. 77-97.

[2] Nowell, D., Hills, D. A., and Moobola, R. "Length scale considerations in fretting fatigue," *Fretting Fatigue: Current Technology and Practices, ASTM STP 1367*, D. W. Hoeppner, V. Chandrasekaran, and C. B. Elliott, Eds., American Society for Testing and Materials, West Conshohocken, PA, 2000, pp. 141-153.

[3] Edwards, P. R., Ryman, R. J., and Cook, R., "Fracture Mechanics Prediction of Fretting Fatigue under Constant Amplitude Loading," *RAE Technical Report 77056*, 1977.

[4] Hattori, T., Nakamura, M., Sakata, H. and Watanabe, T. "Fretting Fatigue Analysis using Fracture Mechanics," *JSME International Journal, Series 1*, 31, 1, 1988, pp. 100-107.

[5] Paris, P. C., Gomez, M. P., and Anderson, W. P. "A Rational Analytic Theory of Fatigue," *The Trend in Engineering*, 13, 1961, pp. 9-14.

[6] Xi, N. S., Zhong, P. D., Huang, H. Q., Yan, H., and Tao, C. H. "Failure Investigation of Blade and Disk in First Stage Compressor," *Engineering Failure Analysis*, 7, 200, pp. 385-392.

[7] Fellows, L. J., Nowell, D., and Hills, D. A., "Analysis of Crack Initiation and Propagation in Fretting Fatigue: the Effective Initial Flaw Size Methodology," *Fatigue & Fracture of Engineering Materials & Structures*, 20, 1, 1997, pp. 61-70.

[8] Nowell, D., and Araújo, J. A., "Analysis of Pad Size Effects in Fretting Fatigue using Short Crack Arrest Methodologies," *International Journal of Fatigue*, 21, 1999, pp. 947-956.

[9] Chan, K. S., Li, Y., Davidson, D. L., and Hudak, S. J. Jr, "A Fracture Mechanics Approach to High Cycle Fretting Fatigue based on the Worst Case Fret Concept: Part 1 – Model Development," *International Journal of Fracture*, 112, 2001, pp. 299-330.

[10] Kitagawa, H., & Takahashi, S., "Applicability of Fracture Mechanics to Very Small Cracks or the Cracks in the Early Stage," *Proceedings of the 2nd International Conference on Mechanical Behaviour of Materials*, American Society for Metals, 1976, pp. 627-631.

[11] Szolwinski, M. P., Harish, G., McVeigh, P. A., and Farris, T. N., "Experimental Study of Fretting Crack Nucleation in Aerospace Alloys with Emphasis on Life Prediction," *Fretting Fatigue: Current Technology and Practices, ASTM STP 1367*, D. W. Hoeppner, V. Chandrasekaran, and C. B. Elliott, Eds., American Society for Testing and Materials, West Conshohocken, PA, 2000, pp. 267-281.

[12] Neu, R. W., Pape, J. A., and Swalla-Michaud, D. R., "Methodologies for Linking Nucleation and Propagation Approaches for Predicting Life under Fretting Fatigue," *Fretting Fatigue: Current Technology and Practices, ASTM STP 1367*, D. W. Hoeppner, V. Chandrasekaran, and C. B. Elliott, Eds., American Society for Testing and Materials, West Conshohocken, PA, 2000, pp. 369-388.

[13] Fouvry, S., Kapsa, P., and Vincent, L., "A Multiaxial Analysis of Fretting Contact taking into account the Size Effect" *Fretting Fatigue: Current Technology and Practices, ASTM STP 1367*, D. W. Hoeppner, V. Chandrasekaran, and C. B. Elliott, Eds., American Society for Testing and Materials, West Conshohocken, PA, 2000, pp. 167-182.

[14] Dang Van, K., Griveau, B., and Message, O., "On a New Multiaxial Fatigue Limit Criterion: Theory and Application," *Biaxial and Multiaxial Fatigue*, M. W. Brown and K. J. Miller, Eds., MEP, London, 1989, pp. 479-496.

[15] Bramhall, R. "Studies in Fretting Fatigue," D.Phil. thesis, University of Oxford, 1973.

[16] Nowell, D., "An Analysis of Fretting Fatigue," D.Phil. Thesis, University of Oxford, 1988.

[17] Nowell, D., and Hills, D. A. "Crack Initiation Criteria in Fretting Fatigue", *Wear*, 136, 1990, pp. 329-343.

[18] Araújo, J. A., and Nowell, D., "The Effect of Rapidly Varying Contact Stress Fields on Fretting Fatigue," *International Journal of Fatigue*, 24, 2002, pp. 763-775.

[19] Johnson, K. L., "Contact Mechanics", Cambridge University Press, 1985.

[20] Ruiz, C., and Chen, K.C., "Life Assessment of Dovetail Joints between Blades and Discs in Aero-Engines", *Proceedings of the International Conference on Fatigue, Sheffield, 1986*, I.Mech.E., London, 1986.

[21] Smith, K. N., Watson, P., and Topper, T. H., "A Stress-Strain Function for the Fatigue of Metals," *Journal of Materials*, 5, 4, 1970, pp. 767-778.

[22] Fatemi, A., and Socie, D. F., "A Critical Plane Approach to Multiaxial Fatigue Damage Including Out of Phase Loading," *Fatigue and Fracture of Engineering Materials and Structures'*, 11, 1988, pp. 149-165.

[23] Mindlin, R.D., "Compliance of Elastic Bodies in Contact," *Journal of Applied Mechanics*, 16, 1949, pp. 259-268.

[24] Nowell, D., and Hills, D. A., "Mechanics of Fretting Fatigue Tests," *International Journal of Mechanical Sciences*, 29, 5, 1987, pp. 355-365.

[25] Hills, D. A., and Nowell, D., "Mechanics of Fretting Fatigue", Kluwer, Dordrecht, 1994.

[26] Fellows, L. J., Nowell, D., and Hills, D. A., "Contact Stresses in a Moderately Thin Strip (with Particular Reference to Fretting Fatigue Experiments)", *Wear*, 185, 1995, pp. 235-238.

[27] Hills, D. A., and Nowell, D., "A Discussion of : "Peak Contact Pressure, Cyclic Stress Amplitudes, Contact Semi-Width and Slip Amplitude: Relative Effects on Fatigue Life" by K. Iyer," *International Journal of Fatigue*, 23, 2001, pp. 747-748.

[28] Araújo, J. A., "On the Initiation and Arrest of Fretting Fatigue Cracks," D.Phil. Thesis, University of Oxford, 2000.

Andrew Mugadu,[1] David A. Hills,[2] and Ludwig Limmer[3]

A Theoretical and Experimental Procedure for Predicting the Fretting Fatigue Strength of Complete Contacts

REFERENCE: Mugadu, A., Hills, D. A., and Limmer, L., "A Theoretical and Experimental Procedure for Predicting the Fretting Fatigue Strength of Complete Contacts," *Fretting Fatigue: Advances in Basic Understanding and Applications, STP 1425*, Y. Mutoh, S. E. Kinyon, and D. W. Hoeppner, Eds., ASTM International, West Conshohocken, PA, 2003.

Abstract: An asymptotic approach to the complete fretting contact problem is presented, with a view to elucidating the crack nucleation process. It is suggested that the characteristic stress state at the edge of the prototypical contact can be matched to a test having the same local contact angle and coefficient of friction. A generalized stress intensity factor is then defined, to scale both fields, hence matching the magnitudes of the stress state. Experimental investigations of this approach has prompted the need for some modifications to our original test apparatus, in order to accommodate complete contacts, and these are briefly outlined. Last, the limitations of this approach are discussed.

Keywords: fretting fatigue, crack nucleation, complete contacts, asymptotic analysis

Introduction

Much of the experimental work in fretting fatigue carried out recently at Oxford has been aimed at understanding the basic mechanics of the fretting fatigue problem, using a two-actuator test apparatus, and employing Hertzian contacts [1] and [2]. This approach has been taken because it is possible to separate out the contact problem from the crack initiation and propagation problems easily. Furthermore, the contact problem itself is amenable to a solution in closed form, both for the contact pressure distribution and the effects of complex shearing tractions with attendant stick and slip zone regimes. It has therefore proved possible to elucidate a great deal about the nature of the partial slip contact problem and the crack propagation problem, in a compelling way, but, at the same time, there are still some unanswered questions. The principal

[1]DPhil. student, sponsored by the Rhodes Trust, Department of Engineering Science, Oxford University, Parks Road, OX1 3PJ, U.K.
[2]Professor of Engineering Science, Department of Engineering Science, Oxford University, Parks Road, OX1 3PJ, U.K.
[3]Department of Engineering Science, Oxford University, Parks Road, OX1 3PJ, U.K.

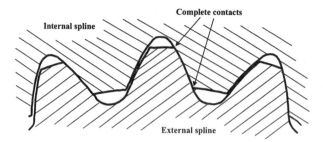

Figure 1 - *Schematic diagram of the cross-section of a spline tooth pair, looking in the axial direction*

one concerning the results of Hertzian tests is the "size effect," whereby we find that, for two identical tests sustaining the same stress state, the one involving a small contact will exhibit an infinite life, while the one involving a large contact will show a finite life. This has never been fully resolved, although various tentative explanations have been presented. For example, it has been suggested that a critical volume of material at the initiation point needs to experience a critical plane parameter before nucleation can proceed. A second problem concerns the development of a life-prediction approach when the contact pressure is singular. Practical examples of this abound: the conformal contact between involute spline teeth (Figure 1), various types of bolted and clamped joints, and certain designs of dovetail fixings at fan-blade roots. In all these cases, because the pressure at the edge of the contact as implied by an elasticity solution is singular, any method involving the simulation of the stress state using a Hertzian type specimen would obviously be inappropriate.

An alternative approach to this problem is to use ideas borrowed from fracture mechanics (and sharp notch mechanics), viz. to attempt to simulate the process zone in which the crack initiates, at the contact edge. The strategy to be developed is as follows: first, the state of stress near the edge of a slipping contact has been studied in the literature. It is known that there is a power order singularity (usually less than square root in strength), related to both the local contact angle and coefficient of friction. In the experimental apparatus, a tensile test specimen with contacting pads pressed onto the surface is utilized [1,3]. However, in contrast to the Hertzian contacts employed in previous experiments, the pads have flat faces with edges inclined at the same contact angles as those present in the prototypical problem. Thus, with the coefficient of friction matched in both problems, the spatial distribution of stresses is the same and hence, providing small scale yielding conditions exist, so is the nature of the strain history of material present in the process zone. There is a strong analogy between this problem and that of two cracks, each suffering say, mode I loading. In the case of the crack, to make the stress fields equal in magnitude as well as spatial variation, it would be necessary to match the stress intensity factors. The same is true of the two fretting problems; we define generalized stress intensity factors to scale the asymptotic fields in the two problems and arrange for these to be matched so that the

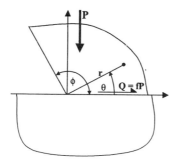

Figure 2 - *A wedge sliding on an elastic half-plane*

stress states are, identical in every respect. To ensure that the test truly represents the prototypical conditions further matching is needed, in the form of ensuring that the two components have the same surface finish characteristics, and also, that the same slip displacement is employed. It is then proposed that the nucleation/no nucleation conditions, and the number of cycles to nucleate a crack in the former case, will be the same for the two problems.

Asymptotic Analysis

The fundamental asymptotic solution required is that of the stress state adjacent to the corner of a component sliding on a half-plane (Figure 2). Although a full understanding of the macroscopic contact of (Figure 1) would require formulation in a different domain, in an asymptotic solution we focus in on a very small process zone at the contact corner. In these circumstances the remote shape of the contact domain is unimportant. This problem has been studied in detail before [4] and [5], and here, only the salient points will be presented. The stress state in the neighborhood of the wedge corner is of the form

$$\sigma_{ij} = K^* r^{\lambda - 1} g_{ij}(\theta; f, \phi) \quad i, j = r, \theta \quad \text{for } \lambda \text{ real and } 0 < \lambda < 1, \quad (1)$$

where (r, θ) are polar coordinates centered on the contact corner, λ and $g_{ij}(\theta; f, \phi)$ are, respectively, the eigenvalue and eigenfunctions of the solution, f, the coefficient of friction, ϕ, the contact angle, and K^*, the generalized stress intensity factor defined as

$$K^* = \sigma_{\theta\theta} r^{1-\lambda} \quad Lt \quad r \to 0 \quad \text{on} \quad \theta = 0. \quad (2)$$

It is not possible, here, to provide a complete description of the method of solution, and we simply show, in Figure 3, the eigenvalue, λ, as a function of f and ϕ. It will be noted that, when $\lambda - 1 > 0$ the state of stress at the apex is bounded, and the procedure being developed does not apply; the figure therefore focuses on the singular region. Note that the curves are annotated with the coefficient of friction, which is given a

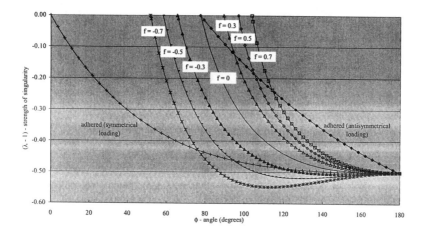

Figure 3 - *Plot of the order of singularity,* $(\lambda - 1)$, *against the frictional wedge angle,* ϕ, *for various coefficients of friction,* f

sign. A positive sign indicates that the frictional force is directed away from the wedge corner, and gives rise to the weaker singularity, for a given wedge angle, and vice versa. This figure displays the strongest singularity which can be found and, in contrast to the monolithic wedge problem (Figure 4), which is conveniently split into a symmetric and antisymmetric solution, the corresponding eigenfunctions here, $g_{ij}(\theta; f, \phi)$, contains neither inherent symmetry nor anti-symmetry. The order of singularity associated with these solutions (symmetric and antisymmetric loading) is also included in (Figure 3) where $\phi = 2\alpha - \pi$. The majority of the region depicts a singularity which is weaker than square root singular, although, there is a range of parameters for which the order of the singularity is greater than $-1/2$. This rather surprising result is, in fact, physically correct, and represents conditions when the coefficient of friction in the wedge problem is greater than the "naturally occurring" traction ratio, $\tau_{r\theta}/\sigma_{\theta\theta}$, in a monolithic wedge of the same geometry i.e. $\phi = 2\alpha - \pi$.

Calibration

The multiplicative factor scaling the magnitude of the solutions, K^*, has, in general, to be found by a purely numerical procedure, such as the finite element method. This is not without difficulty, as the asymptotic form of the stress field ought to be imposed on the general behavior of the stress state around the contact corner, in order to accelerate convergence of the solution, and this is not easy to achieve. However, it will be noted from Figure 3 that for a contact angle of $\phi = 90°$, if the coefficient of friction is equal to $2/3$, a square root singularity exists (for the shearing

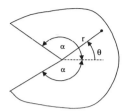

Figure 4 - *A notch of internal angle* 2α

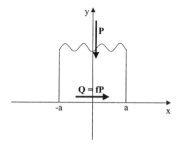

Figure 5 - *Representation of a square-ended rigid punch on an elastic half plane*

force directed towards the wedge corner, negative f), This permits standard crack-tip elements to be used adjacent to the contact corner [6].

A problem that admits an analytical solution is that of a rigid, square-ended pad, of half width a, subject to a normal load, P, and sliding against an elastically dissimilar half-plane (Figure 5). For this case the normal traction distribution, $p(x)$, is given by [7]

$$\frac{ap(x)}{P} = -\frac{\sin \lambda_1 \pi}{\pi}(1 - x/a)^{\lambda_1 - 1}(1 + x/a)^{-\lambda_1} \tag{3}$$

where

$$\tan \lambda_1 \pi = \frac{1}{f\beta} = \frac{2(1 - \nu)}{f(1 - 2\nu)} \qquad 0 < \lambda_1 < 1. \tag{4}$$

Dundurs' parameter, β, varies such that $0 \leq \beta \leq 0.5$ which, from equation (4) above, corresponds to $0.5 \geq \nu \geq 0$. Here we define the generalized stress intensity factor, K_1^*, as

$$K_1^* = \lim_{r \to 0} p(r)r^{\lambda_1}, \tag{5}$$

where r is a polar coordinate measured from the pad corner. Making the substitution

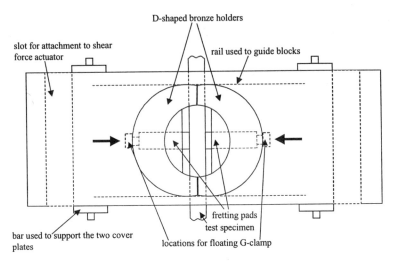

Figure 6 - *Schematic of the carriage showing the D-shaped bronze holders journalled between two rails*

$r = x + a$ in equation (3), we take the limit $x \to -a$, i.e. for the left hand corner of the pad, (Figure 5). This leads to a solution for the normalized generalized stress intensity factor

$$\frac{a^{1-\lambda_1} K_1^*}{P} = -\frac{\sin \lambda_1 \pi}{\pi 2^{1-\lambda_1}}. \tag{6}$$

Experimental

A modified form of the fretting apparatus used in the past to conduct Hertzian type tests has been developed to enable the new type of test pad geometry to be employed. Clearly, the essential difference between a flat-ended pad and a Hertzian cylindrically faced pad is that the alignment of the pad front face against the test specimen face is absolutely critical. Any imperfection here will materially change the form of the stress state at the pad corner, and so, a self-aligning arrangement is highly desirable. At the same time, the shearing force must be applied strictly along the plane of the contact, rather than above it, so that no rocking moment is developed. These stringent and, to some extent, opposing requirements have been achieved by means of the pad holding assembly shown schematically in Figure 6. The essential feature of the fixture is the use of D-shaped bronze holders for the specimens, having an included angle of about 190°, and arranged so that the front faces of the pads lie exactly along the diameter of the D. The holders are permitted to rotate freely, and so the pads, loaded by a centrally applied contact force, are free to take up a natural orientation such that the pad face is pressed uniformly along the test piece face. At the same time, the shearing force is guaranteed to be transmitted along the correct line of action.

It is clear that, if the arrangement is geometrically correct and the generalized stress intensity factors for the two problems are matched, the state of stress in the neighborhood of the initiation zones will be identical. For complete fidelity between the prototype and the experiment the surface finish and slip displacement must also be matched.

Over 150 tests on super CMV shaft steel have been conducted on this apparatus for Rolls Royce plc. The results obtained are internally consistent and support the methodology described in this paper, but we cannot report the data for reasons of commercial confidentiality.

Nucleation Conditions

It is argued that, if the process zone where crack initiation occurs is sufficiently small for the local strains to be completely dominated by a surrounding zone in which the asymptotic singular field applies, the nucleation conditions are properly represented in the experiment. This is completely analogous to the concept of "small scale yielding" in fracture mechanics terms. For the particular problem of a rigid, square-ended block sliding along an elastic half-plane described above, it is possible to deduce the complete contact stress field, as well as to determine the eigenfunctions corresponding to the singular (dominant) term. Explicitly, for this problem, the singular field associated with $f = 2/3$ in the half plane, $(-\pi \leq \theta \leq 0)$, is given by

$$\begin{bmatrix} \sigma_{rr} \\ \sigma_{\theta\theta} \\ \tau_{r\theta} \end{bmatrix} = K^* r^{\lambda-1} \begin{bmatrix} -2.727\sin(0.44\theta + 0.794) & 1.050\sin(1.56\theta + 0.589) \\ -1.743\sin(0.44\theta + 0.794) & -1.050\sin(1.56\theta + 0.589) \\ -0.492\sin(0.44\theta - 0.777) & -1.050\sin(1.56\theta - 0.982) \end{bmatrix}$$
(7)

where K^* is defined in equation (2). The complete stress field is found from the Muskhelishvili potential for this problem, which is given by

$$\Phi(z) = (i + f)\frac{P}{2a\pi}\frac{1}{(z - 1)^{1-\lambda_1}(z + 1)^{\lambda_1}}$$
(8)

where $z = x + iy$, and $i = \sqrt{-1}$. When once the potential is known, the interior stress field may easily be found from standard results, by differentiation alone.

We can therefore find the discrepancy between the two solutions. This could be done by using any stress component, but, it would seem most appropriate to use a composite stress parameter such as the von Mises' yield parameter, J_2. Contours of the percentage difference between the full field solution and the singular term only solution are given in Figures 7(a and b). The quantity plotted is $\left[\left(\sqrt{J_2}_{full} - \sqrt{J_2}_{asymp}\right)/\sqrt{J_2}_{full}\right]$ for $f = \pm 2/3$. Further, the extent of the process or plasticity zone may be estimated by employing the elasticity solution, and noting the size of the region in which the yield condition is exceeded. This is analogous to the standard Irwin type procedure for estimating the plastic zone size in fracture

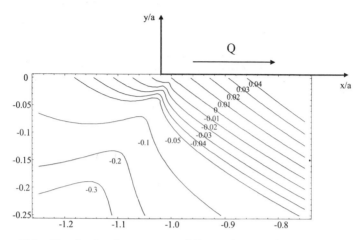

Figure 7(a) - *Plot showing the percentage difference between the full and asymptotic fields, $(\sqrt{J_2}_{full} - \sqrt{J_2}_{asymp})/\sqrt{J_2}_{full}\%)$ for $f = 2/3$*

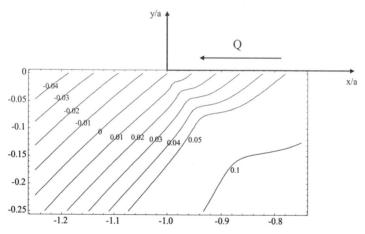

Figure 7(b) - *Plot showing the percentage difference between the full and asymptotic fields, $(\sqrt{J_2}_{full} - \sqrt{J_2}_{asymp})/\sqrt{J_2}_{full}\%,$ for $f = -2/3$*

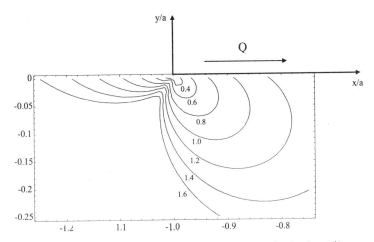

Figure 8(a) - *Plot of the normalized punch load, Pa/k, for $f = 2/3$*

mechanics. Figures 8(a and b) show the extent of the implied plasticity zone for a given normal load, P. Here a is the half-width of the pad, and k is the yield stress in pure shear. Just as in fracture mechanics, judging the size of the acceptable plasticity zone so that it might reasonably be assumed to exist under conditions of small scale yielding is not something which has an exact outcome. If, for example, a maximum of 3% difference between the full-field and singular solution is deemed acceptable, the largest load which may be tolerated, and still come within the scope of small scale yielding is about 0.8 for $f = 2/3$ (Figure 8(a)), and 0.6 for $f = -2/3$ (Figure 8(b)).

A further aspect of the initiation process which warrants attention is the effect of reversing the direction of the shearing force which, by definition, is a feature present in all fretting problems. It should be noted that this does not imply that the slip displacement need be the same in each direction, although, if this were not so, creep would result: and it should also be noted that, in a slipping region, the shear tractions can *only* be of magnitude $\pm fp(x)$, and not lower. We have already shown that, for some values of the contact angle, ϕ, the state of stress will be singular, regardless of whether the shearing tractions are directed towards or away from the pad corner. These problems present a particular challenge, as both the exponent in the asymptotic solution, and the multiplicative factor, the generalized stress intensity factor, will vary from one half cycle to the other. It is therefore not immediately clear whether it is the half cycle which generates the larger stress intensity factor, or the one which generates the stronger singularity, if different, or a combination of the two, which controls crack nucleation conditions. These cases will require further analysis, and the behavior will, to some extent, be controlled by the extent of the process (plasticity) zone. However, there is a significant range of geometries and coefficients of friction where the stress state is bounded when the shearing tractions act away from the corner, and is singular when the shearing tractions act towards it. In these cases, providing small scale

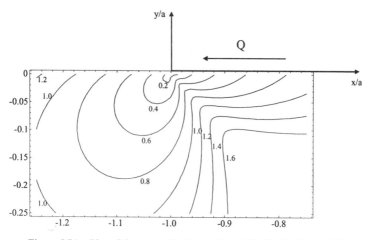

Figure 8(b) - *Plot of the normalized punch load, Pa/k, for $f = -2/3$*

yielding obtains, it is reasonable to assume that it is the characteristics of the solution (strength of singularity and generalized stress intensity factor), which will govern the nucleation conditions. They will control both the nucleation/no nucleation conditions, and the number of cycles taken to nucleate in the former case.

Conclusion

The procedure outlined in this paper provides a self-contained procedure for design against fretting fatigue crack initiation in cases where the prototypical contact is "complete" in nature. Arguments for its rigorous application can be made when the interfacial contact characteristics (surface finish and slip displacement) can be reproduced in the experiment, and where the stress state induced is singular when the shearing force acts towards the contact corner, but bounded when acting away from it. A modified form of the fretting apparatus which permits flat-faced pads to be employed has been developed and described.

References

[1] Hills, D. A. and Nowell, D., *Mechanics of Fretting Fatigue*, Kluwer Academic Publishers, Dordrecht, The Netherlands, 1994.

[2] Fellows, L. J., Nowell, D., and Hills, D. A., "Analysis of Crack Initiation and Propagation in Fretting Fatigue: The effective initial flaw size methodology," *Fatigue and Fracture of Engineering Materials and Structures*, Vol. 20, No. 1, 1997, pp. 61-70.

[3] Hills, D. A. and Nowell, D., "The Development of a Fretting Fatigue Experiment

with Well Defined Characteristics," *Standardization of Fretting Fatigue Test Methods and Equipment, ASTM STP 1159*, M. H. Attia and R. B. Waterhouse, Eds., American Socicty for Testing and Materials, West Conshohocken, PA, 1992, pp. 69-84.

[4] Gdoutos, E. E. and Theocaris, P. S., "Stress Concentration at the Apex of a Plane Indenter Acting on an Elastic Half Plane," *Journal of Applied Mechanics*, Vol. 42, 1975, pp. 688-692.

[5] Comninou, M., "Stress Singularity at a Sharp Edge in Contact Problems with Friction," *Journal of Applied Mathematics and Physics* (ZAMP), Vol. 27, 1976, pp. 493-499.

[6] Mugadu, A., Hills, D. A., and Limmer, L., "An Asymptotic Approach to Crack Initiation in Complete Contacts," *Journal of Mechanics and Physics of Solids*, Vol. 50, No. 3, 2002, pp. 531-547.

[7] Hills, D. A., Nowell, D., and Sackfield, A., *Mechanics of Elastic Contacts*. Pub. Butterworth Heinemann, Oxford, 1993.

FRETTING FATIGUE PARAMETER EFFECTS

Toshio Hattori,[1] Masayuki Nakamura,[2] and Takashi Watanabe[3]

Improvement of Fretting Fatigue Strength by Using Stress-Release Slits

REFERENCE: Hattori, T., Nakamura, M., and Watanabe, T., **"Improvement of Fretting Fatigue Strength by Using Stress-Release Slits,"** *Fretting Fatigue: Advances in Basic Understanding and Applications, ASTM STP 1425,* Y. Mutoh, S. E. Kinyon, and D. W. Hoeppner Eds., ASTM International, West Conshohocken, PA, 2003.

ABSTRACT: A new estimation method for fatigue crack initiation criteria using plain fatigue limit, threshold stress intensity factor range, and specific depth is proposed. Using this, new design method for improving fretting fatigue strength have been developed by using stress-release slits on the pads near the contact edges. This design has analyzed on the basis of stress singularity parameters at the contact edges, and the analysis results were used to optimize the slit shapes, such as slit depth and slit position. This optimization takes into account the fretting fatigue crack initiation strength at the contact edge and the fatigue strength at the slit root. We confirmed that the fretting fatigue crack initiation strength of the optimized design is about twice that of a conventional (i.e., slitless) design.

KEYWORDS: crack initiation, stress singularity parameters, stress-release slit, knurling,

Introduction

Fretting can occur when a pair of structural elements are in contact under a normal load, while cyclic stress and micro-slippage are forced along the contact surface. This condition can be seen in bolted or riveted joints [1,2], in shrink-fitted shafts [3,4], and in the blade dovetail region of turbo machinery [5,6], among others. When fretting occurs the fatigue strength decreases to less than one-third of that without fretting [7,8]. The strength is reduced because of the concentration of contact stresses such as contact

[1]Chief Researcher, Mechanical Engineering Research Laboratory, Hitachi Ltd., Tsuchiura, Ibaraki, Japan.

[2]Researcher, Mechanical Engineering Research Laboratory, Hitachi Ltd., Tsuchiura, Ibaraki, Japan.

[3]Chief Engineer, Hitachi Works, Hitachi Ltd., Hitachi, Ibaraki, Japan.

pressure and tangential stress at the contact edge, where cracks from fretting fatigue form and propagate.

Previously, the authors presented new fretting fatigue strength estimation methods using fracture mechanics for the fretting fatigue limit [7,9] and stress singularity parameters for the fretting fatigue crack initiation strength [10]. Using these methods, many fretting fatigue improvement designs such as grooving or knurling the contact surface [11], placing a low stiffness spacer on a contact interface, and reducing the wedge angle [10] were introduced. In this paper the authors introduce another fretting fatigue strength improvement method using stress-release slits on the pads near the contact edges. This design was analyzed on the basis of stress singularity parameters at the contact edges, and the analysis results were used to optimize the slit shapes, such as slit depth and slit position. This optimization takes into account the fretting fatigue crack initiation strength at the contact edge and fatigue strength at the slit root. It was confirmed that the fretting fatigue crack initiation strength of the optimized design is about twice that of a conventional (i.e., slitless) design.

Traditional fretting fatigue strength improvement methods

The authors have already presented many fretting fatigue strength improvement

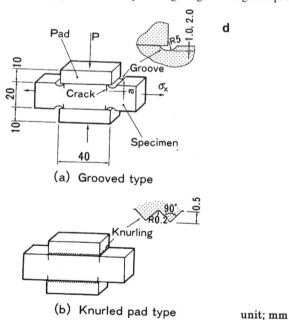

(a) Grooved type

(b) Knurled pad type   unit; mm

Figure 1-Fretting fatigue strength improvement method using grooving or knurling.

design methods. Firstly, design methods involving grooving or knurling a contact surface were introduced as shown in Fig. 1.By using stress analysis, fracture mechanics analysis, and fretting fatigue tests, the improvement of fretting fatigue limit about 70-80% have been confirmed as shown in Fig. 2. Then designs having a low stiffness spacer on a contact interface and a reduction of the wedge angle were introduced, and by using the stress and stress singularity parameter analyses, the improvement of fretting fatigue crack initiation strength was estimated as shown in Figs. 3 and 4. But these designs created many problems, such as complicating the manufacturing process and reducing safety. In this paper therefore another fretting fatigue strength improvement design using stress-release slits on the pads near the contact edges was presented.

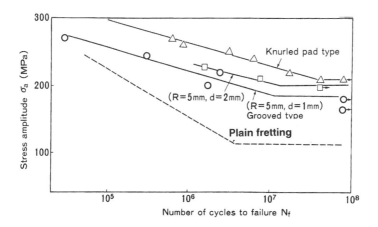

Figure 2-Fretting fatigue test results for grooving or knurling type specimens.

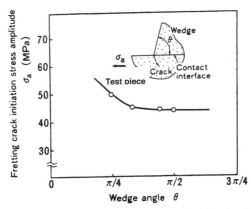

Figure 3-Estimation results of fretting fatigue crack initiation limits for each wedge angle.[10]

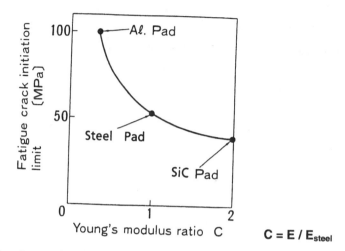

Figure 4-Estimation results of fretting fatigue crack initiation limits for each pad stiffness.[10]

Contact model with stress-release slits

The stress-release slit type contact model is shown in Fig. 5. In this model, the uniform contact pressure P_0 is applied to the outermost edge of the pad and uniform nominal stress σ_0 is applied to the specimen. The stress-release slits are formed on the pads near the contact edges. By using these slits, the concentration of contact pressures and tangential stresses at the contact edges are decreased and fretting fatigue crack initiation strength can be increased. The validity of this fretting fatigue crack initiation strength improvement method is confirmed by the stress and stress singularity parameters analyses that follow.

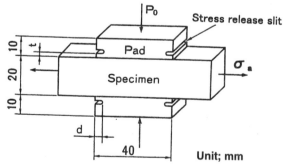

Figure 5-Stress-release slit type contact model.

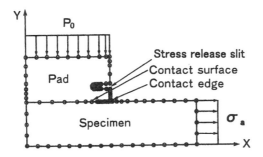

Fig. 6-Boundary element mesh of stress release slit type model.

Stress and stress singularity parameters analyses

To analyze the stress and stress singularity parameters of the stress-release slit contact model as shown in Fig. 5, the boundary element method for contact problems was used. The boundary element mesh for analyzing the stress and stress singularity parameters is shown in Fig. 6. This is a quarter section of the analytical model   shown in Fig. 5. The frictional coefficient μ on the contact surface was set at 0.7 by assessing the damage to the contact surface by repeated micro slippage [3].

To evaluate the fretting fatigue crack initiation strength of this model we used the stress singularity parameters as follows. The stress fields near the contact edges show singularity behavior as shown in Fig. 7.   These stress distributions near the contact edges can be expressed using the two stress singularity parameters, H and λ , as follows,

$$\sigma(r) = H/r^{\lambda} \qquad\qquad (1)$$

where σ is the stress (MPa), r is the distance from the singularity point (mm), H is the intensity of the stress singularity, and λ is the order of the stress singularity. Parameters λ and H are calculated as follows.   The order of the stress singularity (λ) for the contact edge can be calculated analytically [12] using wedge angles θ_1 and θ_2 , Young's moduli E_1 and E_2 , Poisson's ratios ν_1 and ν_2 the frictional coefficient μ (Fig. 8).   However, in this paper λ is calculated by finding the best fit of Eq.(1) to the stress distribution near the contact edge.   The stress distribution can be calculated by numerical stress analysis methods, such as the finite element or boundary element methods.   The intensity of the stress singularity (H) is calculated by the best fit of Eq.(1)

to the numerically analyzed stress distribution.   By comparing these calculated parameters (H and λ) with the crack initiation criterion Hc, fretting crack initiation conditions can be estimated for each wedge angle and pad stiffness.

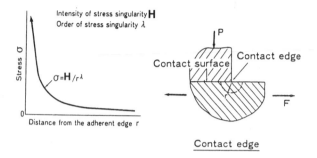

Fig. 7-*Stress distributions near a contact edge and stress singularity parameters.*

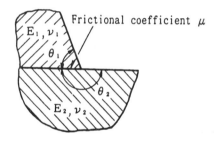

Fig. 8-*Geometry of contact edge.*

Fatigue strength analysis

In this paper three points (A: contact edge, B: inner - contact interface, and C: root of stress-release slit) were selected as the targets for the fatigue strength evaluation of this stress-release slit model (Fig. 9). The fatigue strength of these points was evaluated by using the stress at a certain depth from the surface or singular point. For instance, in the case of stress singular points, the fretting fatigue crack initiation criterion Hc was derived from the plain fatigue limit (σ_{wo}) and the threshold stress intensity factor range (ΔK_{th}) of the specimen material Ni-Mo-V steel. On both critical conditions the stress distributions cross at a point 0.012 mm in depth from the singular point as shown in Fig. 10. We expect that for the critical conditions for each stress concentration case, the stress distributions cross at this specific depth point $r_C = 0.012$ mm. From this assumption we can estimate the critical intensity value of stress singularity H_C for each order of stress singularity λ

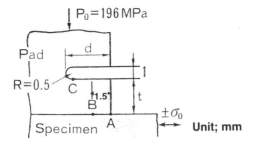

Fig. 9-Fatigue strength evaluation targets on stress-release slit type model.

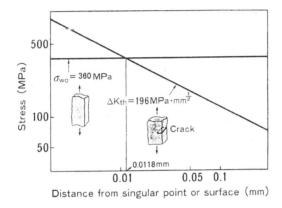

Fig. 10-Stress distributions near singular point on crack initiation conditions and specific length.

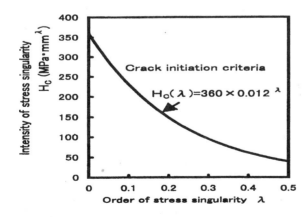

Fig. 11-Critical intensity value of stress singularity for Ni-Mo-V steel.

as shown in Fig. 11. First, the fretting fatigue crack initiation limits at contact edge A and contact interface B can be estimated as follows. The stress distributions near points A and B with slit depth d = 3.8 mm, slit root radius R = 0.5 mm, and slit position t = 3 mm, and on the loading conditions of nominal contact pressure P_0 = 196 MPa and nominal axial repeated stress σ_a = ±98 MPa are shown in Fig. 12. From these calculated results the order of stress singularity can be calculated as λ = 0.281 and intensity of stress singularity H_a = $\Delta H/2$ = 100 MPa·mm$^\lambda$ for contact edge A, and λ = 0.0 and H_a = 230 MPa·mm$^\lambda$ for contact interface B. By comparing these calculated results of stress singularity parameters H_a with the critical intensity value of stress singularity H_C, the fretting fatigue crack initiation limits at contact edge A and contact interface B can be estimated. Then the fatigue crack initiation limit at slit root C can be estimated by the assumption that the fatigue cracks initiate when the stress at the specific depth point r_C = 0.012 mm reaches the plain fatigue limit of σ_{w0} = 360 MPa as shown in Fig. 13. Using these comparisons of analyzed stress at specific depth r_C = 0.012 mm near each point with the plain fatigue limit σ_{w0} = 360 MPa the fatigue crack initiation limit for each point can be estimated as Fig. 14. From these estimated results we can see that the fretting fatigue crack initiation limit at contact edge A increases in accordance with the increase of slit depth, but on the contrary the fatigue crack initiation limit at slit root C and on contact interface B decrease in accordance with the increase of slit depth. And finally the relation between slit position and fatigue crack initiation strength for each optimized slit shape can be obtained as Fig. 15. From these results we can see that the fatigue crack initiation limit with stress-release slits of the optimum design can be increased to about 100 MPa, which is more than twice that of a conventional (i.e., slitless) design.

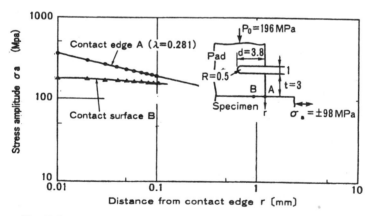

Fig. 12-Stress distributions near contact edge and contact interface.

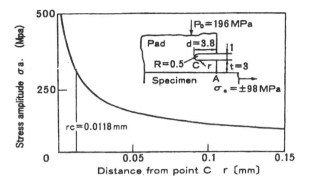

Fig. 13-Stress distribution near a slit root.

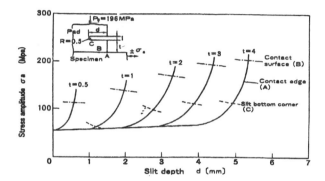

Fig. 14-Relation between crack initiation stress amplitude and slit shapes.

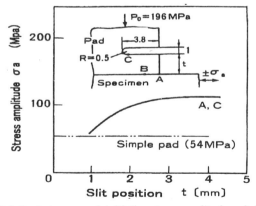

Fig. 15 Relation between crack initiation stress amplitude and slit position.

Conclusions

1. A new estimation method for fatigue crack initiation criteria using plain fatigue limit, threshold stress intensity factor range, and specific depth is proposed.

2. Using these fatigue crack initiation criteria, the stress-release slit design is optimized.

3. By optimizing the slit depth and slit position, the fretting fatigue crack initiation strength can be increased to about 100 MPa, which is about twice that of a conventional (i.e., slitless) design.

References

[1] Gassner, E., "The value of surface-protective media against fretting corrosion on the basis of fatigue strength tests," *Laboratorium fur Betriebsfestigkeit TM19/67,* 1967.

[2] Buch, A., "Fatigue and fretting of pin-lug joints with and without interference fit," *Wear,* 1977, 43, p. 9.

[3] Hattori, T., Kawai, S., Okamoto, N. and Sonobe, T., "Torsional fatigue strength of a shrink- fitted shaft," *Bulletin of the JSME,* 24, 197, 1981,   p. 1893.

[4] Cornelius, E. A. and Contag, D., "Die Festigkeits-minderung von Wellen unter dem Einflu β von Wellen-Naben- Verbindungen durch Lotung, Nut und Pa/feder, Kerbverzahnungen und Keilprofile bei wechselnder Drehung," *Konstruktion,* 14, 9, 1962,   p. 337.

[5] Hattori, T., Sakata, S. and Ohnishi, H., "Slipping behavior and fretting fatigue in the disk/blade dovetail region," *Proceedings, 1983 Tokyo Int.   Gas Turbine Cong.,* 1984, p. 945.

[6] Johnson, R. L. and Bill, R. C., "Fretting in aircraft turbine engines," *NASA TM X-71606,* 1974.

[7] Hattori, T., Nakamura, M. and Watanabe, T., "Fretting fatigue analysis by using fracture mechanics," *ASME Paper No.84-WA/DE-10,* 1984.

[8] King, R. N.and Lindley, T. C., "Fretting fatigue in a 3 $^1/_2$ Ni-Cr-Mo-V rotor steel," *Proc. ICF5,* 1980, p. 631.

[9] Sakata, H., Hattori, T. and Hatsuda, T., "An application of fracture mechanics to fretting fatigue analysis," *Role of Fracture Mechanics in Modern Technology,* Elsvier Science Publications B. V., 1987, p. 303.

[10] Hattori, T. and Nakamura, M., "Fretting fatigue evaluation using stress singularity parameters at contact edges," *Fretting Fatigue, ESIS Publication 18,* 1994, pp. 453.

[11] Hattori, T., Nakamura, M., and Watanabe, T., "Fretting fatigue analysis of strength improvement models with grooving or knurling on a contact surface," *ASTM STP 1159,* 1992, pp. 101.

[12] Dempsey, J. P. and Sinclair, G. B., "On the singular behavior at the vertices of a bi-material wedge," *J. Elasticity, 11,* 1981, P. 317.

Kozo Nakazawa,[1] Norio Maruyama,[1] and Takao Hanawa[1]

Effect of Contact Pressure on Fretting Fatigue in Type 316L Stainless Steel

Reference: Nakazawa, K., Maruyama, N., and Hanawa, T., "**Effect of Contact Pressure on Fretting Fatigue in Type 316L Stainless Steel,**" *Fretting Fatigue: Advances in Basic Understanding and Applications, ASTM STP 1425*, Y. Mutoh, S. E. Kinyon, and D. W. Hoeppner, Eds., ASTM International, West Conshohocken, PA, 2003.

Abstract. The effect of contact pressure on fretting fatigue in solution-treated austenitic stainless steel was studied under load control at a stress amplitude of 180 MPa and a stress ratio of 0.1. With an increase in contact pressure, fretting fatigue life was almost unchanged at contact pressures between 15 and 45 MPa, but it decreased drastically at contact pressures beyond 60 MPa. Frictional stress amplitude at the fretted area increased smoothly with contact pressure. It was impossible to explain the contact pressure dependence of life by the change in frictional stress amplitude. At low contact pressures, stress concentration due to fretting damage occurred at the middle portion of fretted area and the main crack responsible for failure was initiated there. At high contact pressures, deep concavity associated with plastic deformation of the specimen under the contact of fretting pad was formed without accompanying heavy wear. The main crack was initiated at the outer edge corner of the concavity which probably acted as a notch. It was suggested that the stress concentration at the concavity edge corner played an important role in fretting fatigue at high contact pressures.

Keywords: fretting fatigue, contact pressure, concavity, notch effect, austenitic stainless steel

Introduction

In fretting fatigue, contact pressure is one of the most important factors. The effect of contact pressure has been studied by many researchers. Most of the researchers found that fretting fatigue life or strength decreased with an increase in contact pressure [1-9]. A few researchers found that the fretting fatigue life showed a minimum at a certain contact pressure [10-14]. The present authors found a singular phenomenon in high strength steel that under a certain test conditions, the fretting fatigue life exhibited a minimum and a maximum with an increase in contact pressure [15, 16]. Waterhouse

---

[1]Leader, Senior Researcher, and Team Leader, respectively, Biomaterials Research Team, National Institute for Materials Science, 1-2-1 Sengen, Tsukuba, 305-0047 Japan.

reported the effect of normal load on fretting fatigue life under the test of cylinder pad on flat specimen and a fixed breadth of contact using alpha brass and aluminum alloy [10]. A minimum in life was observed at a certain normal load in aluminum alloy, but not in alpha brass. The contact pressure dependence of life depended on materials used even if the test conditions were the same. It is necessary to obtain a fundamental and consistent understanding of the phenomenon as to the effect of contact pressure on fretting fatigue. In the present study, the contact pressure dependence of life in type 316L austenitic stainless steel was studied. It was found that contact pressure dependence of life was different from that observed so far and that at high contact pressures, deep concavity was formed at the fretted surface without accompanying heavy wear. This concavity was suggested to significantly affect fretting fatigue judging from fracture behavior and contact pressure dependence of life. This paper aims to make clear the effects of contact pressure and the concavity on fretting fatigue life in type 316L steel.

Experimental Procedures

The material used was a type 316L austenitic stainless steel bar 25 mm in diameter. Table 1 shows the chemical composition of the steel. The steel was solution-treated at 1323K for 10 min then water quenched. Its austenitic grain size was about 15 μm. Mechanical properties of the steel are shown in Table 2. The yield strength and the ultimate tensile strength of the steel are about 330 and 600 MPa, respectively. Dimensions of the fatigue specimen and the fretting pad are shown in Fig. 1. Bridge-type fretting pads of the same material as the fatigue specimen were used. The gage parts of the fatigue specimens and the fretting pads were polished with 0-grade (#600) emery paper and then degreased with acetone.

The fretting fatigue tests were performed on a 100 kN capacity closed loop electro-hydraulic fatigue testing machine. A constant normal pad load was applied by a small hydraulic actuator. The contact pressure, calculated by dividing a normal load by the apparent contact area for one foot of the fretting pad ($2x6=12$ mm^2), was maintained

Table 1 - *Chemical composition of 316L stainless steel (mass%).*

C	Si	Mn	P	S	Ni	Cr	Mo
0.019	0.48	1.18	0.038	0.013	12.10	16.72	2.05

Table 2 - *Mechanical properties of 316L stainless steel used.*

0.2% P. S. (MPa)	U. T. S. (MPa)	El. (%)	R. A. (%)	Vickers H.
328	602	69	83	159

Fretting Fatigue Specimen

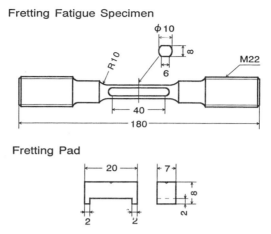

Fretting Pad

Figure 1 - *Shape and dimensions (mm) of the fretting fatigue specimen and the pad.*

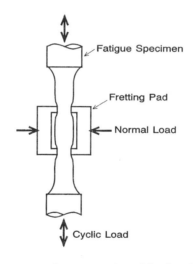

Figure 2 - *Exaggerated representation of fretting damage.*

at a given contact pressure of 15 to 120 MPa. The tests were carried out using a sinusoidal wave form at a frequency of 20 Hz, under load control at a stress ratio of 0.1 in laboratory air of 40 to 70% relative humidity at room temperature. The contact pressure dependence of fretting fatigue life was studied at an axial stress amplitude of 180 MPa. This stress amplitude is equivalent to a maximum stress of 400 MPa. Tangential friction force between the fatigue specimen and the pad was measured using strain gauges bonded to the side of the central part of the pad. Frictional stress amplitude was calculated by dividing a tangential friction force amplitude by apparent contact area

$(2 \times 6 = 12\ mm^2)$ per one foot of the pad.

When the fretting fatigue test was done at high contact pressures, deep concavity was formed at the fretted surface of the specimen due to the contact of fretting pad. No heavy wear was observed macroscopically by naked eyes. The maximum stress applied was higher than the yield strength of the steel. The formation of the concavity is closely related to plastic deformation of the specimen during early stage of fretting fatigue test as described later in discussion. An exaggerated representation of fretting damage is shown in Fig. 2. This concavity was suggested to affect fretting fatigue life judging from fracture behavior and contact pressure dependence of life as described later. So, another set of fretting fatigue test was carried out, where special pre-treatment prior to fretting fatigue test was given to the fatigue specimens to reduce the depth of the concavity. Firstly, the fatigue specimens were subjected to fatigue for 10^4 cycles without using the fretting pads under the same test conditions as the fretting fatigue tests. Then, fretting fatigue test were continued using the fretting pads at a given contact pressure until specimen failure. In this fretting fatigue test, the number of fatigue cycles, 10^4, is not included in the number of cycles to failure. This set of fretting fatigue test is hereafter termed F&FF (Fatigue & Fretting Fatigue) test. The conventional fretting fatigue test is hereafter termed FF (Fretting Fatigue) test. The profile of concavity along cyclic load axis was measured at the central portion of the contact surface of the specimen with a surface roughness analyzer. The depth of concavity, D was defined as an averaged value of two height differences between the bottom surface of fretted area and the upper surfaces that were 1 mm apart from both bottom corners as shown in Fig. 3. The plain fatigue life data were obtained with round bar fatigue specimens whose diameter and length of the gauge section were 8 and 20 mm, respectively.

Results

Fretting fatigue

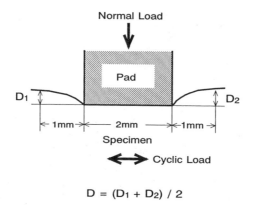

$$D = (D_1 + D_2)\ /\ 2$$

Figure 3 - *Definition of the depth of concavity.*

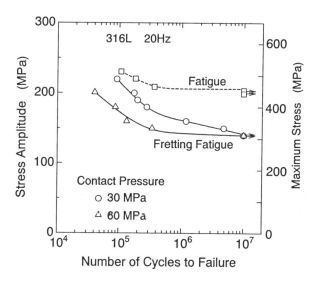

Figure 4 - *S-N curves of fatigue and fretting fatigue for contact pressures of 30 and 60 MPa.*

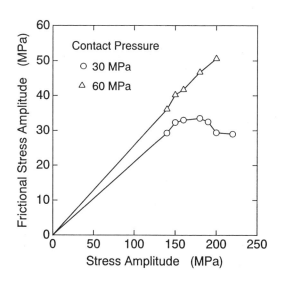

Figure 5 - *Relation between stress amplitude and frictional stress amplitude around 10⁴ cycles for contact pressures of 30 and 60 MPa.*

The S-N curves of fatigue and fretting fatigue at contact pressures of 30 and 60 MPa are shown in Fig. 4. Plain fatigue strength at 10^7 cycles is approximately 200 MPa. At stress amplitudes of more than 150 MPa, fretting fatigue life at a contact pressure of 60 MPa is less than one third of that at a contact pressure of 30MPa. However, fretting fatigue strength at 10^7 cycles at a contact pressure of 60 MPa is almost equal to that at a contact pressure of 30 MPa. Relations between axial stress amplitude and frictional stress amplitude for contact pressures of 30 and 60 MPa are shown in Fig. 5. The frictional stress amplitude at a contact pressure of 60 MPa is larger than that at a contact pressure of 30 MPa.

The effect of contact pressure on fretting fatigue lives of FF and F&FF tests are shown in Fig. 6. The number of cycles to failure at a contact pressure of 0 means plain fatigue life and is beyond 1×10^7 cycles. In the specimens of FF test, the fretting fatigue life decreases sharply to 2×10^5 cycles at a contact pressure of 15 MPa. With an increase in contact pressure, the life remains $(2-3) \times 10^5$ cycles until at a contact pressure of 45 MPa, then decreases drastically at contact pressures of more than 60 MPa. At contact pressures of 90 to 120 MPa, the life is 5×10^4 cycles. In the specimens of F&FF test, the fretting fatigue life decreases sharply at a contact pressure of 15 MPa just like in the specimens of FF test. The life remains $(2-3) \times 10^5$ cycles at contact pressures of 15 to 90 MPa and then decreases to about 1×10^5 cycles at a contact pressure of 120 MPa. At a contact pressure of 15 to 45 MPa, the fretting fatigue lives in the specimens of both FF and F&FF tests are almost the same, but the former is shorter than the latter at contact pressures of more than 60 MPa. Relations between contact pressure and frictional stress

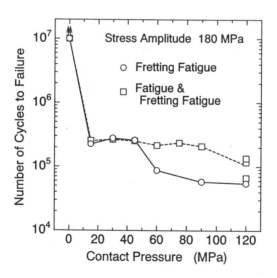

Figure 6 - *The effect of contact pressure on fretting fatigue lives of FF and F&FF tests.*

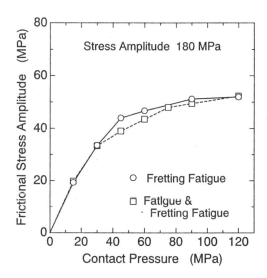

Figure 7 - *Relations between contact pressure and frictional stress amplitude around 10^4 cycles for FF and F&FF tests.*

amplitude for FF and F&FF tests are shown in Fig. 7. Frictional stress amplitudes of both tests increase smoothly with an increase in contact pressure. There is almost no difference between the frictional stress amplitudes for both tests.

Fretting damage

The initiation sites of the main cracks responsible for failure depend on the contact pressure. Figure 8 shows fretted surfaces near the initiation sites of the main cracks in the fractured specimens of both FF and F&FF tests for contact pressures of 30 and 120 MPa. At a contact pressure of 30 MPa, the main cracks are initiated at a middle portion of the fretted area for both specimens of FF and F&FF tests. At a contact pressure range of 15 to 45 MPa, the main cracks occurred similarly. At a contact pressure of 120 MPa, however, the main cracks are initiated at the outer edge of the fretted area for the specimens of both tests. At contact pressures beyond 60 MPa, the main cracks occurred almost at the outer edge of fretted area for both tests. Especially in the specimens of FF test, deep concavity is formed at the fretted area and the main crack occurs just at the outer edge corner of the concavity as shown in Fig. 8(b).

Relations between depth of fretted area, D and stress amplitude at contact pressures of 30 and 60 MPa in the specimens of FF test shown in Fig. 4 are shown in Fig. 9. The depth of fretted area increases with an increase in stress amplitude, but the rate of increase in the depth at a contact pressure of 60 MPa is higher than that at a contact pressure of 30 MPa. Relations between depth of fretted area and contact pressure at a stress amplitude of 180 MPa in the specimens of both FF and F&FF tests shown in Fig. 6

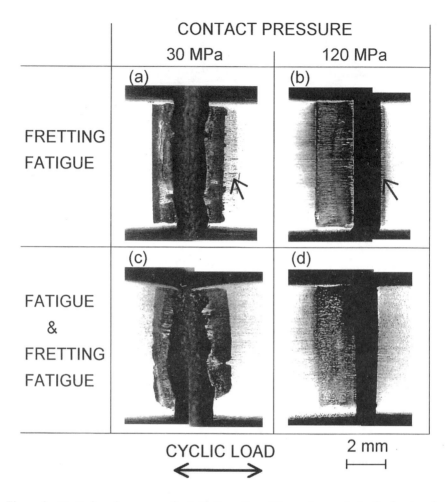

Figure 8 - *Fretted surfaces near the initiation sites of the main cracks in the fractured specimens of FF and F&FF tests for contact pressures of 30 and 120 MPa. Arrows indicate fretting scar produced in the early stage of test.*

are shown in Fig. 10. The depth of fretted area increases almost linearly with an increase in contact pressure. The rate of the increase in the depth for the specimens of FF test is about five times higher than that for the specimens of F&FF test.

Discussion

Contact pressure dependence of life

With an increase in contact pressure, the fretting fatigue lives decrease sharply at a

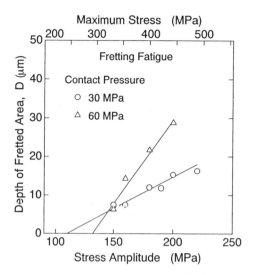

Figure 9 - *Relations between depth of fretted area and stress amplitude for contact pressures of 30 and 60 MPa in the specimens of FF test.*

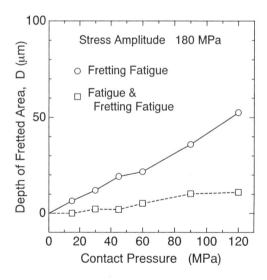

Figure 10 - *Relations between depth of fretted area and contact pressure in the specimens of FF and F&FF tests.*

contact pressure of 15 MPa in both specimens of FF and F&FF tests as shown in Fig. 6. At contact pressures of 15 to 45 MPa, the fretting fatigue lives in the former specimens are almost equal to those in the latter specimens. This means that the application of plain fatigue cycles of 10^4 prior to fretting fatigue test has almost no effect on subsequent fretting fatigue life. With an increase in contact pressure from 15 to 45 MPa, both of the lives remain almost unchanged, although the frictional stress amplitudes in both specimens increase from 20 to about 40 MPa as shown in Fig. 7. It is well known that the frictional stress amplitude is closely related to fretting fatigue life. In general, the fretting fatigue life becomes shorter as the frictional stress amplitude increases [1]. There exists no relationship between the change in fretting fatigue life and that in frictional stress amplitude at contact pressures of 15 to 45 MPa.

   The contact pressure and the frictional stress amplitude defined in the present paper are nominal values, that is, normal load and tangential friction load per one foot of the bridge-type pad were divided by apparent contact area ($2 \times 6 = 12 \, mm^2$). Hence, it is assumed that their distribution at the fretted area is uniform. It is impossible to explain a sharp decrease in life at a contact pressure of 15 MPa, directly by apparent contact pressure and frictional stress amplitude. The decrease in life at low contact pressures was probably caused by the local stress concentration at the fretted area proposed by Nakazawa et al [15]. Under a certain testing condition, there exists a stick region in the middle portion of a fretted area and slip regions on either side of it. This situation depends on the contact pressure, relative slip amplitude and number of cycles [17, 18]. The higher the contact pressure, the wider the stick region. When the contact pressure is low and the stick region is narrow, net values of contact pressure and frictional stress amplitude acting in the stick region must be higher, while those in slip region must be lower than nominal values due to the existence of wear debris and its removal from the fretted area. The decrease in life at a contact pressure of 15 MPa was probably caused by the concentration of contact pressure and frictional stress amplitude at narrow stick region. Their net values must have been considerably larger than the apparent ones. At low contact pressures, the main cracks responsible for failure occurred at the middle portion of fretted area as shown in Fig. 8. This result also supports an existence of both narrow stick region and stress concentration there, since the crack is sometimes initiated at slip region near the boundary between the stick region and the slip region [15, 18].

   In the specimens of F&FF test, the fretting fatigue life at contact pressures of 60 to 90 MPa, is almost the same as that at low contact pressures. It decreases slightly at a contact pressure of 120 MPa. In the specimens of FF test, however, the fretting fatigue life decreases down to about one third of the life in F&FF test at contact pressures of more than 60 MPa. The shorter fretting fatigue lives in FF test at contact pressures of more than 60 MPa compared with that in F&FF test cannot be explained by the frictional stress amplitude shown in Fig. 7, since the frictional stress amplitudes in both tests are almost the same. When the contact pressure is high, the depth of concavity at the fretted area is large in FF test, but small in F&FF test as shown in Fig. 10. Furthermore, the main crack initiation in the specimens of FF test occurred just at the outer edge corner as shown in Fig. 8(b). These results suggest that the outer edge of fretted area acts as a source of stress concentration. The shorter fretting fatigue lives were probably related to stress concentration arising from profile of the outer edge rather than from so-called fretting damage. The fretting fatigue lives of both tests at contact pressures of 15 to 45

MPa are probably not affected by the concavity, since its depth is small.

At stress amplitudes of more than 150 MPa, fretting fatigue lives at a contact pressure of 60 MPa are extremely shorter than that at a contact pressure of 30 MPa as shown in Fig. 4. It is suggested that the fretting fatigue life at a contact pressure of 60 MPa is probably affected rather by the stress concentration due to the concavity than by the frictional stress amplitude shown in Fig. 5, since the life in F&FF test at a contact pressure of 60 MPa is almost the same as that at a contact pressure of 30 MPa as shown in Fig. 6. The fretting fatigue strength at 10^7 cycles for a contact pressure of 30 MPa is 140 MPa and is equal to that for a contact pressure of 60 MPa. At a stress ratio of 0.1 used in the present study, the stress amplitude of 140 MPa corresponds to the maximum stress of about 310 MPa. This maximum stress value is smaller than the yield strength of 316L steel used (Table 2). Hence, the depth of concavity is small as described later and there is no effect of concavity on the fretting fatigue strength.

Formation of concavity

In fatigue or fretting fatigue test under load control, parallel gauge section of the specimen is deformed plastically and elongated during loading when the maximum stress is exceeded beyond yield stress. In 316L steel used in the present study, work hardening occurs with an increase in strain. After unloading, parallel gauge section remains as being uniformly elongated to the loading axis due to the plastic deformation. Figure 11 shows a

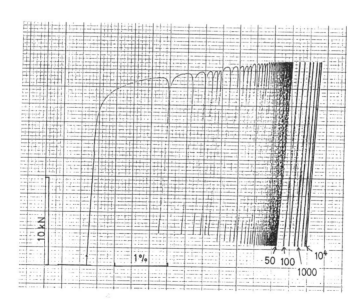

Figure 11 - *Load-elongation response of the fretting fatigue specimen subjected to plain fatigue of 10^4 cycles without fretting pads at a stress amplitude of 180 MPa.*

load-elongation response of the fretting fatigue specimen subjected to plain fatigue of 10^4 cycles without fretting pads under load control at a stress amplitude of 180 MPa at a frequency of 1 Hz. The elongation was measured using an extensometer attached to the parallel gauge section of the specimen. With an increase in number of cycles, the elongation increases, and the maximum load increases and the minimum load decreases. The load change in the early stage of fatigue experiment probably resulted from transient loading state to stable load response under load control in closed loop electro-hydraulic fatigue testing machine. At 50 cycles, the load approaches to a certain value, and permanent elongation is about 3.5%. At 10^4 cycles, the elongation is almost saturated, and the permanent elongation is about 4.2%.

The fretting pad is deformed elastically due to frictional force that develops between two pad feet, although fretting fatigue specimen is deformed plastically and elongated permanently with number of cycles in the early stage of test. Thus, the span length between the two fretting scars of pad feet on the specimen surface along cyclic stress axis is expanded outwards with an increase in number of cycles, but the position of contact area is almost fixed after 10^4 cycles. In Fig. 8 (a) and (b), fretting scar produced in the early stage of test is seen to be adjacent to the main scar as shown by arrows.

In the specimens of FF test, the depth of concavity at the fretted area increases with an increase in contact pressure as shown in Fig. 10. This result was probably caused by the fact that the degree of constraint of specimen deformation along its thickness direction by fretting pad is large when the contact pressure is high. The depth of concavity also increases with an increase in stress amplitude under a given contact pressure as shown in Fig. 9. Furthermore, the maximum stress values at the depth of concavity zero of the extrapolated relationship between stress amplitude and depth of concavity for contact pressures of 30 and 60 MPa are about 250 and 300 MPa, respectively. These stress values are smaller than the yield strength. However, they are roughly equal to the yield strength, and this suggests that the formation of concavity is closely related to plastic deformation. The deep concavity observed in the present study is probably associated with low yield strength of 316L steel used.

On the other hand, in the specimens of F&FF test, cyclic strain hardening occurs due to plain fatigue of 10^4 cycles prior to fretting fatigue test. By this plain fatigue, Vickers hardness number of the specimens is increased from 160 to 200. Consequently, the depth of concavity is small even though fretting fatigue test is started successively after the plain fatigue, since the plastic deformation of the specimen during fretting fatigue test is slight. The situation of concavity formation in a cold worked 316L steel may be different from that in a solution-treated one used in the present study.

Conclusions

The effect of contact pressure on fretting fatigue in type 316L stainless steel has been studied at a stress amplitude of 180 MPa. The conclusions obtained are as followings.
(1) With an increase in contact pressure, fretting fatigue life is almost unchanged at contact pressures of 15 to 45 MPa, but it decreases drastically at contact pressures beyond 60 MPa. It is impossible to explain the contact pressure dependence of life by the change in frictional stress amplitude since the frictional stress amplitude increases

smoothly with contact pressure.

(2) The concavity associated with plastic deformation of the specimen under the contact of fretting pad is formed at the specimen surface. The depth of the concavity becomes larger, as the contact pressure increases.

(3) At low contact pressures, stress concentration due to fretting damage occurs at the middle portion of fretted area, and the main crack responsible for failure is initiated there. The existence of concavity has almost no effect on the life since the depth of concavity is small.

(4) At high contact pressures, deep concavity is formed without accompanying heavy wear. The main crack is initiated at the outer edge corner of the concavity which probably acts as a notch. This suggests that the stress concentration there plays an important role in the life.

References

[1] Nishioka K. and Hirakawa K., "Fundamental Investigation of Fretting Fatigue (Part 6, Effects of Contact Pressure and Hardness of Materials)," *Bulletin of the JSME*, Vol. 15, 1972, pp. 135-144.

[2] Endo K. and Goto H., "Initiation and Propagation of Fretting Fatigue Cracks," *Wear*, Vol. 38, 1976, pp. 311-324.

[3] Gaul D. J. and Duquette D. J., "The Effect of Fretting and Environment on Fatigue Crack Initiation and Early Propagation in a Quenched and Tempered 4130 Steel," *Metallurgical Transactions*, Vol. 11A, 1980, pp. 1555-1561.

[4] Sato K. and Fujii H., "Fretting-Fatigue Strength and Fracture Morphology of Carbon Steel S45C," *Journal of Japan Society of Strength and Fracture of Materials*, Vol. 18, 1984, pp. 98-113 (in Japanese).

[5] Nagata K., Matsuda T., and Kashiwaya H., "Effect of Contact Pressure on Fretting Fatigue Strength," *Transactions of the Japan Society of Mechanical Engineers*, Vol. 53, 1987, pp. 196-199 (in Japanese).

[6] Mutoh Y., Nishida T., and Sakamoto I., "Effect of Relative Slip Amplitude and Contact Pressure," *Journal of the Society of Materials Science, Japan*, Vol. 37, 1988, pp. 649-655 (in Japanese).

[7] Satoh T., Mutoh Y., Yada T., Takano A., and Tsunoda E., "Effect of Contact Pressure on High Temperature Fretting Fatigue," *Journal of the Society of Materials Science, Japan*, Vol. 42, 1993, pp. 78-84 (in Japanese).

[8] Adibnazari S. and Hoeppner D. W., "A Fretting Fatigue Normal Pressure Threshold Concept," *Wear*, Vol. 160, 1993, pp. 33-35.

[9] Satoh T., Mutoh Y., Nishida T., and Nagata K., "Effect of Contact Pad Geometry on Fretting Fatigue Behavior," *Transactions of the Japan Society of Mechanical Engineers*, Vol. 61, 1995, pp. 1492-1499 (in Japanese).

[10] Waterhouse R. B., "The Effect of Clamping Stress Distribution on the Fretting Fatigue of Alpha Brass and Al-Mg-Zn Alloy," *Transactions of American Society for Lubrication Engineers*, Vol. 11, 1968, pp. 1-5.

[11] Switek W., "Fretting Fatigue Strength of Mechanical Joints," *Theoretical Applied Fracture Mechanics*, Vol. 4, 1985, pp. 59-63.

[12] Fernando U. S., Farrahi G. H., and Brown M. W., "Fretting Fatigue Crack Growth

Behaviour of BS L65 4 percent Copper Aluminum Alloy under Constant Normal Load," *Fretting Fatigue, ESIS 18* (Edited by R. B. Waterhouse and T. C. Lindley), 1994, Mechanical Engineering Publications, London, pp. 183-195.

[13] Del Puglia A., Pratesi F., and Zonfrillo G., "Experimental Procedure and Parameters Involved in Fretting Fatigue Tests," *Fretting Fatigue, ESIS 18* (Edited by R. B. Waterhouse and T. C. Lindley), 1994, Mechanical Engineering Publications, London, pp. 219-238.

[14] Lee S.-K., Nakazawa K., Sumita M., and Maruyama N., "Effects of Contact Load and Contact Curvature Radius of Cylinder Pad on Fretting Fatigue in High Strength Steel," Fretting Fatigue: *Current Technology and Practices, ASTM STP 1367,* 2000, pp. 199-212.

[15] Nakazawa K., Sumita M., and Maruyama N., "Effect of Contact Pressure on Fretting Fatigue of High Strength Steel and Titanium Alloy," *Standardization of Fretting Fatigue Test Methods and Equipment, ASTM STP 1159,* 1992, pp. 115-125.

[16] Nakazawa K., Sumita M., and Maruyama N., "Effect of Relative Slip Amplitude on Fretting Fatigue of High Strength Steel," *Fatigue and Fracture of Engineering Materials and Structures,* Vol. 17, 1994, pp. 751-759.

[17] Vingsbo O. and S derberg S.,"On Fretting Map," *Wear,* Vol. 126, 1988, pp. 131-147.

[18] Waterhouse R. B. and Taylor D. E., "The Initiation of Fatigue Cracks in A 0.7% Carbon Steel by Fretting," *Wear,* Vol. 17, 1971, pp. 139-147.

Chung-Hyun Goh,[1] Richard W. Neu,[1] and David L. McDowell[1]

Influence of Nonhomogeneous Material in Fretting Fatigue

REFERENCE: Goh, C.-H., Neu, R. W., and McDowell, D. L., "Influence of Nonhomogeneous Material in Fretting Fatigue," *Fretting Fatigue: Advances in Basic Understanding and Applications, STP 1425,* Y. Mutoh, S. E. Kinyon, and D. W. Hoeppner, Eds., ASTM International, West Conshohocken, PA, 2003.

Abstract: Since fretting fatigue often leads to catastrophic failure in components clamped together with relatively small amplitude displacements, several analysis methods have been developed to quantify fretting fatigue damage. Fretting damage analyses using crystal plasticity have potential to address the issues of accounting for inherent material microstructure heterogeneity as well as more realistic treatment of crystallographic slip in fretting fatigue. The primary focus of this study is to explore the influence of microstructure as well as coefficient of friction within the fretting fatigue boundary layer during fretting fatigue process. Crystal plasticity theory is used for nonhomogeneous finite element simulations and the results are compared to those developed by an initially homogeneous J_2 plasticity theory with nonlinear kinematic hardening flow rule as well as experiments.

Keywords: Fretting fatigue, coefficient of friction (COF), stick/ slip, crystal plasticity

Introduction

Fretting is a damage process that is associated with local adhesion and oscillatory displacements between contact surfaces, subjected to a cyclic, relative motion of extremely small amplitude. Since fretting fatigue damage accumulation occurs over relatively small volumes, the subsurface-stress state is expected to be rather non-uniformly distributed in polycrystalline materials. The scale of the cyclic plasticity and the damage process zone is often on the order of microstructural dimensions and the resolution of the local stick/slip behavior along the interface may play a key role in fretting fatigue damage process. Thus, the role of the microstructure as well as the coefficient of friction (COF) within the fretting boundary layer is quite significant in fretting fatigue analysis. The latter influence is found to play a first order role in the distribution of contact stresses and plastic strain fields, yet cannot be directly controlled in the experiment and evolves during testing.

The heterogeneity of discrete grains and their crystallographic orientation is accounted for using continuum crystallographic cyclic plasticity models coded in an ABAQUS User MATerial (UMAT) subroutine [1]. To provide credence for the non-homogeneous crystal plasticity modeling approach, simulations of fretting fatigue were compared with the results developed by an initially isotropic J_2 plasticity model as well as experiments when Ti-6Al-4V is fretted against itself. This study explores the nature of subsurface cyclic plasticity and contact stress/strain fields considering different pad loads and different

[1]Graduate Research Assistant, Associate Professor, and Professor, The George W. Woodruff School of Mechanical Engineering, Georgia Institute of Technology, Atlanta, GA 30332-0405.

values of COF. In this work, we primarily address the latter issue by conducting the finite element (FE) simulations based on crystal plasticity theory, comparing to J_2 plasticity theory with nonlinear kinematic hardening flow rule as well as experimental results and observations.

J_2 Nonlinear Kinematic Hardening Plasticity Algorithm

In this study, a pure kinematic hardening rule is considered to represent a cyclically stable response. The model is valid for arbitrarily finite elasto-plastic deformations. We assume the rate-independent plasticity with a simple J_2 yield surface:

$$f = \bar{\sigma} - \sigma_0 = \sqrt{3J'_2} - \sigma_0 = \sqrt{\frac{3}{2} \underset{\sim}{s}' : \underset{\sim}{s}'} - R \qquad (1)$$

where $\bar{\sigma}$ is the effective stress, $\underset{\sim}{s}'$ is the reduced deviatoric stress tensor, σ_0 is the initial yield strength, and R is the radius of the yield surface. The reduced deviatoric stress tensor is given by

$$\underset{\sim}{s}' = \left(\underset{\sim}{s} - \underset{\sim}{x} \right) \qquad (2)$$

where $\underset{\sim}{s}$ and $\underset{\sim}{x}$ are the deviatoric Cauchy stress and the deviatoric backstress tensors, respectively: $\underset{\sim}{s} = \underset{\sim}{\sigma} - \frac{1}{3} tr\left(\underset{\sim}{\sigma}\right) \underset{\sim}{I}$ with $\underset{\sim}{\sigma}$ the Cauchy stress tensor and $\underset{\sim}{I}$ the second order identity tensor.

For an associated flow rule, the unit normal vector can be represented as

$$\underset{\sim}{n} = \frac{\partial f / \partial \underset{\sim}{\sigma}}{\left\| \partial f / \partial \underset{\sim}{\sigma} \right\|} = \frac{\underset{\sim}{s}'}{\left\| \underset{\sim}{s}' \right\|} \qquad (3)$$

The associated plastic flow rule is given by

$$\underset{\sim}{D}^p = \left\langle \frac{1}{H} \dot{\underset{\sim}{\sigma}} : \underset{\sim}{n} \right\rangle \underset{\sim}{n} = \dot{p} \underset{\sim}{n} \qquad (4)$$

where H is the plastic modulus, $\dot{\underset{\sim}{\sigma}} = \underset{\approx}{C} : \left(\underset{\sim}{D} - \underset{\sim}{D}^p \right)$, $\dot{p} = \left(\dot{\varepsilon}^p : \dot{\varepsilon}^p \right)^{1/2}$, and the Macauley bracket denotes that $\langle Z \rangle = Z$ if $Z \geq 0$, else $\langle Z \rangle = 0$. Here, $\underset{\approx}{C}$ is the fourth order elastic stiffness tensor and $\underset{\sim}{D}$ is the rate of deformation tensor; the additive decomposition of the rate of deformation tensor $\underset{\sim}{D}$ into elastic and plastic components is assumed, i.e., $\underset{\sim}{D} = \underset{\sim}{D}^e + \underset{\sim}{D}^p$. For a pure kinematic hardening, the consistency condition for the yield criterion is represented as

$$\dot{f} = \frac{\partial f}{\partial \underset{\sim}{\sigma}} : \dot{\underset{\sim}{\sigma}} + \frac{\partial f}{\partial \underset{\sim}{x}} : \dot{\underset{\sim}{x}} = 0 \tag{5}$$

The plastic modulus with N decoupled backstress components is given by the normality (or associated) flow rule based on a kinematic hardening Mises yield surface, i.e., for $\dot{f} = 0$:

$$H = \sum_{\alpha=1}^{N} \left(h^{\alpha} - h_D^{\alpha} \, \underset{\sim}{x}^{\alpha} : \underset{\sim}{n} \right) \tag{6}$$

where N represents the number of backstress components and h and h_D are direct hardening and dynamic recovery coefficients, respectively. In this study, two components of the backstress are used and the constants for duplex Ti-6Al-4V, based on correlation to the cyclic stress-strain response, are: $\sigma_0 = 765 \, MPa$, $h^1 = 7484 \, MPa$, $h^2 = 725 \, MPa$, $h_D^1 = 486$, and $h_D^2 = 37$. The reader is referred to the thesis of Marin [2] and related paper (Marin and McDowell [3]) for more details.

Crystal Plasticity Algorithm

The crystal plasticity algorithm follows that described in Bennett [4], with the numerical implementation described by McGinty and McDowell [5], building on the fully implicit method outlined by Cuitiño and Ortiz [6]. The crystal plasticity algorithm begins with the multiplicative decomposition (see Figure 1). The total deformation gradient is given by

$$\underset{\sim}{F} = \underset{\sim}{F}^e \underset{\sim}{F}^p \tag{7}$$

where $\underset{\sim}{F}^e$ is the elastic deformation gradient representing elastic stretch (or compress) and rotation of the lattice, as well as the rigid body rotation, while $\underset{\sim}{F}^p$ is the plastic deformation gradient describing the collective effect of dislocation motion along the active slip planes relative to a fixed lattice in the reference configuration. The kinematics of elastic-plastic deformation of crystals is shown in Fig. 1. In the figure, the grids represent the crystal lattice; $\underset{\sim}{s}_0^{\alpha}$ and $\underset{\sim}{n}_0^{\alpha}$ denote the initial unit vectors in the slip direction and the slip plane normal direction, respectively, for the αth slip system in the undeformed configuration. $\underset{\sim}{s}^{\alpha}$ and $\underset{\sim}{n}^{\alpha}$ are not generally unit vectors, but they remain orthogonal to the slip direction since $\underset{\sim}{s}_0^{\alpha} \cdot \underset{\sim}{n}_0^{\alpha} = \underset{\sim}{s}^{\alpha} \cdot \underset{\sim}{n}^{\alpha} = 0$. As the material plastically deforms, the lattice undergoes elastic deformation and rigid rotations.

The resolved shear stress (or the Schmid stress) on the αth slip system is given by

$$\tau^{\alpha} = \underset{\sim}{\sigma} : \left(\underset{\sim}{s}^{\alpha} \otimes \underset{\sim}{n}^{\alpha} \right) \tag{8}$$

where $\underline{s}^{\alpha} \otimes \underline{n}^{\alpha}$ denotes the tensor product of two vectors $\underline{s}^{\alpha}$ and $\underline{n}^{\alpha}$. The power-law relation on the αth slip system is described by

$$\dot{\gamma}^{\alpha} = \dot{\gamma}_0 \left| \frac{\tau^{\alpha} - x^{\alpha}}{G} \right|^{m} \text{sgn}\left(\tau^{\alpha} - x^{\alpha}\right) \qquad (9)$$

where G represents the isotropic drag strength, m is the flow exponent (or inverse strain-rate sensitivity exponent), x^{α} is the backstress on the αth slip system, and $\dot{\gamma}_0$ is the reference shearing rate. Here, $\alpha = 1, 2, 3$ for the planar triple slip systems. The backstress evolves according to a pure nonlinear kinematic hardening rule of the Armstrong-Frederick type, i.e.,

$$\dot{x}^{\alpha} = h\dot{\gamma}^{\alpha} - h_D x^{\alpha} \left| \dot{\gamma}^{\alpha} \right| \qquad (10)$$

The crystal plasticity algorithm is coded into an ABAQUS UMAT subroutine [4-5].

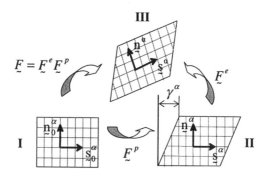

Figure 1. *Kinematics of elastic-plastic deformation of crystalline solid deforming by crystallographic slip (I: Undeformed, II: Intermediate, and III: Deformed Configuration).*

Planar Triple Slip Model

The microstructure of the Ti-6Al-4V alloy consists of two phases, 60% primary α and 40% lamellar colonies of α+β phase, with a Young's modulus of 118 *GPa* and a Poisson's ratio of 0.349 [7]. In the Ti-6Al-4V alloy with duplex microstructure, slip in the HCP α phase typically dominates and prismatic slip is the most active deformation mode at room temperature [8].

As shown in Fig. 2, the crystal plasticity model with the planar triple slip idealization has been shown to accurately represent the macroscopic cyclic stress-strain behavior of the Ti-6Al-4V, which exhibits a nearly elastic-perfectly plastic response, while the planar double slip idealization, commonly used for FCC materials, doesn't exhibit this relationship with any combination of constants. This is due to the fact that with only two slip systems in the 2D idealization, any arbitrary 2D plastic deformation must involve both slip systems activated to the same order of shearing rate. Accordingly, the hardening within each grain occurs at a higher rate during plastic straining. Adding a third slip system gives a redundant system (more systems than required to accommodate arbitrary 2-D deformation), and leads to less work hardening. Therefore, the third slip system in the planar triple slip model is essential for describing the nearly elastic-perfectly plastic cyclic behavior of this material. Furthermore, it is quite sensible in view of the room temperature dominance of the prismatic slip mode for α, which has three systems. We believe that this is a general characteristic of two-phase systems with strong heterogeneity and relatively weak matrix strain hardening. These systems tend to promote shear localization. The constants for the planar triple slip model describing duplex Ti-6Al-4V are: $\dot{\gamma}_0 = 0.001$ s^{-1}, $G = 404$ MPa, $m= 63$, $h = 500$ MPa, $h_D = 100$. The method of determining the constants is described in Ref [9].

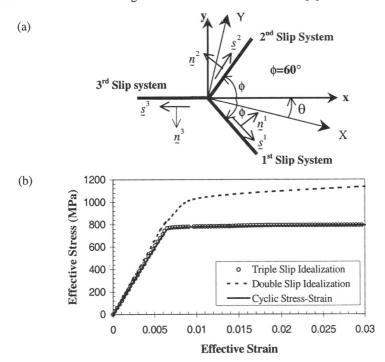

Figure 2. *(a) Planar triple slip system idealization and (b) comparison of planar double and triple slip idealization for Ti-6Al-4V (G=404 MPa, m=63, h=500 MPa, h$_D$=100, $\dot{\gamma}_0 = 0.001 s^{-1}$, E=118 GPa, and v=0.349).*

Coulomb Friction Model

The frictional behavior is usually expressed in terms of the interfacial stresses of the contact. Coulomb friction is the standard friction model implemented in ABAQUS [1].

The Coulomb friction model assumes that sticking (no relative motion) occurs if the maximum shear stress (τ_{max}) is less than a critical stress (τ_{crit}) and the contacting bodies will slide when $\tau_{max} = \tau_{crit}$. The critical stress (τ_{crit}) is given by

$$\tau_{crit} = \mu p \tag{11}$$

where μ and p are the COF and the normal (or direct) contact pressure, respectively. Two methods exist in ABAQUS to implement Coulomb friction: a Lagrange multiplier method and a penalty method (see Fig. 3). The advantage of a Lagrange multiplier method is the exact nature it captures the sticking conditions, where the relative motion is zero. However, the Lagrange multiplier implementation tends to slow or sometimes prevent convergence of the solution due to the presence of rigid constraints. In the penalty method, the sticking condition is approximated with penalty stiffness (G) to facilitate convergence. Instead of $G = \infty$, G is adapted continuously by the ratio of critical shear stress and elastic slip, i.e. $G = \tau_{crit} / \gamma_{crit}$ in the penalty method. Even with the convergence difficulties, the Lagrange multiplier formulation is attractive in fretting fatigue analysis since the resolution of stick/slip behavior along the interface is considerably important in fretting.

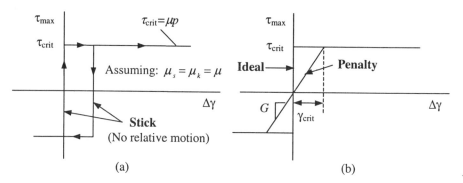

Figure 3. *Diagram of the Coulomb friction model [1]: (a) Lagrange multiplier method and (b) penalty method, where τ_{max} and τ_{crit} are the maximum and critical shear stresses, respectively. Also, γ_{crit} and $\Delta\gamma$ are the critical elastic slip and the slip (or relative motion) increment, respectively, and G is penalty stiffness.*

Finite Element Implementation

A finite element analysis is conducted by modeling a symmetric section of the experimental configuration consisting of the fretting pad and fatigue specimen as shown in Fig. 4(a). The fretting pad is cylindrical with a radius of curvature (R) of 50 mm and the fatigue specimen has dimensions of 5.0 mm (z-direction) by 12.7 mm (y-direction). In this configuration, the trailing edge is on the left side of the contact zone for a tensile loading. The pad and specimen are fixed in the x- and y-directions along planes of symmetry. The elements of the spongy layer are constrained in both the x- and y-directions along the very top of the layer. The normal load, W, is first imposed and then three fatigue loading cycles are prescribed to the specimen in the x direction to account for cumulative history effects (Fig. 4(b)). The fatigue load (F_{sp}) applied to the specimen is given by

$$F_{sp} = \sigma\left(w_s \times t_s\right) \tag{12}$$

where σ is the bulk applied stress to the specimen, and w_s and t_s are the width and thickness of the specimen, respectively. The maximum and minimum values of F_{sp} are determined by stress ratio ($R_\sigma = \sigma_{min}/\sigma_{max}$) and the fatigue specimen bulk stress amplitude ($\sigma_a = (\sigma_{max} - \sigma_{min})/2$) for a given dimension. For all the cases considered in this paper, $R_\sigma = 0.1$ and $\sigma_a = 150 \ MPa$. Therefore, the applied stress to the specimen (σ) is linearly varied from the minimum of 33.33 MPa to the maximum of 333.33 MPa with time during the loading cycle.

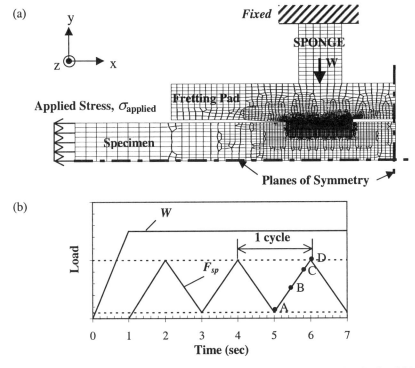

Figure 4. *(a) Finite element mesh of experimental configuration and (b) the load history for $R_\sigma = 0.1$, where W and F_{sp} denote the normal load and the fatigue load applied to the specimen, respectively, and Points A, B, C, D denote time steps to represent interface stress evolutions in the later section.*

For the elements under the contact zone, a mapped mesh is used with the element size of 10 μm. To improve efficiency, only the elements in the region of contact, 4 mm (length) × 0.5 mm (depth), employ the crystal plasticity constitutive model, while the remaining elements are either specified as J_2 plastic or purely elastic depending on the location relative to the contact region. The pad is elastic and the spongy layer is assigned a modulus of $10^{-5}E$.

Eight-noded biquadratic quadrilateral 2-D solid plane strain elements with reduced integration (ABAQUS CPE8R) are used and plane strain conditions are assumed in the analysis. Contact pair elements are used with the small sliding option within ABAQUS [1] to represent fretting contact between the fretting pad and the specimen. A Lagrange multiplier implementation is used to enforce the sticking constraints along the interface between two contacting surfaces since the resolution of the stick/ slip behavior is very important for simulating fretting contact between two bodies.

Assignment of discrete grains (size and geometry) is achieved by grouping elements with the same prescribed crystallographic orientation prior to running the simulation. The angle of orientation (θ) of the bisector axis of each grain relative to the global x-direction is randomly assigned in the ABAQUS UMAT subroutine. To facilitate comparison among the different loading cases, the same grain orientation assignment is used for each case. The grain is assumed as a square of 50 μm by 50 μm (i.e., 5 elements by 5 elements); no distinction is made between α grains and α-β lamellar grains in the duplex microstructure.

Results

Since fretting fatigue is a surface-related damage process, the role of the coefficient of friction (COF) along the interface is quite significant in fretting fatigue analysis. To study the influence of the COF, FE simulations have been conducted for the following conditions: stress ratio (R_σ) of 0.1, a bulk fatigue stress amplitude (σ_a) of 150 MPa, two normal fretting pad loads, $W = 1350\ N$ ($\approx 0.3 P_y$) and $W = 4050\ N$ ($\approx 0.9 P_y$), and three different values of COF, $\mu = 0.75$, $\mu = 1.0$, and $\mu = 1.5$; P_y is the critical peak load for initial subsurface yield [10]. Simulations using completely elastic material as well as J_2 plasticity with a nonlinear kinematic hardening rule are also used to "benchmark" the results of the simulations using crystal plasticity.

Contact Stress Fields along the Interface of Contact

The contact stress fields along the interface may be first investigated in order to understand the effects caused by the local surface stresses during the fretting fatigue process. The fretting damage processes are not only associated with the microslip at the contact surface but also the cyclic fretting contact stresses [11]. The stress distributions along the interface at the end of 3rd loading cycle, developed from various material models, are compared in Fig. 5. The width of the stick region found experimentally for this loading condition is more closely represented by the $\mu = 1.5$ case [12]. For all material models, the stress distributions are almost the same except near the trailing edge in the σ_{xx} distributions (see Fig. 5 (a) and (b)). At the trailing edge, particularly for the

$\mu = 1.5$ case, both the J_2 and crystal plasticity simulations predict a lower stress than the elastic simulations due to stress redistribution resulting from plastic deformation. The stress distributions of the crystal plasticity and J_2 plasticity simulations are essentially the same except for small fluctuations at the trailing edge due to crystallographic grains when crystal plasticity is used.

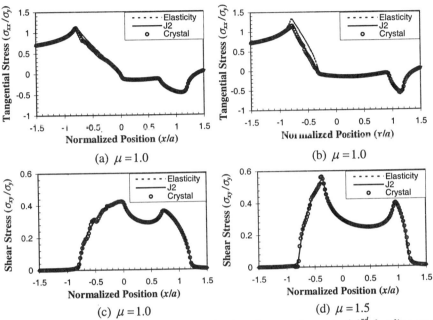

(a) $\mu = 1.0$

(b) $\mu = 1.0$

(c) $\mu = 1.0$

(d) $\mu = 1.5$

Figure 5. *The stress distributions along the interface at the end of 3rd loading cycle, comparing material models and COF: tangential stress (σ_{xx}) distributions and shear stress (σ_{xy}) distributions for the case: $R_\sigma = 0.1$, $\sigma_a = 150$ MPa, and $W = 1350$ N ($\sigma_y = 765$ MPa and $a = 514$ μm).*

Subsurface Distributions at the Trailing Edge of the Contact

Subsurface distributions of the cumulative effective plastic strain ($\overline{\varepsilon}^p = \int \sqrt{\frac{2}{3} \dot{\underline{\varepsilon}}^p : \dot{\underline{\varepsilon}}^p} \, dt$) at the end of the 3rd loading cycle at the trailing edge of the contact using two different values of COF, $\mu = 1.0$ and $\mu = 1.5$, are shown in Figs. 6 and 7, respectively. The crystal plasticity model has the tendency to predict greater shear strain localization than when a homogeneous J_2 plasticity theory is used. This shear strain localization leads to nonhomogeneous distributions of plastic deformation in the subsurface field of the crystal plasticity simulations. The peak values of cumulative plastic strain and the penetration depth are both greater for crystal plasticity. Both the maximum value of cumulative plastic strain and the depth of its penetration increase with increasing COF. The penetration depths of the localized plastic strain for crystal plasticity are similar to the depth of the non-catastrophic cracks experimentally observed in the subsurface layer

[13]. When μ=0.75 is used (not shown), the extent of the cumulative effective plastic strains is very small in the crystal plasticity simulation and no plastic strain accumulation is found in J_2 plasticity simulation. Even with crystal plasticity, elastic shakedown is reached by the third cycle, clearly missing the extent of deformation experimentally observed in the subsurface layers for the same fretting fatigue condition.

Also, the μ=0.75 case predicts a gross slip condition but partial slip conditions clearly prevail in this loading case from experimental observations [12]. Our FE simulations using crystal plasticity and $\mu = 1.5$ offer the closest description of the experimentally observed cracking behavior for specimens with the same loading conditions.

When two materials in contact experience relative sliding motion under cyclic loading, plastic strain accumulation increases with the cycles and this plastic deformation induces crack nucleation. Although the maximum plastic deformation occurs at the surface, the depth of nucleation process volume may contribute significantly to whether crack of a critical length is able to nucleate. Note that there are no significant differences in the stress distributions between the various plasticity models as shown in Fig. 5, yet the cumulative plastic strain fields are significantly different. Therefore, the localization of plastic deformation cannot be represented well by the contact stress distribution or subsurface stress field. Subsurface distributions of the cumulative effective plastic strain as well as the plastic strain-based multiaxial fatigue criteria based on the crystal plasticity model exhibit more realistic representation of the localization of plastic deformation, comparable to experimental observations, than when J_2 plasticity theory is used [13]. Of course, purely elastic calculation renders little information relevant to microstructure-scale plastic deformation and damage processes.

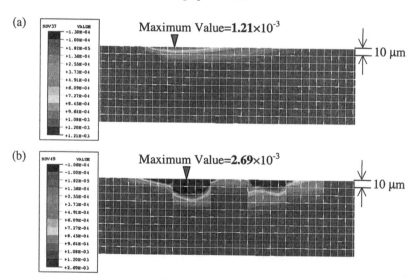

Figure 6. *Cumulative effective plastic strain distribution in the subsurface region at the trailing edge at the end of 3rd loading cycle for the case: $R_\sigma = 0.1$, $\sigma_a = 150$ MPa, $W = 1350$ N, and $\mu = 1.0$, showing influence of material model (a) J_2 plasticity and (b) crystal plasticity.*

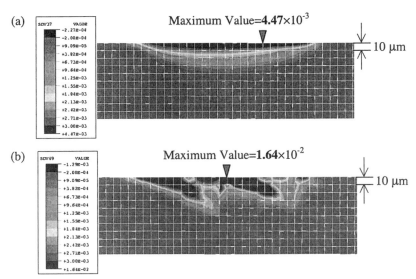

Figure 7. *Cumulative effective plastic strain distribution in the subsurface region at the trailing edge at the end of 3rd loading cycle for the case: $R_\sigma = 0.1$, $\sigma_a = 150$ MPa, $W = 1350$ N, and $\mu = 1.5$, showing influence of material model (a) J_2 plasticity and (b) crystal plasticity.*

Stick and Slip Regions on the Contact Surface

In fretting, the contact area generally consists of two regions, stick and slip (or micro-slip) zones. A modification of surface topography may contribute locally to increasing the COF and give way to stick in some micro-slip regions. In sticking zone, adhesion can occur but little surface modification is anticipated in the sense that no relative motion will occur within this area. On the other hand, plastic deformation and some surface modification are inevitable in the relative slip zone [*14*].

The relative slip range along the interface is shown in Fig. 8. The peak value of the relative slip range at the edge of contact on the trailing edge side is around 2.2 μm for both COF cases. Interestingly, near the boundary between the stick and slip regions crystal plasticity predicts a region where alternating stick and nanoslip occurs, whereas both elasticity and J_2 plasticity predict that the transition between stick and slip occurs at a single point (see Figs. 8(b) and (d)). The mixed stick/slip region of the lower COF case (Fig. 8(b)) extends wider and has smaller magnitude than the higher COF case (Fig. 8(d)).

The mixed stick/slip region can be also seen by plotting the ratio of local shear traction, $q(x)$, to normal traction, $p(x)$ as shown in Fig. 9. Fig. 9 shows the variation of $q(x)/p(x)$ over the contact area at the end of 3rd loading cycle for the cases of $\mu = 1.0$ and $\mu = 1.5$. The successive stick/ slip regions lead to multiple stress concentration sites along the interface. As a consequence, the plastic deformation from the crystal plasticity theory naturally leads to a surface roughness from an initially smooth surface without the need to employ any ad hoc asperity size and spacing algorithms. Thus, the surface may locally experience the sticking conditions in the slip region. In the Figures 8 and 9, the dashed boxes represent a stick zone in the mixed stick/slip regions; the locations correspond to each other between the two figures.

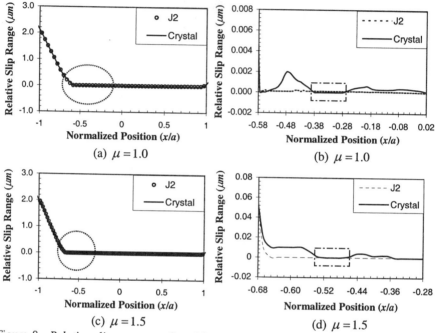

(a) $\mu = 1.0$

(b) $\mu = 1.0$

(c) $\mu = 1.5$

(d) $\mu = 1.5$

Figure 8. *Relative slip range predicted by the FE simulation along the interface of the fatigue specimen during the 3rd cycle and a magnification of the circled region showing the mixed stick/slip region for conditions:* $R_\sigma = 0.1$, $\sigma_a = 150$ *MPa, and* $W = 1350$ *N* ($a = 514$ μm).

The ratio of the frictional force to normal contact load, Q/P, is computed from the FE simulations throughout three cycles using

$$\frac{Q}{P} = \frac{\int q(x)\,dA}{\int p(x)\,dA} \qquad (13)$$

The ratio of Q/P starts from 0.01 in the beginning of the 1st loading cycle and at the end of the loading cycle reaches a value of 0.88 ($\mu = 1.0$) and 0.9 ($\mu = 1.5$). On the other hand, the ratio of the local $q(x)/p(x)$ varies from 0.65 to 1.0 for $\mu = 1.0$ and from 0.55 to 1.5 for $\mu = 1.5$ along the interface during the 3rd loading cycle with the peak values being in the slip region (see Fig. 9). An estimate of the COF in the steady-state slip region based on the size of the stick and slip zones found experimentally gave average values of COF in the slip region ranging from 0.71 to 1.06 [*12*]. The analysis suggests that very local stress contractions near the stick/slip regions lead to high local effective COF values.

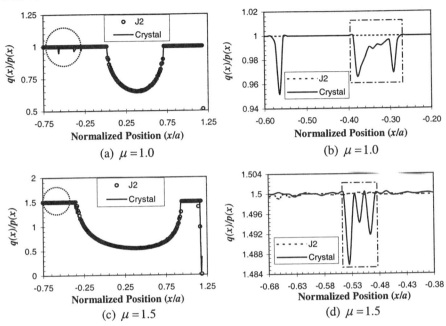

Figure 9. *The ratio of local shear traction ($q(x)$) to normal traction ($p(x)$) predicted by the FE simulation along the interface of the fatigue specimen at the end of 3rd loading cycle and a magnification of the circled region showing the mixed stick/slip region for conditions: $R_\sigma = 0.1$, $\sigma_a = 150$ MPa, and $W = 1350$ N ($a = 514$ μm).*

Interface Shear Stress Evolution during Fretting

To better understand the evolution of the surface stresses and stick/slip conditions during the fretting fatigue process, four points in the loading portion of the 3rd cycle, noted as A, B, C, and D in Fig. 4(b), are examined. The interface shear stress distributions are shown in Fig. 10. In the figure, Q denotes frictional force (F) per unit length (L), i.e. $Q = F/L$, and σ represents the applied stress to the specimen during cyclic loading. The shear stress (σ_{xy}) is normalized by the cyclic yield stress ($\sigma_y = 765$ MPa) and the applied stress (σ) is normalized by the product of COF (μ) and the maximum normal pressure ($p_0 = 349$ MPa). The frictional forces (F) are computed from FE analysis by summing the reaction forces in the pad along the y-axis plane of symmetry. In this component model, these reaction forces are not prescribed, but are coupled with the fatigue load (F_{sp}) applied to the specimen. The frictional force per unit length ($Q = F/L$) is then normalized by COF (μ) and the normal load per unit length

$(P = W/L)$ of 1350 N/mm. The nature of the stick and slip regions along the interface at these four points in the loading portion of the 3rd cycle for the cases of $\mu = 1.0$ and $\mu = 1.5$ are illustrated in Fig. 11.

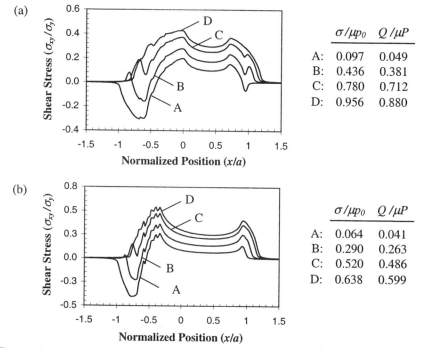

Figure 10. *Interface shear stress* (σ_{xy}) *evolution during the 3rd loading cycle when (a)* $\mu = 1.0$ *and (b)* $\mu = 1.5$ *for the case:* $R_\sigma = 0.1$ *and* $\sigma_a = 150$ *MPa* ($\sigma_y = 765$ *MPa,* $p_0 = 349$ *MPa, $P = 1350$ N/mm, and $a = 514$ μm).*

For μ=1.5, at the minimum load just after the load reversal (Point A), there exists only a stick region with no slip. At Point B, slip occurs at the trailing edge while a mixed stick/slip region is present at the leading edge. The negative shear stress exists at the trailing edge until the load reaches the value of $Q/\mu P = 0.49$ and $\sigma/\mu p_0 = 0.52$ (Point C) (see Fig. 10). As the load increases, the slip zone size increases while the stick zone size decreases. Interestingly, the mixed stick/ slip zone is observed at the boundary between the slip and stick zones at the trailing edge near the peak of the loading. Also, the contact zone is shifted to the right by 0.2a mm at the final state of the 3rd loading cycle (Point D: $Q/\mu P = 0.6$ and $\sigma/\mu p_0 = 0.64$). During the unloading of the 3rd cycle, the order of phenomena is reversed. The lower COF case, μ=1.0, shows similar tendency as the μ=1.5 case though the mixed stick/slip region near the leading edge is observed at a later time step (Point C); it is present from Point B for μ=1.5 case (see Fig. 11).

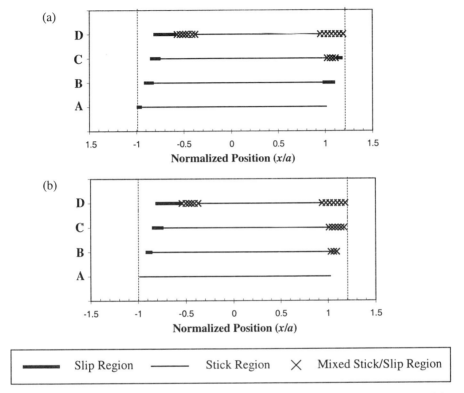

Figure 11. *Stick and slip regions along the interface based on crystal plasticity model at four points in the 3rd loading cycle, as noted in Fig. 4(b), (a) $\mu = 1.0$ and (b) $\mu = 1.5$ for the case: $R_\sigma = 0.1$ and $\sigma_a = 150$ MPa (W = 1350 N and a = 514 μm).*

Contact Strain Fields along the Interface of Contact

The total and plastic strain distributions predicted using the different material models and COF are compared in Figs. 12 and 13, respectively. In the figures, the strain components are normalized by the yield strain ($\varepsilon_y = 0.0065$). The strain distributions show similar tendency for all material models but there exist significant differences among the material models at the trailing edge where significant plasticity occurs, particularly in shear strain distribution with the higher COF (Fig. 12(d)). In the plastic strain distributions, crystal plasticity theory predicts nonhomogeneous distributions, which likely results from the shear localization due to the heterogeneity of the microstructure whereas the J_2 plasticity model predicts smooth distributions as shown in Fig.13. Crystal plasticity simulations exhibit fluctuations in stress and strain distributions along the contact surface due to alternating hard and soft orientations of surface grains leading to a natural length scale for roughness, i.e. in these simulations, the grain scale.

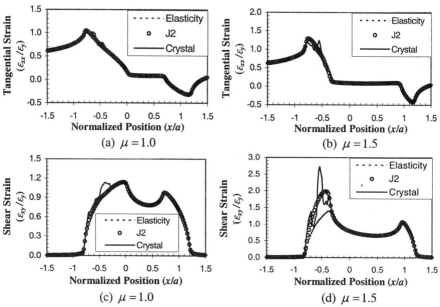

Figure 12. *The total strain distributions along the interface at the end of 3rd loading cycle, comparing material models and COF: tangential strain (ε_{xx}) and shear strain (ε_{xy}) distributions for the case:* $R_\sigma = 0.1$, $\sigma_a = 150\,MPa$, *and* $W = 1350\,N$ ($\varepsilon_y = 0.0065$ *and* $a = 514\,\mu m$).

Influence of Normal Contact Load in Fretting

The subsurface distributions of the cumulative effective plastic strain based on the J_2 and crystal plasticity simulations for an increase in contact load, $W=4050\,N$, with $\mu = 1.5$ are shown in Fig. 14. These results should be compared to the lower normal load case ($W = 1350\,N$) shown in Fig. 7. Comparing the two cases, the heterogeneity of the cumulative plastic strain field for crystal plasticity appears to be magnified in the lower normal load case, when the extent of plastically deforming material at the surface is of the same magnitude as the grain size or smaller. To better understand the influence of contact load on the subsurface cumulative plastic strain fields, the distributions of stress and strain along the interface at the end of 3rd loading cycle are shown in Figs. 15 and 16. For clarity, only the stress and strain distributions from J_2 plasticity model are shown; the crystal plasticity distributions are similar though with some fluctuation near the trailing edge similar to what is shown in Figs. 5 and 12. With increasing COF, the maximum shear stress increases particularly at the trailing edge in both normal loading cases (Fig. 15). The distributions in both tangential (σ_{xx}) and shear (σ_{xy}) stresses significantly vary with increasing COF over the entire contact region in case of the lower normal load, $W = 1350\,N$ while there are no significant differences due to variation of COF except near the trailing edge (left side of the curve) in the higher normal load case, $W=4050\,N$.

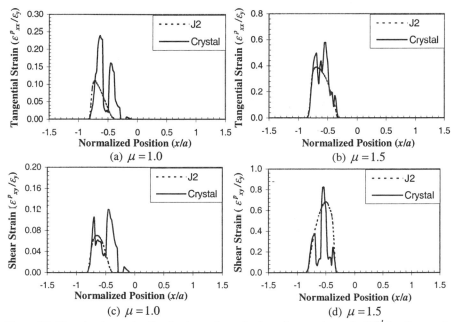

Figure 13. *The plastic strain distributions along the interface at the end of 3rd loading cycle, comparing plasticity models and COF: tangential strain and shear strain distributions for the case:* $R_\sigma = 0.1$, $\sigma_a = 150$ MPa, $W = 1350$ N, $\varepsilon_y = 0.0065$, and $a = 514$ μm.

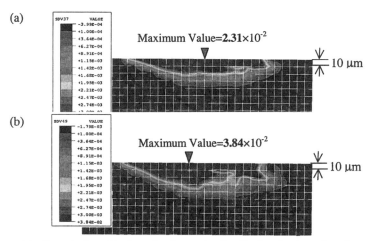

Figure 14. *Cumulative effective plastic strain distribution in the subsurface region at the trailing edge at the end of 3rd loading cycle for the case:* $R_\sigma = 0.1$, $\sigma_a = 150$ MPa, $W = 4050$ N, and $\mu = 1.5$, *showing influence of material model (a) J_2 plasticity and (b) crystal plasticity.*

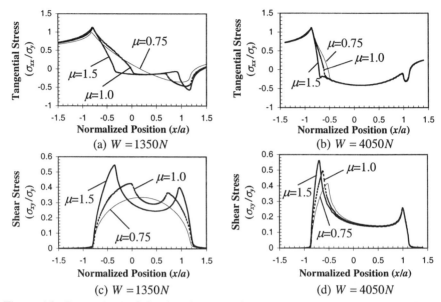

Figure 15. *Comparison of the distributions of tangential stress* (σ_{xx}) *and shear stress* (σ_{xy}) *along the interface at the end of 3rd loading cycle when various COFs are used with conditions:* $R_\sigma = 0.1$ *and* $\sigma_a = 150\ MPa$ ($\sigma_y = 765\ MPa$, $a = 514\ \mu m$ *for* $W = 1350\ N$, *and* $a = 889\ \mu m$ *for* $W = 4050\ N$).

The influence of COF is even more significant when considering the distributions of the strain components as shown in Fig. 16. It is observed that the shear strain (ε_{xy}) is largest for the higher COF ($\mu = 1.5$), which is about twice as large than when $\mu = 1.0$ is used, and the general tendency follows in the same manner as the distributions of stress components with increasing COF in both loading conditions, $W = 1350\ N$ and $W = 4050\ N$ (Fig. 15). The decrease in the apparent heterogeneity of the cumulative plastic strain field for the higher load case may be partially attributed to the narrower peak in the shear strain along the interface (see Fig. 16 (d)). When the peak is more distributed as in the lower contact load case (Fig. 16 (c)), the crystal plasticity model samples a greater number of microstructural variations leading to multiple stress concentrations and resulting in greater heterogeneity of the plastic strain field.

A History-Dependent COF Evolution in Fretting

The influence of COF in the hysteresis of frictional force versus the applied fatigue stress (which is proportional to the nominal relative displacement [12]) is shown in Fig. 17. The hysteresis curves represent near stick, partial slip, or gross slip condition, depending on the COF. A lower COF manifests gross sliding conditions, whereas a higher COF promotes partial slip conditions with very little hysteresis. As the COF increases, the hysteresis curve shifts up and shows some discrepancy with the curve obtained from the experiments, which does not exhibit this shift. A higher COF yields no gross slip conditions even in the first cycle and thus it always leads to a positive mean stress (see Fig. 17). However, there are gross sliding conditions during the first few cycles in real contacts, as shown in the experimental results (Fig. 18(a)). This implies that the COF should evolve in the simulations to capture this transient effect. When the COF is incremented after each cycle from 0 to 1.5, the steady-state mean frictional force is better predicted (Fig. 18(b)). However, there is still some difference between the magnitude of the predicted frictional force range (ΔQ) and the experimental results.

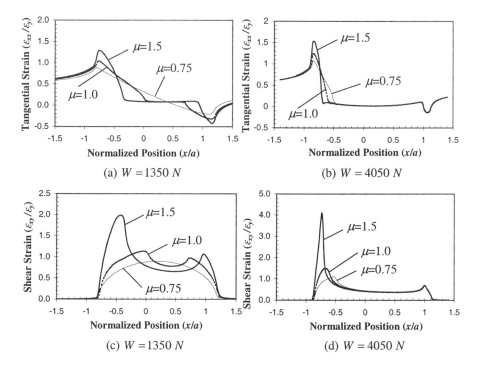

Figure 16. *Comparison of the distributions of tangential strain (ε_{xx}) and shear strain (ε_{xy}) along the interface at the end of 3rd loading cycle when various COFs are used with conditions: $R_\sigma = 0.1$ and $\sigma_a = 150$ MPa ($\varepsilon_y = 0.0065$, $a = 514$ μm for W = 1350 N, and $a = 889$ μm for W = 4050 N).*

Discussion

The prescribed COF plays a key role in the distribution of contact stress and plastic strain fields, with increasing COF leading to intensification of subsurface plasticity. The

COFs of 1.0 and 1.5 at first glance may seem unreasonably high. However, it is likely that the local value of COF, defined as $q(x)/p(x)$, may be considerably higher than what is measured macroscopically, Q/P, which is simply the integral of all the local interactions, as noted in Eq. (13). The COF is *not* a given material property since it also depends on several factors such as the surface interactions and environment. As the local normal load decreases, $q(x)/p(x)$ may go to the infinite at the edge of contact (or at edge of asperities when considering rough interfaces) in the ideal case when no plasticity occurs. Therefore, surface interactions at the microscopic scale play a significant role in determining the COF under partial slip conditions.

From fretting wear tests on three titanium alloys such as Ti-6Al-4V, Fayeulle et al. [15] observed that the experimentally measured COF, in the steady state after 10^3 cycles, was always very high, greater than unity, and the tangential load contributes to plastic deformation of the substrates and formation of friction ridges. Also, recent work by Swalla et al. [16] suggests that when prescribing the macroscopic fretting fatigue conditions under high cycle fretting fatigue conditions, a higher COF has the effect of accounting for the locally elevated shear stress due to asperity-asperity contacts, albeit in an average sense. Here, the experimental deformation fields and widths of the stick/slip regions are better predicted when a higher COF is used. The deformation field is developed due to the local behavior and not the smeared average behavior collectively accounting for several asperity-asperity interactions.

Since the nature of the surface interaction between two surfaces is determined by the real area of contact, shear forces as well as normal forces acting on junctions may play a significant role in determining the value of COF. The COF is also related to the surface topography. These mechanisms may result in local stress and plastic strain concentrations and thus produce a much higher value of COF than that measured in an average sense. Surface irregularities, such as roughness and waviness, also induce surface displacement in normal direction resulting in a dynamic normal force that can further result in fluctuation of local COF, $q(x)/p(x)$ [17].

The interfacial stick-slip process is caused by the fact that the frictional force varies as a function of displacements, time, or velocity, rather than a constant [18]. Initially, the COF rapidly increases until reaching a steady-state value in less than 20 cycles in experiments and partial slip conditions prevail [12]. Based on the overall tangential traction of the contact, Q, the ratio of Q/P is considerably less than the value of $\mu_{prescribed}$ with the COF being statically indeterminate within the stick zones. Hence, our method for estimating the value of COF for the partial slip case involves comparing computational crystal plasticity and experimental results. Purely experimental methods such as progressively increasing the slip amplitude until partial slip gives way to gross slip over the entire contact are likely no less subject to idealization and perhaps even moreso since the COF is dependent upon the history of sliding. Therefore, gross slip measures cannot reflect very well the behavior within the relatively small slip zones where fretting wear and damage are concentrated in the partial slip case and locally the pressure at which slip occurs is comparably lower [16]. The frictional force may undergo significant changes during the early stages of sliding before reaching the steady-state conditions. Thus, a variable history-dependent COF in the slip and mixed slip-stick zones may be more appropriate.

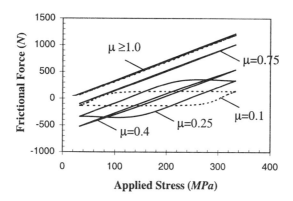

Figure 17. *Comparison of the hysteresis curves of frictional force versus the applied fatigue stress to the specimen for a constant COF throughout the 3rd cycle ($R_\sigma = 0.1, \sigma_a = 150$ MPa, and W $= 1350$ N).*

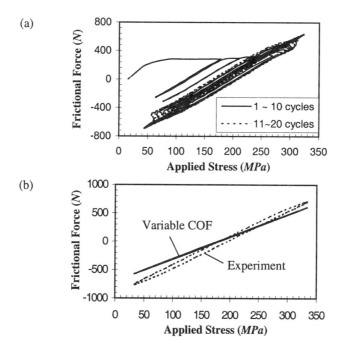

Figure 18. *Comparison of frictional force versus the applied fatigue stress to the specimen: (a) the experimental results obtained from first 20 cycles [12] and (b) the hysteresis curves obtained from FE analysis using a variable COF with cycles ($R_\sigma = 0.1, \sigma_a = 150$ MPa, and W $= 1350$ N).*

Concluding Remarks

Fretting fatigue simulations using the nonhomogeneous crystal plasticity exhibit important features that are not manifested by the initially homogeneous J_2 theory: first, crystal plasticity simulations predict highly heterogeneous plastic strain distributions, particularly at lower contact loads, with greater cumulative plastic strain magnitude as well as deeper penetration of cyclic plastic strain. Second, regions of mixed stick/slip, leading to local stress/strain concentrations, are predicted when crystal plasticity is used.

Finite element simulations using crystal plasticity theory with $\mu = 1.5$ more closely represent the local plasticity behavior along the interface and in the subsurface field observed in the fretting fatigue experiments under partial slip conditions.

The value of the prescribed coefficient of friction (COF) particularly affects the interfacial strain distributions, especially in the component of shear strain. The stress distributions along the interface are not significantly different comparing J_2 and crystal plasticity. As the contact pressure is reduced, the influence of COF in the stick/slip conditions is great.

Since the COF depends on the history of sliding, assuming a constant COF throughout the cycling is an idealization. Under partial slip conditions, the initial COF is very low and it rapidly increases with fretting cycles. Hence, a variable history-dependent COF in the slip and mixed stick-slip regions may provide a more reasonable description of the transient response leading to better prediction of the steady-state response.

Acknowledgments

This work is supported under AFOSR Grant No. F49620-01-1-0034, Length Scale Considerations in the Formation of Attachment Fatigue Cracks. The grant is jointly funded by the Metallic Materials program (Dr. Craig Hartley, program manager) and the Structural Mechanics program (Dr. Dan Segalman, program manager).

References

[1] ABAQUS, version 5.8, Hibbitt, Karlsson and Sorensen, Inc., Pawtucket, RI, 1998.

[2] Marin, E.B., "A Critical Study of Finite Strain Porous Inelasticity," Ph.D Thesis, Georgia Institute of Technology, Atlanta, GA, 1994.

[3] Marin, E.B. and McDowell, D.L., "A Semi-Implicit Integration Scheme for Rate-Dependent and Rate-Independent Plasticity," *Computers and Structures*, Vol. 63, No. 3, 1997, pp. 579-600.

[4] Bennett, V., "A Study of Microscale Phenomena in Small Crack Propagation under Multiaxial Fatigue," Ph.D. Thesis, Georgia Institute of Technology, Atlanta, GA, 1999.

[5] McGinty, R.D. and McDowell, D.L., "Multiscale Polycrystal Plasticity," *Journal of Engineering Materials and Technology*, 1999, Vol. 121, pp. 203 209.

[6] Cuitiño, A.M. and Ortiz, M., "Computational Modeling of Single Crystals," *Modeling and Simulation in Materials Science and Engineering*, 1992, Vol. 1, pp. 225-263.

[7] Eylon, D, "Summary of the Available Information on the Processing of the Ti-6Al-4V HCF/LCF Program Plates," University of Dayton, Dayton, OH, 1998.

[8] Russo, R.A. and Seagle, S.R., "Deformation and Recrystallization of Titaniumand Its Alloys," *ASM International*, Course 27, Lesson, Test 5, 1994.

[9] Goh, C.-H., Wallace, J.M., Neu, R.W., and McDowell, D.L., "Computational Crystal Plasticity Applied to Attachment Fatigue Problems," *proc. 5th National Turbine Engine High Cycle Fatigue Conference* (HCF '00), Chandler, Arizona, 7-9 March 2000.

[10] Goh, C.-H., Neu, R.W., and McDowell, D.L., "Shakedown, Ratchetting, and Reversed Cyclic Plasticity in Fretting Fatigue of Ti-6Al-4V based on Polycrystal Plasticity Simulations," *proc. 6th National Turbine Engine High Cycle Fatigue Conference* (HCF '01), Jacksonville, Florida, 5-8 March 2001.

[11] Nakazawa, K., Sumita, M., and Maruyama, N., "Effect of Contact Pressure on Fretting   Fatigue of High Strength Steel and Titanium Alloy," *ASTM STP* 1159, M. Helmi Attia and R.B. Waterhouse, Eds., 1992, pp. 115-125.

[12] Wallace, J.M. and Neu, R.W., "Fretting Fatigue Crack Nucleation in Ti-6Al-4V," *Fatigue and Fracture of Engineering Materials and Structures*, submitted for review, 2001.

[13] Goh, C.-H., Wallace, J.M., Neu, R.W., and McDowell, D.L., "Polycrystal Plasticity Simulations of Fretting Fatigue," *International Journal of Fatigue*, in press.

[14] Hills, D.A. and Nowell, D., *Mechanics of Fretting Fatigue*, Kluwer Academic Publishers, Dordrecht, 1994.

[15] Fayeulle S., Blanchard P., and Vincent L., "Fretting Behavior of Titanium Alloys," *Tribology, Transactions*, 1993, Vol. 36, No. 2, pp 267-275.

[16] Swalla, D.R. and Neu, R.W., "Influence of Coefficient of Friction on Fretting Fatigue Crack Nucleation Prediction," *Tribology International*, in press.

[17] Yoon, E.-S., Kong, H.S., Kwon, O.-K., and Oh, J.-E., "Evaluation of Frictional Characteristics for a Pin-On-Disk Apparatus with Different Dynamic Parameters," *Wear*, Vol. 203-204, 1997, pp 341-349.

[18] Williams, J.A., *Engineering Tribology*, Oxford University Press, Oxford, 1994.

Marie-Christine Dubourg [1]

Local Fretting Regime Influences on Crack Initiation and Early Growth

---

REFERENCE: Dubourg, M.-C., "**Local Fretting Regime Influences on Crack Initiation and Early Growth**," *Fretting Fatigue: Advances in Basic Understanding and Applications, STP 1425,* Y. Mutoh, S. E. Kinyon, and D. W. Hoeppner, Eds., ASTM International, West Conshohocken, PA, 2003.

ABSTRACT: An investigation of fatigue crack initiation and propagation under fretting conditions is described in which both experimental and theoretical approaches are used. Fretting tests were conducted on an aerospace aluminium alloy. The approach presented here aims to describe the propagation of a fretting crack through a step-by-step understanding of the different stages of its life (1) Stage I crack initiation, i.e., sites, growth directions and fracture mode, (2) Stage I-to-Stage II transition as a result of the competition between failure modes, i.e the branching depth and the growth direction after branching. This methodology was successfully applied to fretting fatigue data obtained under the Partial Slip Regime (PSR). Pertinent parameters were proposed to deal with items (1) and (2) depending on the ranges of shear and tensile stresses and their respective roles in crack growth. This step-by-step methodology is used here to analyze the influence on crack initiation and propagation of the local fretting regime and the associated mechanical stress-strain state.

KEYWORDS: crack initiation, propagation, fretting, friction, crack branching

Nomenclature

a Contact half-width
c Radius of the stick zone
f, f_s, f_d Coefficient of friction, static (partial slip condition) and dynamic (gross slip condition) values of the coefficient of friction
P, Q Normal and tangential loads
R Radius of the sphere
x Abscissa at the contact surface, x = 0 corresponds to contact center
z sample depth
α Angle of the initiation plane with respect to the surface contact, i.e, Stage I propagation plane

---

[1]Research scientist, Laboratoire de Mécanique des Contacts UMR CNRS 5514, INSA de Lyon, Bât. J. D'Alembert, 20 Av. A. Einstein, 69621 Villeurbanne Cédex, France.

β Angle of the Stage II propagation plane with respect to contact surface

$\Delta\delta$ Displacement amplitude

σ_m average value of normal stress σ_{nn} along plane of length 20 μm at an angle α

$\sigma_{m,max}$ Maximum value of σ_m over a whole loading cycle

$\sigma_{m,min}$ Minimum value of σ_m over a whole loading cycle

$\Delta\sigma_m$ $(\sigma_{m,max} - \sigma_{m,min}) / 2$, range of tensile stress

$\Delta\sigma_m*$ Effective range of tensile stress σ_m

τ_m Average value of shear stress σ_{nt} along plane of length 20 μm at an angle α

$\tau_{m,max}$ Maximum value of τ_m over a whole loading cycle

$\tau_{m,min}$ Minimum value of τ_m over a whole loading cycle

$\Delta\tau_m$ $(\tau_{m,max} - \tau_{m,min}) / 2$, range of shear stress

Introduction

Fretting is one of the plagues of modern industrial machinery. It is sometimes responsible for premature fatigue failures and often limits component life. Fretting occurs whenever a junction between components is subjected to cyclic sliding, with small relative displacements at the interface of the contacting surfaces. Further cyclic bulk stresses may be superimposed to one or both components. Fretting is divided into fretting wear and fretting fatigue [1,2] and in the literature they are often related to the total degradation response. In these cases fretting wear is considered to be a wear problem, i.e. a loss of material, and fretting fatigue to be a cracking problem causing a reduction of the fatigue limit of the material. Numerous studies have shown that wear and cracking often coexist at the same contact and that the development of long cracks is not restricted to fretting fatigue but also occurs under fretting wear conditions.

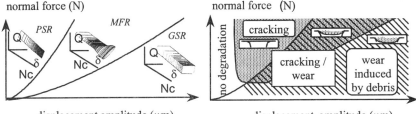

Figure 1- *A Running Condition Fretting Map (RCFM) and its associated Material Response Fretting Map (MRFM). Here PSR= Partial Slip Regime characterized by quasi-closed Q-δ loops, MFR= Mixed Fretting Regime characterized by quasi-closed, parallelepipedic and elliptic Q-δ loops and GSR= Gross Slip Regime characterized by parallelepipedic Q-δ loops [3].*

The mapping concept proposed by Vincent et al [3] is based on two sets of fretting maps. These maps describe on the one hand the local fretting regime i.e Running Condition Fretting Maps (RCFM), with corresponding contact kinematics (adhesion, partial slip, gross slip) while on the other hand is the main initial fretting damage i.e. Material

Response Fretting Maps (MRFM) involving non-degradation, cracking, particle detachment for combinations of normal load and relative displacement (see Figure 1). They establish a close link between the main initial damage response and the contact sliding regime. For the case of very small displacement amplitudes, partial slip conditions prevail within the contact area and induce crack nucleation, while greater displacement amplitudes leading to full sliding over the contact area favor wear. Cracks initiate at a very early stage and most of the component life incorporates crack growth. Cracking is thus related through these maps to both the Partial Slip Regime (PSR) and the Mixed Fretting Regime (MFR), but nevertheless there is a difference, since partial slip conditions prevail during all the tests under PSR, and the wear of the surface within the sliding zone is negligible. Under MFR, there is an evolution from one sliding condition to another, generally from gross slip to partial slip. As can be seen in Fig.1 these conditions give rise to cracking and wear as the main damage processes with the amount of wear increasing as the displacement amplitude increases.

A combined experimental and theoretical analysis of fretting crack nucleation and propagation has therefore been undertaken in order to improve our understanding of the cracking mechanisms. This analysis has been conducted for the following conditions:

- a sphere is repeatedly fretted against a flat surface of the same material, the normal load being applied first and then a sequence of cyclic oscillations is imposed. This then allows a constant bulk stress to be superimposed on the planar specimen,
- normal elastic load ranges can be applied,
- displacement amplitudes δ between the contacting surfaces, which provide a PSR fretting regime, can be recorded.

This latter condition guarantees negligible wear of the contacting surfaces which remain smooth. These specific experimental conditions correspond to idealized conditions and these perfectly fit with the assumptions of the model. Two crack types have been observed during Stage I, corresponding to specific mode I and mode II conditions. Transition from Stage I to Stage II for both crack types is characterized by a crack branching towards a new propagation direction of $\beta \approx 65°$ to the specimen surface [4]. Then during Stage II, both crack types propagate along the 65° direction. Issues such as (a) the location of crack nucleation and crack directions during Stage I propagation, (b) parameters controlling crack branching during the Stage I-Stage II transition, (c) crack direction during stage II have been theoretically addressed, respectively in references [5], [6] and [7]. Specific macroscopic parameters linked to the propagation driving forces were proposed and the predicted results are in good agreement with experimental observations.

The aim of this paper is to analyze the influence of the fretting regime on the initiation and the early growth of fretting cracks. Different displacement amplitudes, leading either to PSR or MFR conditions are considered when the normal load is constant and provides elastic behavior. It was observed that although Stage I shear and tensile crack initiation coexist at the same time under experimental conditions at the boundary of PSR and MFR, shear crack initiation prevailed in the MFR whilst tensile crack growth was manifest in the PSR. Fretting tests conducted on specimens of 7075 aluminium alloy under spherical-plane contact under PSR for larger displacement amplitudes and MFR conditions conclude the analysis.

Results Under PSR and Small Displacement Amplitude

Tests Results

These results have been presented elsewhere [5] and are only briefly recalled here. Multiple cracks initiate in the annulus microsliding zone of the contact area (See Figure 2) and these are symmetrical about the centerline x = 0. Two main cracks develop from this network while most of the other cracks self arrest. At the contacting surface, cracks propagate along a semi-elliptical trajectory further below the surface, while their growth occurs along two distinct macroscopic directions during, respectively, Stage I and Stage II. The transition between these two propagation Stages is revealed by crack branching along the Stage II propagation direction. Two types of crack are observed during Stage I.

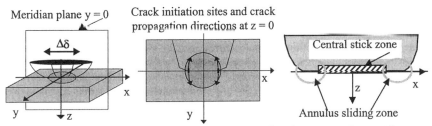

Figure 2 - *Sphere – plane configuration, Top view at z=0 and section in the meridian plane y = 0 .*

- Type I cracks grow initially at a shallow angle to the specimen surface with a propagation direction ranging from 15° to 35° to the surface. These cracks mainly occur in the middle of the annulus sliding zone.
- Type II cracks grow along a direction approximately perpendicular to the surface with a direction ranging from 75° to 90° to the surface. They appear near the edge of the contact area in the annulus sliding zone, inside and/or outside the contact patch.

Finally, it was observed that, for both crack types, the nearer to the contact center they initiate, the shallower to the surface is their initial growth direction. Then, during Stage II, both types of crack propagate along a direction of approximately 65° to the surface. The branching depth corresponds roughly to a maximum crack length of 160 μm.

Fretting Crack Prediction

The contact problem between the contacting surfaces is solved, using Coulomb's Law, as a unilateral contact problem with friction. An incremental description of the loading is used to account for hysteresis phenomena. The subsurface stress field is calculated within a zone close to the contact surface. This stress field is multiaxial, non-proportional, and exhibits very sharp gradients [8]. Fatigue crack initiation during Stage I and early propagation up to the Stage I- Stage II transition have been analyzed as a continuum mechanics problem. The range of shear and tensile stresses have been computed for all material locations within the sliding zone during the cyclic fretting loading for all possible plane directions. Average stresses are used [6], calculated along planes of length L=20μm. The existence of macroscopic cracks [9] and their influence on the stress and strain fields is accounted for in the analysis of the fatigue crack process during Stage II growth.

Some observations on Stage I:

- A type I crack is a shear mode fatigue crack. Two specific parameters have been proposed to predict both the initiation site and a single growth direction. The growth occurs macroscopically in the direction given by an angle α along which the value of the shear driving force, i.e the amplitude of the average shear stress, $\Delta\tau_m$ is maximum $\Delta\tau_m = \Delta\tau_{m,max}$ and such that the average value σ_a of the average tensile stress perpendicular to that direction, σ_m ($\sigma_a = 0.5*(\sigma_{m,max} + \sigma_{m,min})$) is minimum (an absolute value of 40 MPa is considered here for this minimum). The corresponding crack initiation domain is the zone where both conditions are fulfilled, being assumed that the risk is the highest where σ_a is zero.

- A type II crack is a tensile mode fatigue crack. The propagation driving force is assumed to be the maximum amplitude of crack opening. The crack extension angle α is therefore defined by the direction along which the *effective* amplitude of the average stress normal to crack trajectory $\Delta\sigma_m*$ is maximum, $\Delta\sigma_m*=\Delta\sigma_m*,_{max}$. Furthermore the amplitude of the average shear stress tends to a minimum value along that direction. A Type II crack location is therefore near the edge of the contact area, inside or just outside, where a high level of tensile stress occurs. The angle α thus ranges from 90° (outside the contact patch) to 70° (inside the contact zone).

The conditions governing crack branching from Stage I to Stage II are identified. Both initial crack types (I and II) branch as a consequence of:

- the decrease of the Stage I propagation driving force, gradually leading to a mixed I+II mode, (Note that, crack locking effects gradually hinder type I crack propagation)
- at the same time, the effective amplitude of the tensile stress increases perpendicular to the plane that is 65° to the contact surface, where the shear mode driving force is at a minimum.

Hence, branching and initial propagation during Stage II are governed by the maximum amplitude of the crack opening and not by its maximum value.

The Influence of the Local Fretting Regime

The fretting crack mechanisms described above were theoretically investigated under PSR and a displacement amplitude of 15 μm. Various relative displacement amplitudes are considered for a given normal load and bulk stress. This allows an analysis of the effects of the contact kinematic conditions for a local fretting regime and then of the local fretting regime itself on crack initiation and growth. Three contact conditions were simulated and these are presented in Table 1. The first two conditions belong to the PSR while the third condition belongs to the MFR. An analysis of the stresses induced in the vicinity of the contact zone has been conducted as in previous work for those 3 cases. Looking at the initiation zones, the growth directions, and the stress levels versus the contact conditions (see Figure 3) that:

Table 1 - *Fretting Conditions*

Local fretting regime	Experimental conditions	Modeling data
PSR, Number 1	P = 1000 N Q = +/- 500 N R = 0.3 m f = 1. $\Delta\delta$ = +/- 25 μm	P = 420 N/mm Q = +/- 210 N/mm R = 0.239 m f = 1. c/a=0.71
PSR, Number 2	P = 1000 N Q = +/- 1000 N R = 0.3 m f = 1.2 $\Delta\delta$ = +/- 35 μm	P = 420 N/mm Q = +/- 420 N/mm R = 0.239 m f = 1.2 c/a=0.41
MFR, Number 3	P = 1000 N Q = +/- 800 N R = 0.3 m f_s = 0.8 f_d = 0.7 $\Delta\delta$ = +/- 50 μm	P = 420 N/mm Q = +/- 335 N/mm R = 0.239 m f_s = 0.8 , f_d = 0.7 c/a=0

Local Fretting Regime	Degradations

PSR (1)

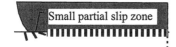

- No damage exept cracking
- Few cracks

PSR (2)

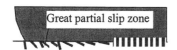

Larger potential cracking zone
- More cracks
- Particle detachment through crack coalescence
- Particle expulsion dependant on relative displacements

MFR (3)

- Particle detachment through crack coalescence
- Accelerated expulsion of particles
- Surface state modification
- Overstresses, initiation of new cracks

‖‖‖ Central stick zone within the contact	＼Crack	▼ Detached particle	◄─► Relative movement

- for type I cracks, the greater the slip zone within the contact area, the larger the potential zone for crack initiation. This leads to an increase in the number of cracks that may initiate and a greater variety of crack inclinations. The risks of cracking are linked to the shear stress amplitude which in turn is dependent on the friction coefficient,
- for type II cracks, the initiation process is controlled either by a tensile-compressive or a tensile fatigue mechanism. Since the normal load and the bulk stress are kept constant, then the variations of the propagation driving force are only induced by the variations of the tangential tractions. The fatigue limit is equal to 230 MPa for R = $\sigma_{min}/\sigma_{max} = 0.1$. It is clear that each material point within the sliding zone is subjected to a different maximum tensile stress amplitude and a different mean tensile stress. There are no fretting fatigue data available for these pairs. Only qualitative or rough estimations of the risks of cracking can be predicted. Nevertheless the stresses are the highest in the zone close to the contact boundary.
- The growth direction is nearly unchanged, except that, for a high friction coefficient, cracks are more inclined towards the specimen surface,
- For a given risk of initiation, as quantified by the maximum shear and the effective tensile stress amplitude, the damage occurring at the contacting surfaces depends on the "contact dynamics" and their effect in trapping the wear particles generated from crack coalescence. This efficiency is, among other things, linked to the amplitude of the relative displacement $\Delta\delta$ between the contacting bodies. From a certain relative displacement amplitude$\Delta\delta$, here around 35 μm, a ridge is observed at the contact edge and favors particle entrapment inside the contact. Furthermore, the initiation sites and initial growth directions may be different to those predicted without accounted for the presence of this ridge.

The influence of a ridge

The ridge that forms at the contact boundary is subjected during the contact life to a continuous development in its mechanical properties and its geometry. We are interested in its influence on contact parameters i.e., contact pressure, tangential tractions and contact patch dimensions, which govern, in turn, the internal stresses and finally the crack initiation process.

A ridge is considered as an asperity and its mechanical properties are assumed to be the same as those of the specimens. Its dimensions are indicated in Fig. 4, and correspond to experimental observations. In the analysis PSR(2) conditions are simulated with and without this ridge.

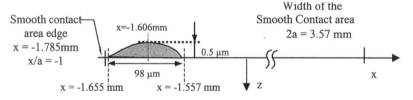

Figure 4 - *Schematic representation and dimensions of a ridge at the contact edge*

The normal and tangential tractions are indicated in Fig. 5. The variations of the specific parameters for the (smooth = no ridge) and (with a ridge) cases for type I and II cracks are presented respectively in Figs 6 and 8.

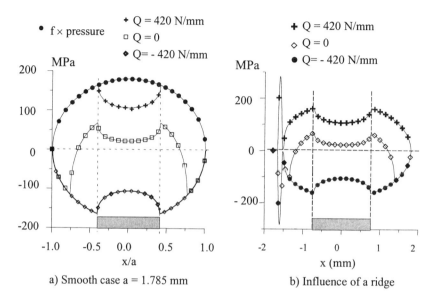

a) Smooth case a = 1.785 mm   b) Influence of a ridge

Figure 5 – *Tangential tractions for PSR(2) Conditions*

In the smooth case, Po, the maximum hertzian pressure and 2a, the contact width are respectively 150 MPa and 3.57 mm. The presence of the ridge (i) induces a sharp peak of pressure reaching 236 MPa, (ii) shifts the left contact edge from x=-1.785 mm to x=-1.665 mm and (iii) reduces slightly the contact area to 3.5 mm.

The stress state at the near surface is strongly influenced by the ridge which induces localized stress concentrations. This has the following consequences for:
• type I shear cracks:

a considerable increase in $\Delta\tau_{m,max}$ is obtained at x=-1.78 mm, i.e from 296 to 573 MPa, see Figs 6(a) and 6(b), while a decrease is observed for points situated inside the contact area; for instance at x=-1.60 mm, from 324 to 278 MPa. Looking at the crack nucleation sites, it appears that the condition σ_a close to zero is satisfied at two different locations, x= -1.78 mm and x= -1.60 mm, while in the smooth case only the latter point was obtained. Furthermore, $\Delta\tau_{m,max}$ has increased by a factor of nearly 2. This large increase may accelerate the crack initiation process and increases the risks. Isovalues of τ_{max} are shown in Fig. 7 for one load step, at Q = 420 N/mm. A confined plastic zone appears behind the ridge which will enhance fretting crack nucleation. It can be seen that in front of and immediately beneath this zone τ_{max} decreases very rapidly. At x = -1.78mm, the initial crack propagation direction is equal to 35° while in the smooth case it is about 31-

32°. The ridge thus influences the crack direction during Stage I, leading to shallower cracks.

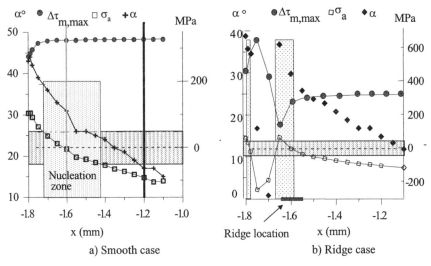

a) Smooth case

b) Ridge case

Figure 6 - *Variations in α, $\Delta\tau_{m,max}$, σ_a over part of the relative sliding zone for a Shear crack.*

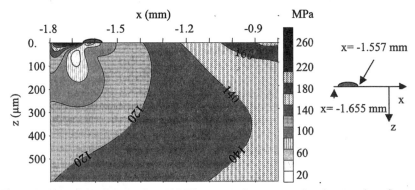

Figure 7 - *Isovalues of $\Delta\tau_m$ at $Q = 420$ N/mm near the contact edge showing the influence of the ridge.*

- type II tensile crack; see Figs 8(a) and 8(b):

The variations of $\sigma_{m,max}$, $R = \sigma_{m,min} / \sigma_{m,max}$ and α in the smooth case and of $\sigma_{m,max}$, $\sigma_{m,min}$ and α in the ridge case over part of the relative sliding zone are shown in the figures. In the smooth case, $\sigma_{m,max}$ decreases continuously with respect to the contact center. The ridge induces localized stress concentrations and especially an increase in the tensile stress at the contact edge together with a large compressive effect. There is a maximum and a relative maximum in values of $\sigma_{m,max}$. Though the effects of these complex contact stresses leave the crack directions almost unchanged, the crack nucleation sites change significantly. The locations corresponding respectively to the maximum and the relative maximum value, $\sigma_{m,max}$, are shifted towards the contact edge for the former and towards the contact center for the latter.

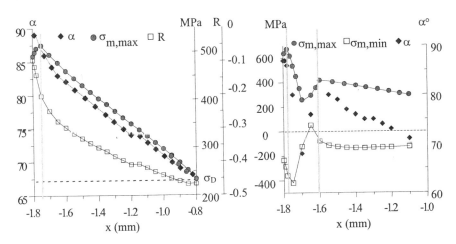

a) Smooth case: $\sigma_{m,max}$, α and R variations.

b) Ridge case: $\sigma_{m,max}$, $\sigma_{m,min}$ and α variations.

Figure 8 - *Variations of specific parameters over part of the relative sliding zone for a tensile crack.*

• Stage I / Stage II transition,

Type I and II cracks branch as a consequence of the decrease in the Stage I propagation force which occurs concomitantly with an increase in the effective tensile stress perpendicular to the plane at 65° to the surface [6]. This branching is likely to occur between z_{min} and z_{max}. The compressive part of the amplitude of the axial stresses increases continuously with depth z, until $\Delta\sigma_m$ becomes compressive during the whole cycle for a characteristic depth, z_{max}. Here z_{max} is the maximum depth at which branching may occur, as the driving forces for type I and II cracks, the relative sliding and the opening amplitude respectively, are inhibited. The minimum depth at which branching may occur, z_{min} is not quantitatively defined as it is a result of a competition between the shear and the tensile driving forces. For the present, we suggest that a criterion of a 50% reduction in $\Delta\tau_{max}$ and this correlates well with experimental observations of crack branching.

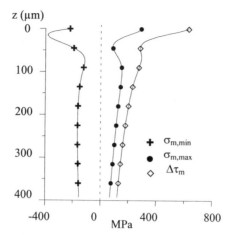

Figure 9 - *Influence of the ridge on* $\sigma_{m,min}$, $\sigma_{m,max}$ *and* $\Delta\tau_m$ *variations versus depth when* $x = -1.78$ mm.

Figure 9 shows $\sigma_{m,min}$, $\sigma_{m,max}$ and $\Delta\tau_m$ variations versus depth at $x = -1.78$ mm, along $\alpha=35°$ which is the initial growth direction for the tensile crack. Values of $\Delta\tau_m$ and $\sigma_{m,max}$ decrease very rapidly through a zone of depth 50 μm, close to the contact surface, the former being divided by nearly a factor 2 and the latter by more than a factor 2. Then from a depth of 100 μm, they decrease very slowly and $\sigma_{m,max}$ does not become negative. It is thus not possible to define a z_{max} value as previously. However it is clear that the relative sliding along the crack faces are rapidly limited due to the strong reduction in both $\Delta\tau_m$ and the tensile stress.

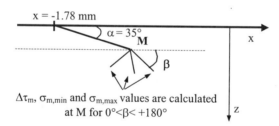

Figure 10 - *Different branching direction for the crack tip M.*

Ranges of tensile and shear stresses are computed at the crack tip situated at varying depths from $z = 21$ μm to $z = 99$ μm along the initial crack growth direction $\alpha=35°$ (see Fig. 10) for all possible branching directions β. Figure 11(a) shows the values of $\sigma_{m,min}$ and $\sigma_{m,max}$ versus β. The value of $\Delta\sigma_m{}^*{}_{,max}$ is obtained for β ranging from 75° to 80°. Figure 11(b) presents $\Delta\tau_{m,min}$ variations along the direction defined by β. It should be noted that the Stage I/Stage II transition zone is likely to be bounded between 50 and 70 µm.

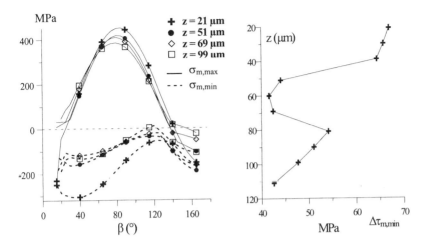

(a) Variations of $\sigma_{m,min}$ and $\sigma_{m,max}$ versus β for different depths

(b) Variations of $\Delta\tau_{m,min}$ versus depth along β

Figure 11 - *Stress state variations at x=-1.78 mm for a shear crack inclined initially at $\alpha = 35°$ for the PSR(2) condition and showing the effect of a ridge at the contact edge*

Conclusions

A theoretical model for fretting contact fatigue has been developed. Stage I crack initiation and Stage I/ Stage II transition processes were analyzed through an approach based on continuum mechanics. The computational predictions were first validated for the Partial Slip Regime (PSR) with a   small displacement amplitude between the contacting surfaces.

This paper highlights the influence of the local fretting regime on the crack initiation and early growth processes. The effects of the displacement amplitudes between the contacting surfaces and of a ridge are analyzed. On the basis of the results presented, the following conclusions can be made:

- The displacement amplitude $\Delta\delta$ between the contacting surfaces, is a key parameter,
  - the greater the value of $\Delta\delta$, the larger the fretting crack initiation zone giving an increase in the density of secondary cracks which favors crack coalescence and particle detachment,
  - particle expulsion or trapping is dependant on $\Delta\delta$ and the contact geometry,
  - the crack direction is nearly independent on $\Delta\delta$,

- For values of $\Delta\delta$ greater than about 35 µm, a ridge is formed at the contact edge. In comparison to a smooth surface tested under similar conditions, localized stress concentrations are obtained that modify the Stage I crack initiation and Stage I/Stage II processes. The transition zone is closer to the contact surface and the Stage II direction of growth is changed;

Short crack growth behavior is thus clearly linked to the contact conditions and the local fretting regime. The predicted results correlate well with the experimental ones if the surface evolution is taken into account in the analysis of the fretting fatigue mechanism. Above a certain displacement amplitude, cracking is no more the dominant fatigue mechanism and competes with wear processes, which may be a consequence of a great density of small cracks that coalesce leading to particle detachment and expulsion from the contact. Pits and scales appear at the contacting surfaces and induce localized stress concentrations which may accelerate or decelerate the growth of secondary cracks, which in turn may modify the growth of the predominant cracks situated at the contact edges.

Acknowledgements

The author is grateful to V. Lamacq for a fruitful collaboration.

References

[1] Waterhouse R.B., "Fretting Fatigue", Elsevier, Applied Science, London, 1981.

[2] Waterhouse R.B., "Fretting Corrosion", Pergamon, Oxford, 1972.

[3] Vincent L., Berthier Y., Godet M., " Testing Methods in Fretting Fatigue : A Critical Appraisal ", ASTM STP 1159 on *'Standardization of Fretting Fatigue Test Methods and Equipment'*, American Society for Testing and Materials, West Conshohocken, PA, 1992, pp. 33-48.

[4] Forsyth P.J.E., " A Two-Stage process of fatigue crack growth ", in Symposium on Crack propagation, Cranfield, HMSO London, 1971, pp. 76-94.

[5] Lamacq V., Dubourg M.-C., Vincent L., " A theoretical model for the prediction of initial growth angle and sites of fretting fatigue cracks", *Tribology International*, Vol. 30, N°6, 1997, pp. 391-400.

[6] Lamacq V., Dubourg M.-C., " Modeling of initial fatigue crack growth and crack branching under fretting conditions", *Fatigue Engineering Material Structure*, Vol. 22, 1999, pp. 535-542.

[7] Dubourg M.-C., Lamacq V., " Stage II crack propagation direction determination under fretting fatigue loading. A new approach in accordance with experimental observations", ASTM STP 1367 on *'Fretting Fatigue: Current Technology and*

Practices', Ed. D. W. Hoeppner, V. Chandrasekaran, C.B. Elliot III, Salt Lake City, 2000, pp. 436-450.

[8] Brown M.W., Miller K.J., Fernando U .S. Yates J.R., Suker D.K., " Aspects of Multiaxial fatigue crack propagation", in *'Multiaxial and Fatigue Design'* ESIS Publication 21, Edited by Pineau A., Cailletaud G., and Lindley T.C., Mechanical Engineering Publications, London, 1996, pp. 317-334.

[9] Dubourg M.-C., Godet M., Villechaise B., " Analysis of multiple cracks – Part II : Results", *American Society of Mechanical Engineers, Journal of Tribology*, Vol. 114, 1992, pp. 462-468.

[10] Lamacq V., "Amorçage et propagation de fissures de fatigue sous conditions de fretting", Thèse : Doctorat, INSA, 1997, 251 p.

Yasuo Ochi,[1] Yohide Kido,[2] Taisuke Akiyama, [2] and Takashi Matsumura [1]

Effect of Contact Pad Geometry on Fretting Fatigue Behavior of High Strength Steel

---

Reference: Ochi, Y., Kido, Y., Akiyama, T., and Matsumura, T., "**Effect of Contact Pad Geometry on Fretting Fatigue Behavior of High Strength Steel,**" *Fretting Fatigue: Advances in Basic Understanding and Applications, ASTM STP 1425*, Y. Mutoh, S. Kinyon, and D. W. Hoeppner, Eds., ASTM International, West Conshohocken, PA, 2003.

Abstract: The effect of contact pad geometry on fretting fatigue behavior of high strength steel was examined by using the clamping double bridge pads system. In order to investigate the effects of bridge pad shapes, the fretting fatigue tests were carried out by using four types of bridge pad with different leg height h^*. The fretting fatigue strength decreased with decreasing in the h^*. Also, the stick area increased as increasing of the h^*. The contact pressure distribution was measured by strain gauge and calculated by using a finite element method in order to investigate the effect of the pads leg height h^*. The contact pressure concentrated in the inside on the contact surface where the main crack initiated. Two-stage tests were carried out in order to investigate the fretting fatigue damage. The two-stage test was performed to a certain cycles under fretting condition and then plain tests without fretting as removing the pads were continued.

Keywords: fretting fatigue, contact pad geometry, contact pressure distribution, strain gauge measurement, FEM analysis, relative slip amplitude, two-stage test, high strength steel

Introduction

It has been known that the fretting fatigue strength is affected by the shapes of specimen and contact pad. However, there have been a few studies, which have investigated the effect systematically. Also, the experimental method and the shapes of specimen and contact pad have been selected differently by each study. As a result, it is difficult to compare the data published [1]. Therefore, it is important that the effecting factors on the fretting fatigue property should be clarified by systematic fretting fatigue tests with varying testing parameter.

---

[1]Professor and Associate Professor, Department of Mechanical Engineering and Intelligent Systems, University of Electro-Communications, Tokyo 1-5-1, chofugaoka, chohu, Tokyo, Japan.
[2]Graduate school student, University of Electro-Communications.

In order to elucidate the effect that the fretting fatigue property is affected by the shapes of specimen and contact pad, it is necessary to consider the contact condition in detail. A lot of research by using the analytical method was conducted by now. For example, it is reported that, in case of the plane-to-sphere contact and the sphere-to-sphere contact, it is widely carried out the analysis using the equation of Hertz and Mindlin et al. in search of the contact pressure distribution, and that it shows the good coincidence with the experimental results [2-4]. But, in case of the plane-to-plane contact, it has been shown that the contact pressure has a partial concentration at the stress singular point in the outer edge [5-7], or in the outer and the inner edges [8] of the contact surface. Therefore, it is difficult that the analytical results agree with the experimental results, and it still needs to examine carefully.

In this paper, in order to investigate the effect of shapes of contact pad that influenced fretting fatigue behavior, fretting fatigue tests and two-stage test by using the bridge pad with different leg height were carried out with a high strength steel SNCM439 (Japanese Industrial Standard). And, in order to examine the contact pressure distribution, the contact stress in the vicinity of contact surface was measured by using subminiature 5-element strain gauge. Furthermore, a finite element method (FEM) analysis was conducted to compare with the experimental results.

Material and Experimental Procedure

Material

Nickel chrome molybdenum steel SNCM439 was used as the materials for specimen and bridge pad in the present study. The steel was quenched and tempered in the following sequence: heated to 1123K for 6×10^3 seconds then oil cooled and heated to 893K for 7.2×10^3 seconds then water cooled. Tables 1 and 2 show the chemical compositions and mechanical properties of the materials, respectively. The shape and dimensions of specimen and bridge pad are shown in Figs. 1 and 2, respectively. The contact surfaces of specimen and bridge pad were finished by grinding with a #50 grindstone.

Experimental Procedure

Fretting fatigue tests were carried out by using a servo hydraulic testing machine. The equipment for generation of fretting is shown in Fig. 3 (a). Fretting was produced by clamping bridge pads from both sides of specimen under an axial stress amplitude. The tests were carried out at a frequency of 10 Hz under tension-tension mode with a stress ratio of 0.05 at room temperature. The contact pressure P was 200 MPa, where the contact area of two bridge pad legs was 2×18 mm^2 and the normal force was 7,200 N. During a fretting fatigue test, a relative slip amplitude S was measured by using a clip gauge. Figure 3(b) shows the measuring device of relative slip amplitude. L shape arm was clamped to specimen, and a stand was attached in bridge pad, next, a clip gauge was attached in L shape arm and stand. A frictional force F was measured by using a strain gauge attached to the underside of the bridge pad, also. Four types of bridge pad (leg height h^*: 0.5, 1.0, 3.0 and 5.0 mm) were used as shown in Fig. 2. In order to investigate the effect of the fretting fatigue damage due to the difference of h^*, a two-stage test was conducted. The contact stress in the vicinity of contact surface between the bridge pad and the specimen was measured under the condition for the contact pressure P of 200 MPa. Figure 4 shows the measuring device of contact stress using a subminiature 5-element strain gauge (KFR-120-D19-11-N10C made by KYOWA Electronic Instruments Co.). The gauge pitch was 0.5 mm and the distance from the center of gauge to the contact surface was about 0.5 mm. A specimen with a cross section of rectangle (6×8 mm²) was used for the measurement for the stress distribution near the contact surface.

Table 1–Chemical Composition(wt%).

	C	Si	Mn	P	S	Cu	Ni	Cr	Mo
Specimen	0.37	0.28	0.75	0.021	0.017	0.05	0.174	0.81	0.15
Bridge Pad	0.40	0.26	0.71	0.023	0.017	0.01	0.173	0.77	0.16

Table 2–Mechanical Properties.

	Yield strength MPa	Tensile strength MPa	Reduction of area %	Vickers hardness HV [a]	Young's modulus GPa
Specimen	933	1016	54	288	210.7
Bridge Pad	771	922	62	275	–

[a] Indentation force was 0.98 N for measurement of Vickers hardness.

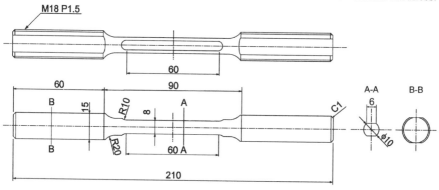

Figure 1–Fretting fatigue specimen (mm).

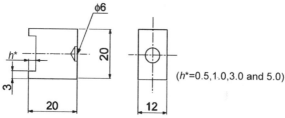

Figure 2–Bridge pad (mm).

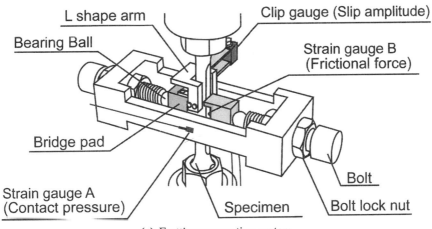

(a) *Fretting generation system.*

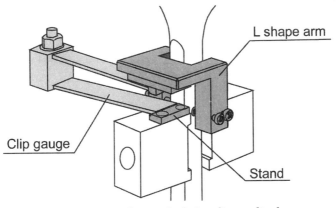

(b) *Measuring device of relative slip amplitude.*
Figure 3–*Fretting equipment.*

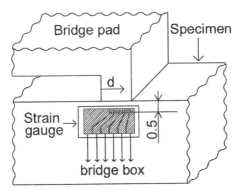

Figure 4–*Measuring device of contact stress* (mm).

Analytical Method

In order to calculate the contact stress distribution and the relative slip amplitude, a finite element method (FEM) was carried out using the program MARC Version K7. The condition of FEM was elastic analysis. The boundary conditions and the load table are shown in Fig. 5. The mesh divisions of the FEM model constituted by plain strain elements are shown in Fig. 6. This mesh division for the leg height h^*=1.0 mm was composed of 3898 elements and 4095 nodes. The Poisson's ratio of SMCM439 steel was 0.28 which was extracted by the reference [9]. The frictional coefficient of 0.5 was obtained from the experimental results and the preliminary analyses.

Results and Discussion

S-N curves

Figure 7 shows the *S-N* curves for plain fatigue and fretting fatigue with different leg heights h^*. In the present study, the plain fatigue strength and the fretting fatigue

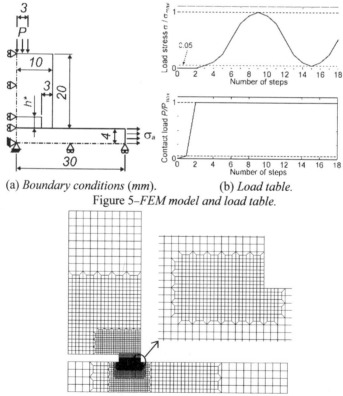

(a) *Boundary conditions (mm)*.   (b) *Load table*.

Figure 5–*FEM model and load table*.

Figure 6–*Mesh division for FEM model (h^* = 1.0 mm model)*.

strength are defined as the strength at $5×10^6$ cycles. The fretting fatigue strengths were as following sequence; $h*$=0.5 and 1.0 mm for about 110 MPa, $h*$=3.0 mm for 125 MPa, $h*$=5.0 mm for 150 MPa, respectively. The fretting fatigue strength remarkably decreased in comparing with the plain fatigue strength that was about 400 MPa. That is, the fretting fatigue strength was only 27 to 37 % of the plain fatigue strength. Difference in the fretting fatigue strength between the conditions of the $h*$ of 0.5 and 1.0 mm was a little. While in the range of the $h*$ of 1.0 to 5.0 mm, the fretting fatigue strength decreased when $h*$ decreased.

Frictional Force

Figure 8 shows the relationship between the frictional force F and the stress amplitude σ_a. The F was in proportion to the σ_a in all leg height $h*$. Difference in the F between the conditions of the $h*$ of 0.5 and 1.0 mm was little, while in the range of the $h*$ of 1.0 to 5.0 mm, the F increased as decreasing of the $h*$. It is known that the frictional force F, with decreasing of the fretting fatigue strength due to the acceleration of crack propagation rate, is the most important factor of the fretting fatigue [10]. Hence, we consider that the $h*$ affects the fretting fatigue strength. It is thought that F decreased as increasing of the $h*$ for the variable form of the bridge pad leg considering the effect of stiffness of bridge pad leg.

Relative Slip Amplitude

Figure 9 shows the relationship between the relative slip amplitude S and the stress amplitude σ_a. The S increased almost linearly as σ_a increased. But the difference in the S by the variation of the $h*$ was not observed at the identical stress amplitude level, and each S was varied within the range of several μm. From observations of fretted damage surface, the roughness of wear scars was severe and the slip area on either side of the fretted area spread with decreasing of the $h*$. Some studies have shown that the same

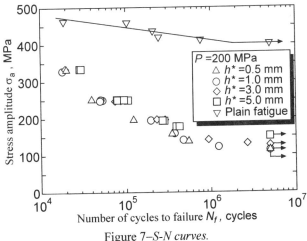

Figure 7–S-N curves.

phenomenon was caused as a result of an increase in the S [11-13]. Assuming that the S is affected by the h^*, the value of the S was calculated by using FEM analysis. The analytical values in Fig. 9 are lower than the experimental values over the whole stress amplitude range. One of the reasons for the results was suggested that because the L shape arm bent by vibration during the fretting fatigue test, the experimental value of the S was amplified and it was not possible to measure the exact value. The increase in the analytic values of the S as decreasing of the h^* was thought to be due to the deformation of the bridge pad leg. The distribution of the S on the contact surface calculated by FEM calculated by FEM is shown in Fig. 10. The stick area (S=0 μm) with the h^* of 3.0 and 5.0 mm exists around the center of the contact surface. The range of stick area increased as increasing of the h^*. The results of FEM coincided with the observation results of the fretted surface on specimen. It is thought that this is due to the effect of stiffness of bridge pad leg.

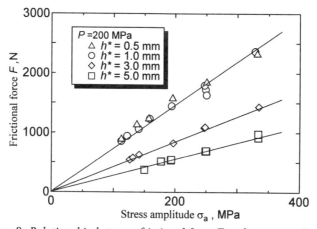

Figure 8–*Relationship between frictional force F and stress amplitude σ_a.*

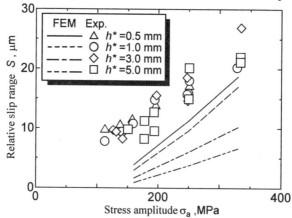

Figure 9–*Relationship between relative slip range S and stress amplitude σ_a.*

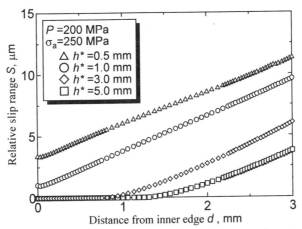

Figure 10–*Relationship between relative slip range S and distance from inner edge d.*

Contact Pressure Distribution

Figure 11 shows the results of the contact stress distribution in the position of 0.5 mm depth from the contact surface measured using 5-element strain gauge and calculated using FEM analysis. In the present experiment, the contact stress on the contact surface could not be measured because the distance of central position of the strain gauge from the contact surface was about 0.5 mm as shown in Fig. 4. And the detailed measurements could not be done, since a gauge pitch was 0.5 mm. In order to evaluate the contact stress, the influence of the position of the strain gauge should be taken into account. Therefore, we calculated the contact stress by using FEM analysis in order to compare it with experimental results. Moreover, it must be considered that the value measured by strain gauge used for this experiment contains the strain by the deformation of specimen by the effect of the frictional force F and the stress amplitude σ_a. The contact stress which is actually measured by the strain gauge was thought to be the value σ_z' that is calculated using the following equation (1).

$$\sigma_z' = \sigma_z + v\sigma_x \qquad (1)$$

Namely, the contact stress σ_z should be added the value multiplied the axial stress σ_x by the Poisson's ratio v. Therefore, in this sturdy, the analytic value to compare the experimental value was defined as the σ_z'.

Fig. 11 (a) and (b) show the relationship between the σ_z' and distance from inner edge d for σ_a=160 MPa, h*=1.0 mm and for σ_a=160 MPa, h*=5.0 mm, respectively. From both figures, at the maximum load of the stress amplitude ($\sigma_{a\ max}$), the σ_z' concentrated inside contact surface and decreased outside where the crack generation side was found. This phenomenon is thought to be due to the deformation of the bridge pad leg pushed out toward outside by tuning the elongation of specimen when stress

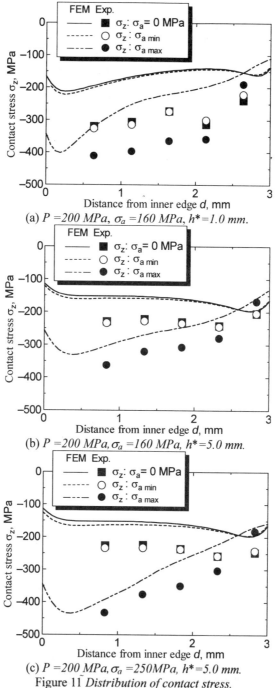

(a) $P = 200$ MPa, $\sigma_a = 160$ MPa, $h^* = 1.0$ mm.

(b) $P = 200$ MPa, $\sigma_a = 160$ MPa, $h^* = 5.0$ mm.

(c) $P = 200$ MPa, $\sigma_a = 250$ MPa, $h^* = 5.0$ mm.

Figure 11 *Distribution of contact stress.*

amplitude was loaded. The difference in the analytical results was little, but the experimental value was higher than analytical value. Especially, at low h^*, the effect was more remarkable. This phenomenon is thought to be originated in the change of the contact area by the abrasion. At the minimum load of stress amplitude ($\sigma_{a\,min}$), the contact pressure distribution was about the same as the result at $\sigma_a=0$ MPa because the stress ratio was 0.05.

Fig. 11 (c) shows the contact stress distribution for $\sigma_a=250$ MPa, $h^*=5.0$ mm. At the $\sigma_{a\,max}$ in the higher σ_a level, the σ_z' increased in all the measurement area and the average contact pressure P increased. Because the σ_z' contains the axial stress σ_x, the σ_z' increased as increasing of the stress amplitude σ_a by the effect of the σ_x. At the $\sigma_{a\,max}$, in the higher σ_a level, the σ_z' was considered to significant increasing as much as the inside on the contact surface, since the deformation of bridge pad leg became larger as the σ_a was higher.

Fig. 12 shows the fretted damage surface observed after the fretting fatigue test for $P=200$ MP, $\sigma_a=160$ MPa at $N=6\times10^3$ cycles. Although the N_{fret} was low cycles, the wear damage at both edges of contact surface was severe and these were shaved off owing to the abrasion. It is thought that because of the change of the contact area by abrasion, the peak of the actual σ_z' was generated inside from the peak point of the σ_z' by FEM.

Fretting Fatigue Damage

The two-stage tests were carried out to investigate the fretting fatigue damage. The tests were performed to the cycles N_{fret} on the process of the fretting fatigue tests, and then to the cycles to failure or to the cycles 2×10^6 with detaching the bridge pads. Fig. 13 shows the results of the two-stage tests. Relations between fretting fatigue cycles N_{fret} and the total cycles N_t which is the sum of N_{fret} and the subsequent cycles are shown in the figure. The fatigue strength is defined as the strength at 2×10^6 cycles in this test. From Fig. 13, the result is divided into three stages as follows; Stage 1: N_t is non-failure period, Stage 2: N_t is longer than N_f, Stage 3: N_t is approximately the same as N_f because of the saturation of the fretting damage. The marks of the solid circle in the Stage 1 show the number of cycles of fretting fatigue limit $N_{fret-lim}$ [14]. Fretting fatigue limit $N_{fret-lim}$ means the minimum cycles of N_{fret} on the Stage 1 required for failure of specimen by plain fatigue without fretting on the Stage 2. The ratio of fretting fatigue limit for fretting fatigue life $N_{fret-lim}/N_f$ were 7% for $h^*=0.5$ and 1.0 mm, 13% for $h^*=3.0$ mm, 20% for $h^*=5.0$ mm. Namely, $N_{fret-lim}/N_f$ decreased as decreasing of the h^*. At low h^*, the crack reached to the critical crack size earlier, because the period of crack initiation and the crack propagation rate was accelerated by the effect of the frictional force [15]. As a result, the fretting fatigue life seems to be decreased as decreasing of the h^*.

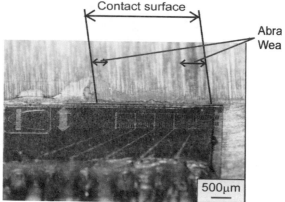

Figure 12–*Contact surface and positions of strain gauge measurement.*

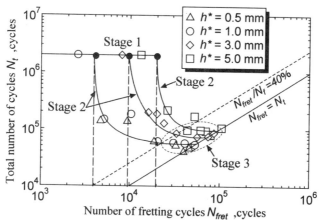

Figure 13–*Result of the two stage tests.*

Conclusions

In order to investigate the effect of shapes of contact pad, the fretting fatigue tests and two-stage tests by using the bridge pad with different leg height were carried out in SNCM439 steel. Then, the contact stress distribution in the vicinity of the contact surface was measured by using a 5-element strain gauge, and the FEM analysis was conducted to compare with the experimental results.

1. Difference in the fretting fatigue strength between the conditions of leg height h^* of 0.5 and 1.0 mm was a little, while in the range of the h^* of 1.0 to 5.0 mm, the fretting fatigue strength decreased as decreasing of the h^*.

2. Difference in the frictional force F between the conditions of the h^* of 0.5 and 1.0 mm was a little, while in the range of the h^* of 1.0 to 5.0 mm, the F increased as decreasing of the h^*.

3. Relative slip amplitude calculated by FEM decreased as increasing of $h*$ because of the deformation of the bridge pad leg, and then, the stick area occurred with the higher $h*$ condition.

4. At the maximum load of stress amplitude, the contact stress σ_z concentrated inside the contact surface and decreased outside where was the crack generation side. And, the contact stress distribution measured by using the subminiature 5-element strain gauge was affected by the axial stress σ_x.

5. As a result of the two-stage test, the fretting fatigue limit decreased as decreasing of $h*$ by the effect of the frictional force.

References

[1] *Standard Method of Fretting Fatigue testing*, The Japan Society of Mechanical Engineers, 2002.

[2] Farris, T., Harish, G., Tieche, C., Sakagami, T., and Szolwinski, M., "Experimental Tools for Characterizing Fretting Contacts," *International Conference on Advanced Technology in Experimental Mechanics*, 1999, pp. 256-263.

[3] Dominguez, J., "Cyclic variations in friction forces and contact stresses during fretting fatigue," *Wear*, Vol.218, 1998, pp. 43-53.

[4] Zhou, Z., and Vincent, L., "Cracking Induced by Fretting of Aluminium Alloys," *Transactions of the ASTM*, American Society for Testing and Materials, 1997, pp. 36-42.

[5] Nagata, K., and Fukakura, J., "Effect of Contact Materials on Fretting Fatigue Strength of 3.5Ni-Cr-Mo-V Rotor Steel and Life-Prediction Method," *Trans. JSME Series A*, The Japan Society of Mechanical Engineers, 58-553, 1992, pp. 1561-1568.

[6] Satoh, T., Mutoh, Y., Nishida, T., and Nagata, K., "Effect of Contact Pad Geometry on Fretting Fatigue Behavior," *Trans. JSME Series A*, The Japan Society of Mechanical Engineers, 61-587, 1995, pp. 1492-1499.

[7] Maruyama, N., and Sumita, M., "Fretting Fatigue Strength Analysis of Ti-6Al-4V in Air," *Tetsu-to-Hagane*, The Iron and Steel Institute of Japan, Vol.76, 1990, pp. 262-269.

[8] Khadem, R., and O'Connor, J., "Adhesive or Frictionless Compression of an Elastic Rectangle between two Identical Elastic Half-spaces, *Int. J. Engng. Sci*, Pergamon Press, Vol.7, 1969, pp. 153-168.

[9] Tanaka, K., Matsuoka, S., Miyazawa, K., "Combined force characteristic of SNCM8 steel," *Trans. JSME Series A*, The Japan Society of Mechanical Engineers, 45-391, 1979, pp. 204-210.

[10] Mutoh, Y., Nishida, T., and Sakamoto, I., "Effect of Relative Slip Amplitude and Contact Pressure on Fretting Fatigue Strength," *Journal of the Society of Materials Science, Japan*, The Society of Materials Science, Japan, 37-417, 1987, pp. 649-655.

[11] Husheng, G., Haicheng, G., and Huijiu, Z., "Effect of slip amplitude on fretting fatigue," *Wear*, Vol.148, 1991, pp. 15-23.

[12] Ochi, Y., Tateno, B., and Kuroki, T., "Fretting Fatigue Strength Properties of Pearlitic Bainitic Steels with Medium Carbon Steel Contact Material," *Journal of the Society of Materials Science, Japan*, The Society of Materials Science, Japan, 47-10, 1998, pp. 1065-1070.

[13] Ochi, Y., Hayashi, H., Tateno, B., Ishii, A., and Urashima, C., "Fretting Fatigue Properties in Rail Steel and Fish Plate Materials (Effects of Contact Pressure and Relative Slip Amplitude)," *Trans. JSME Series A*, The Japan Society of Mechanical Engineers, 63-607, 1998, pp. 453-458.

[14] Wharton, M., Taylor, D., and Waterhouse, R., "Metallurgical Factors in the Fretting-Fatigue Behavior of 70/30 Brass and 0.7% Carbon Steel," *Wear*, Vol.23, 1973, pp. 251-260.

[15] Satoh, T., Mutoh, Y., and Yada, T., "Effect of Contact Pressure on High Temperature Fretting Fatigue," *Journal of the Society of Materials Science, Japan*, The Society of Materials Science, Japan, 42-472, 1993, pp. 78-84.

LOADING CONDITION AND ENVIRONMENT

Jeremy Hooper[1] and Phil E. Irving[2]

Fretting Fatigue Under Block Loading Conditions

Reference: Hooper, J. and Irving, P. E., **"Fretting Fatigue under Block Loading Conditions,"** *Fretting Fatigue: Advances in Basic Understanding and Applications, STP 1425,* Y. Mutoh, S. E. Kinyon, and D. W. Hoeppner, Eds., ASTM International, West Conshohocken, PA, 2003.

Abstract. The effect on fretting fatigue life of applying periodic overloads and two-level block loading has been investigated, and the applicability of Miners damage accumulation under fretting fatigue loading conditions has been tested. The application of single 150% overloads at given intervals had a considerable effect on life, with life extensions on the order of two being observed for overload application intervals of 10^2 to 10^4 cycles. Miner's law was unable to predict the life extensions found. Two-level block loading spectra were applied in which the single overloads were made equivalent to a high-level loading block consisting of a single cycle. The effect of increasing the number of overloads in a high-level block was investigated. It was observed that as the proportion of cycles in the high-level block was increased, the damage accumulation became dominated by the high-level loading. The effect of lower-level constant amplitude loading was to extend life and delay the damage process. A hypothesis based on the contact stress analysis and tangential force hysteresis loops has been proposed and successfully explains all of the variations in fretting fatigue life observed. Miner's law cannot be confidently applied for the loading conditions tested.

Keywords: fretting fatigue, block loading, overloads, variable amplitude loading, Miner

Introduction

Almost all laboratory research in fretting fatigue has been carried out under conditions of constant amplitude loading (CAL). However, real-life loading is invariably variable amplitude by nature. Real-life complex loading spectra are frequently approximated by block loading tests in the laboratory. In order to calculate life under block loading conditions, a damage law such as Miner's rule, given in equation (1), is required. However, as Miner's law is empirical and based on limited fatigue data, it may not be accurate for all conditions. This fact was underlined by Troshchenko et al. [1] who found during an extensive review of the literature, considerable scatter in damage summations.

---

[1] Professor, Damage Tolerance Group, School of Industrial and Manufacturing Science, Cranfield University, Cranfield, MK43 0AL, United Kingdom.
[2] Professor, Damage Tolerance Group, School of Industrial and Manufacturing Science, Cranfield University, Cranfield, MK43 0AL, United Kingdom.

With this in mind, equation (1) can be re-written as equation (2), where a is the actual damage sum for the test. If a is greater than unity then Miner's law gives a conservative prediction for fretting fatigue *life*, and if a is less than unity Miner's law gives a non-conservative prediction of fretting fatigue life.

$$\sum_{i=1}^{s} \frac{n_i}{N_i} = 1 \qquad (1)$$

where n_i is the number of fatigue cycles of stress amplitude σ_i in the ith block and N_i is the number of cycles at σ_i required to induce failure under constant amplitude loading conditions.

$$\sum_{i=1}^{s} \frac{n_i}{N_i} = a \qquad (2)$$

The first significant paper on variable amplitude loading was that of Gassner [2] in 1963, who found that the strength reduction caused by fretting was less marked under VAL conditions than under CAL conditions. In a series of investigations into fretting fatigue under Gaussian-type loading [3, 4, 5], Edwards investigated the effects on fretting fatigue life of applying two types of VAL spectra; Gaussian random loading (GRL) and programmed random loading (PRL). The PRL spectrum was similar to the Gaussian spectrum but with higher peak stresses. It was found that GRL was less damaging than CAL, and PRL was less damaging still. Thus agreeing with Gassner's findings. In contradiction, Mutoh et al [6] found that GRL was actually more damaging than CAL.

Van Leeuwen et al [7] carried out one of the first studies into fretting fatigue under a simplified aero-engine spectrum; namely a low-amplitude, high-frequency "ripple" superimposed on a main high-amplitude low-frequency crenelated waveform. It was concluded that the additional ripple accelerated fretting and crack initiation but did not affect crack propagation. More recently Cortez et al (1999) [8] compared the effect on life of applying similar spectra, where 500 low-amplitude cycles at 200 Hz were superimposed onto a high-amplitude crenelated waveform at 1 Hz, with CAL tests at 200 Hz and at 1 Hz. It was found that the VAL fretting fatigue lives fell within the scatter bands of the 200 Hz CAL results and that Miner's rule provided reasonable, but slightly non-conservative, predictions of VAL fretting fatigue lives. Similar results were found by Szolwinski et al [9], and Iyer and Mall [10].

Mutoh et al [11] studied the effects of two-level block loading on fretting fatigue and investigated the accuracy of Miners summations. It was found that when the number of blocks to failure was high, the linear damage law (LDR) held for fretting fatigue. However, for a low number of blocks to failure, the damage in the first block seemed to significantly influence the cumulative damage value and failure. Troshchenko et al [1] also studied two-level block loading spectra, consisting of greater than ten blocks until failure for fretting fatigue. It was found that at failure, the damage accumulation on the high step remained close to unity, irrespective of the loading regime. However, on the low block step, the damage process was inconsistent with the linear damage rule.

Hooper and Irving [12] investigated the effects of applying single overloads to a CAL spectrum at a range of cycle intervals. It was found that the fretting fatigue life was highly dependent upon the overload application interval and overload amplitude, with life increasing by a factor of 2-3 for the application of 150% overloads every 100-10 000 cycles. Linear damage summations under predicted life by up to a factor of 2. Analysis of the tangential force hysteresis loops showed that a shift in the stick-slip boundaries at the contact patch took place on overload, and that damage was therefore accumulating at more than one location, with consequent increase in life. In the present work, these results have been extended into the two-level block loading regime, and changes in life and accuracy of Miners rule are further investigated.

Experimental

The experimental configuration is shown in Figure 1 and has been described in detail previously [12]. The bulk axial load is applied using a 30 kN servo-hydraulic fatigue machine equipped with digital control allowing variable amplitude spectra to be applied. All tests were carried out at 5 Hz and at an R-ratio (minimum axial load/maximum axial load) of 0.1. Overload tests consisted of constant amplitude loading (CAL) with single 150% overloads [12] applied at specified cycle intervals. The overload starts and finishes at the minimum load of the CAL cycles. In the case of two-level block loading tests, all blocks have $R = 0.1$, to allow application of Miner's law by direct comparison with previous produced CAL stress-life (S-N) data. The normal load P is applied via calibrated spring-loaded bolts. The value of P was chosen to give a peak Hertzian contact pressure of 160 MPa. The selected combination of contact stress, pad radius and sample alternating stress resulted in the contact area being in the partial slip regime. Partial slip is the most damaging slip regime with respect to the specimen life [13].

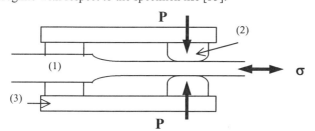

Figure 1 - *Test Configuration and Loading Conditions*

The specimens, (1) on Figure 1, were machined from Al 2618-T6 forgings and had a rectangular gauge length, of cross section 12mm x 7mm. The specified ultimate tensile strength and 0.2% proof strength of the alloy were 430 MPa and 340 MPa, respectively. The steel fretting contact pads (2) were machined with a cylindrical surface, chosen to give a contact stress distribution calculable using well-established expressions. Edge effects and digging-in problems associated with flat and bridge type fretting pads were thus avoided [14]. A radius of curvature $r_c = 220$ mm was chosen for the fretting pad to give a contact semi-width greater than the critical size found for similar materials in the tests of Nowell & Hills [15]. Strain-gauged bridges, (3) on Figure 1, were used to hold the fretting pads in place. The strain-gauge output was used to monitor the tangential force, Q_t, induced at the contact interface due to friction when stress was applied to the specimen. Under sliding conditions, the tangential force equals the frictional force, i.e. the product of the coefficient of friction and normal force. The tangential force was monitored throughout each test.

The extent of slip during fatigue was derived by plotting tangential force against applied bulk axial stress. As sample loading is elastic, load will be directly proportional to displacement at the contact patch. Plots of tangential force against axial stress will be equivalent to plotting tangential force against contact patch displacement. The resulting hysteresis loop shape indicates the slip regime [16]. There are three possible slip regimes in fretting; stick, partial slip and gross slip. The associated hysteresis loop shapes are linear, elliptical and parallelepiped, respectively [16]. The width of the loop provides an indication of partition between slip and stick regions in the partial regime.

Single overload tests

A CAL cycle will be defined as an overload of 100% amplitude. In this series of tests, single overloads of 150% amplitude were applied at intervals. An example overload event is given in Figure 2. The CAL axial stress amplitude, σ_{CAL}, was 60 MPa for all overload tests, i.e. the stress range is 120 MPa for the CAL cycles. Earlier tests by the authors gave the fretting life for a CAL test with σ_{CAL} = 60 MPa to be 4.4×10^5 cycles [12], shown as a dashed line on Figure 3.

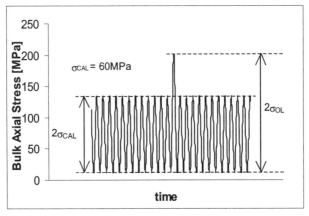

Figure 2 - *Typical Overload Spectrum; 150% Overload*

Effect of Overload Application Frequency

Overloads of 150% amplitude were applied at intervals of 1×10^5, 2×10^4, 1×10^4, 1000, 100, 50, 20, 10 or 5 constant amplitude loading cycles. Figure 3 compares the fretting fatigue lives for each of these tests. It can be seen that fretting fatigue lives arc highly dependent on the frequency of overload application. A maximum life, which is over twice that of the CAL test, exists when 150% overloads are applied every 1000 cycles. At overload application intervals of greater than 2×10^4 cycles, the fretting fatigue lives tend towards that of CAL tests. The fretting fatigue lives are considerably lower than under CAL for tests with overload application intervals of 20 cycles or less. A curve representing Miner's law predictions is also shown in Figure 3.

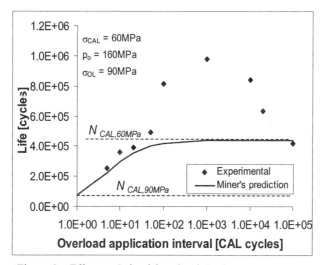

Figure 3 - *Effect on Life of Overload Application Frequency*

From Figure 3 it can be seen that Miner's law is only valid for small overload application intervals (20-100 cycles) and for very large intervals ($>10^5$ cycles). For intermediate intervals Miner's law under estimates life by a factor of approximately 2.

Tangential Force Hysteresis Loops and Effects of Overloads

Figure 4 shows a typical elliptical tangential force hysteresis loop for a CAL test at $\sigma_{CAL} = 60$ MPa. The form of the loop is characteristic of the partial slip regime. The tangential force scale has been normalised to allow direct comparison between hysteresis loops for CAL and overload tests, by setting the $Q_{t,min}$ to zero. The effects on the hysteresis loop shape of applying single 150% overloads is shown in Figure 5. The subscripts 1 and 2 correspond to the CAL cycles immediately before and after the overload application. The important parameters are the tangential force range, ΔQ_t, and the deviation of CAL hysteresis loop position from the pre-overload position. From Figure 5 it can be seen that a shift in the mean tangential force, $\Delta Q_{t,mean}$, value occurs with the application of an overload. This implies a movement of the stick-slip boundaries in the partial slip contact.

The CAL tangential force range, ΔQ_t, undergoes negligible change when an overload is applied.

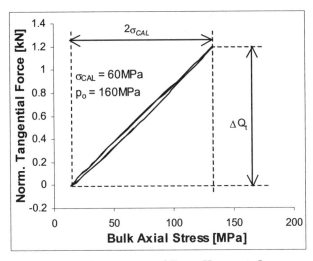

Figure 4 - *CAL Tangential Force Hysteresis Loop*

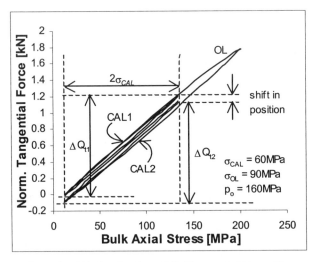

Figure 5 - *Effect of 150% Overload on CAL Tangential Force Hysteresis Loop*

Relaxation Tests

Tests were carried out applying a 150% overload and monitoring the tangential force hysteresis loops over the next 2×10^4 CAL cycles. Figure 5 shows the results of the test for the application of a 150% overload. The tangential force axis has been normalised to zero for comparison purposes. The post-overload CAL hysteresis loop moves back towards the pre-overload CAL hysteresis loop position within approximately 6500 cycles of the 150% overload application (averaged over three tests) and remained in the original equilibrium position throughout the remainder of the 2×10^4 cycles. However, the rate of relaxation is not linear and the size of return shift over the first 1000 post-overload cycles increases with increasing overload size.

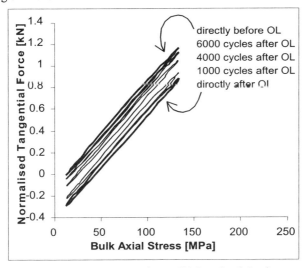

Figure 6 - *Relaxation Test for 150% Overload Application*

Application of Overloads to Non-Fretting Plain Fatigue Tests

Overloads of 60 MPa greater than the CAL range were applied to specimens subjected to simple non-fretting fatigue. A life reduction of 80% was found for the test with overloads every 1×10^3 cycles. This suggests that the *increases* in life due to applying overloads at similar frequencies in fretting fatigue tests were due to a contact phenomenon, and not a material characteristic.

Transition from single overload to block loading

The single overload tests reported in section 3 can be equated to two-level block loading tests where the higher level block consists of a single cycle. An example of a two-level block loading spectrum is given schematically in Figure 7. The spectrum consists of two blocks, of different amplitude and different numbers of cycles, which are repeatedly applied until failure occurs.

The aim of this series of tests was to observe the effect on fretting fatigue life of increasing the number of cycles in the higher level block systematically, from 1 cycle to 1000 cycles. The number of cycles N_A in the lower level block ($\sigma_A = 60$ MPa) was kept constant at 5000 cycles for all tests. The number of cycles N_B in the higher level ($\sigma_B = 90$ MPa) blocks were 1, 10, 100, 250, 500, 750, or 1000 for the relevant test. All blocks had an applied R-ratio of 0.1. Figure 8 shows the variation in total number of cycles to failure N vs. the number of cycles in each high level block N_B.

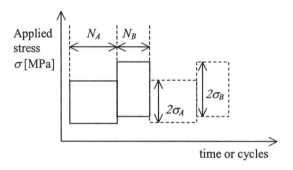

Figure 7 - *Two-Level Block Loading Spectrum*

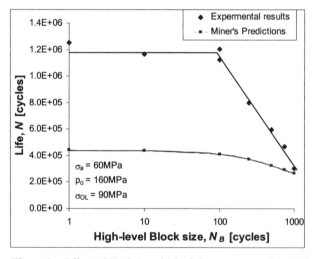

Figure 8 - *Effect of High Level Block Size N_B on Failure Life*

It can be seen from Figure 8 that increasing the number of cycles in the high level block N_B from 1 to 100 cycles does not seem to have an effect on the total number of cycles to failure. However, there is a marked decrease in the total failure life as N_B increases from 100 to 1000 cycles. The predicted failure lives using CAL *S-N* data and

Miner's law is shown by the grey curve. There is an obvious difference between the experimental and predicted lives for most of the tests. A comparison between the experimental results and damage summations for each loading level, $a(\sigma_A)$ and $a(\sigma_B)$, and total damage sum, a_{TOT}, for each of the tests in this series is presented in Table 1.

Table 1 – *Miner's damage summations for "transition" test conditions*

Test	p_o [MPa]	σ_A [MPa]	σ_B [MPa]	$N_A:N_B$ [cycles]	Life, N [cycles]	$a(\sigma_A)$	$a(\sigma_B)$	a_{TOT}
A	160	60	90	5000:1	1.305×10^6	2.966	0.003	2.968
B	160	60	90	5000:10	1.166×10^6	2.645	0.026	2.671
C	160	60	90	5000:100[a]	1.122×10^6	2.501	0.244	2.744
D	160	60	90	5000:100[b]	1.205×10^6	2.685	0.263	2.948
E	160	60	90	5000:250	8.132×10^5	1.760	0.430	2.190
F	160	60	90	5000:500	5.950×10^5	1.229	0.601	1.830
G	160	60	90	5000:750	4.801×10^5	0.949	0.696	1.644
H	160	60	90	5000:1000	3.042×10^5	0.576	0.563	1.139

Tangential Force Hysteresis Loops for Block Loading Tests

From the single overload test results, it was observed that applying a single 150% overload caused a shift in the mean tangential force of the CAL hysteresis loop. These changes have been previously used by the authors [*12*] to explain the life increases shown in Figure 3. The hypothesis proposed is included in the discussion. Therefore it is important to see how effectively applying repeated overloads to make up the high-level block affects the tangential force hysteresis loops.

Figure 9 gives a schematic representation of the two-level block loading spectrum, and will be used to show the variation in tangential force hysteresis loops throughout a test. Figure 10 (a - e) shows the variation in tangential force throughout the transition from the low level block to high level block and back to the low level block for test F, where $N_B =$ 500 cycles.

Figure 10a shows a CAL cycle of $\sigma_A = 60$ MPa. The resulting hysteresis loop is of elliptical form and therefore representative of partial slip. Good agreement with Figure 4 can be seen, although the loop in Figure 10(a) is less uniform. The authors believe that this is likely to be due to the build-up of debris causing damage on the fretting pad surface, as the measurements were taken after 3.95×10^5 cycles. Foundation for this view comes from the fact that the CAL cycle of $\sigma_A = 60$ MPa after the application of the high-level cycles shows similar non-uniformity.

Figure 10b shows the tangential force hysteresis loop for the transition cycle from the low-level block to the high level block, i.e. point B to point C on Figure 9. This cycle is similar to that of the single overload cycle in section 3, but the minimum load at point C is greater than that at point B, hence the loading cycle and therefore the hysteresis loop are incomplete.

Figure 10c shows the first complete high level CAL cycle of $\sigma_B = 90$ MPa. The resulting hystersis loop is also elliptical, although the degree of hysteresis is much larger. This implies that partial slip still results over the contact interface, but the fraction of the interface undergoing slip is greater, as would be expected.

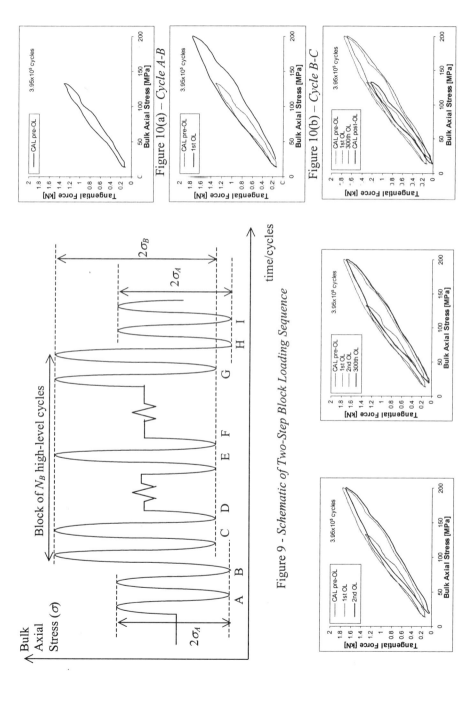

Figure 10(a) – *Cycle A-B*

Figure 10(b) – *Cycle B-C*

Figure 9 - *Schematic of Two-Step Block Loading Sequence*

Figure 10d shows a hysteresis loop representative of any of the other CAL cycles in the high level loading block. There is no further movement in hysteresis loop position, with respect to Q_t, until the transition back to the low-level block occurs.

The transitional loading cycle from the high-level block to the low-level block, i.e. point G to point H on Figure 9, and the respective hysteresis loop are over-complete as the minimum load at point H is lower than that at point G.

Figure 10e shows the first post-overload low-level CAL cycle. A shift in mean tangential force, compared to the pre-overload cycle, similar to that found in the single overload tests (Figure 5) is observed. Subsequent cycling at $\sigma_A = 60$ MPa sees a general "relaxation" of the hysteresis loop back to the pre-overload equilibrium position, similar to Figure 6. A close relationship is therefore observed between the hysteresis loop behaviour of block loading and single overloads.

Discussion

Fretting fatigue endurance is influenced by the stress distribution within the contact region. The localised stresses within this region are well-established for the case of a cylindrical fretting pad in contact with a flat specimen. Applying a normal force to the cylinder will cause a parabolic pressure distribution at the contact interface, as proposed by Hertz [17]. Mindlin [18] developed equations for the shear traction distributions arising in a Hertzian contact. These shear traction distributions were later modified by Nowell and Hills [15] for a typical fretting fatigue situation, considering the effect of applying a bulk axial load to the specimen that is not directly applied to the fretting pads. The effect of this additional load is to cause an eccentricity e of the central stick region, therefore producing a large slip region at the trailing edge of contact and a small, almost negligible, slip region at the leading edge of contact. The associated stress peaks at the stick-slip boundaries are also modified, with a major peak occurring at the trailing edge boundary. Figure 11 shows a simplified approximation of the shear traction distribution for a fretting fatigue test operating under partial slip conditions.

Under conditions of CAL, crack initiation invariably occurs at the trailing edge stick-slip boundary, i.e. at the position of peak contact stress. Therefore it can be assumed that this localised stress peak is the driving force in the initiation and early propagation of fretting fatigue cracks. The current authors have previously used this assumption, along with the changes in tangential force hysteresis loop due to the application of an overload, to explain the considerable variation in fretting fatigue life with overload application interval, as shown in Figure 3 [12]. For completeness, the hypothesis will be reproduced here with the aid of Figure 12, which schematically shows the variation in peak contact stress position due to the application of an overload.

During the application of an overload of amplitude σ_B, the size of the stick region decreases and the eccentricity of the stick region increases [15]. Therefore a smaller stick region exists, centered closer to the leading edge of contact, as shown in Figure 12(a). Referring back to the tangential force hysteresis loops for the application of an overload, Figure 5, it is observed that the overload cycle does not fully complete before the CAL cycling of σ_A re-starts. As the overload cycle is not closed, it implies that the reverse slip back to the original situation has not fully occurred. Hence, for the application of the post-overload CAL cycling, the new stick region will be centered over a similar section of the specimen as for the overload cycle. The eccentricity decreases due to the lower applied axial stress, so that the CAL stress conditions result but over a different area of the specimen surface to the pre-overload CAL cycle, as shown in Figure 12(b).

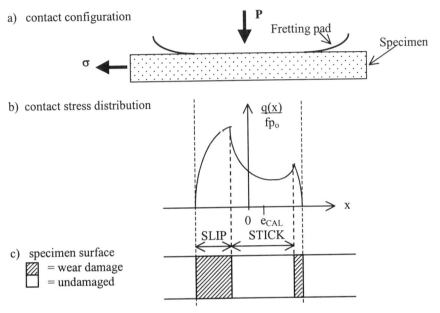

Figure 11 - *Contact Region Stress Analysis and Suggested Stick-Slip Regions from [15]*

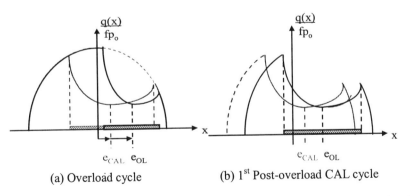

Figure 12 - *Changes in Shear Traction Distribution due to Application of an Overload*

Figure 6 showed that the shift in the CAL hysteresis loop position, due to the application of the overload, gradually returned to the original pre-overload equilibrium position during subsequent cycling at σ_A. For the overload amplitude used in this study, the number of CAL cycles required for complete return was approximately 6500 cycles. This implies that the post-overload CAL peak in Figure 12(b) moves over a region of the specimen surface until it reaches the equilibrium position shown in grey after 6500 cycles. This movement of the peak stress means that damage will be accumulating at a number of spatially separated points and the damage at each site will be less than if there were only a single location. Therefore the number of cycles required for crack initiation and early propagation of a fretting fatigue crack is increased. This hypothesis has previously been used to explain all of the points in Figure 3 [12].

Considering the block-loading test results from section 5. From Figures 9 and 10, it can be seen that the effect on tangential force hysteresis loops of applying a block of overloads σ_B is similar to that of applying a single overload, in that a similar shift in the pre-overload and post-overload CAL (σ_A) hysteresis loop position occurs. This implies that the hypothesis proposed can also be used to explain the effects on life of applying blocks of various numbers of cycles of overloads. From Figures 10(c) and 10(d) it can be seen that applying additional overloads in the high-level block does not cause any further movement in the stick-slip boundaries. Instead the tangential force hysteresis loops are identical for all overload cycles. Implying that the overload block can be assumed to be under similar conditions to a CAL test of σ_B. Therefore if the number of cycles in the high-level block are great enough, a crack nucleation process can occur that would compete with the crack nucleation process of the CAL cycles of σ_A. Table 2 shows the individual contributions to the Miner's damage sum from the low-level and high-level blocks. In all cases, the damage due to the low-level blocks is very high (>unity). However, this does not account for the movement of the peak stress condition. Therefore the majority of the low-level cycles that make up the damage sum have a very minor effect on the actual damage mechanism.

Figure 8 shows that the effect of high-level block size on life can be separated into two regions. There is a negligible effect on life of increasing the number of cycles in each individual high-level block from 1 (single overload) to 100. It can be seen from Table 2 that for this range of numbers of high-level cycles N_B, the damage sum due to the high-level loading is negligible, and therefore is not sufficient to create a competing crack nucleation process. However, when the number of cycles in each high-level block is above 100 cycles, there is an obvious effect of N_B on fretting fatigue life. Increasing the number of σ_B cycles causes life to decrease. This corresponds to an increase in the damage sum due to the high-level loading blocks, as shown in Table 1. Hence it is fair to assume that as the proportion of high-level cycles to low-level cycles increases, the damage mechanism is increasingly due to the high-level cycles because only a fraction of the low-level cycles occur at any given point on the specimen surface. As the number of cycles in the high-level block increases above 100, the damage accumulating at the peak stress location under σ_B becomes increasingly dominant and the total life reduces.

Conclusions

1. In single overload tests, where 150% overloads are applied at given intervals of CAL cycles, the interval of overload application has a large effect on fretting fatigue life.
2. Applying 150% overloads every $1 \times 10^2 - 1 \times 10^4$ cycles causes fretting fatigue life to increase by a factor of two. Miner's law is unable to predict these life variations.
3. As the number of high-amplitude cycles in a two-level block loading test is increased, the fretting fatigue life decreases.
4. The damage accumulated during the high-level loading has the largest influence on fretting fatigue life.
5. A hypothesis has been proposed, based on the shear traction distributions and tangential force hysteresis loops, that successfully explains the life variations found in both single overload and block loading tests.
6. Miner's law cannot be confidently applied for the loading conditions tested.

Acknowledgments

The authors would like to thank Dowty Aerospace Propellers and the EPSRC for jointly funding this research.

References

[1] Troshchenko, V. T., Dragan, V. I., and Semenyuk, S. M., "Fatigue Damage Accumulation in Aluminium and Titanium Alloys Subjected to Block Program Loading under Conditions of Stress Concentration and Fretting," *International Journal of Fatigue*, Vol. 21, 1999, pp. 271–279.

[2] Gassner, E., "On the Influence of Fretting Corrosion on the Fatigue Life of Notched Specimens of an Al-Cu-Mg2 Alloy," *Fatigue of Aircraft Structure*, Pergamon, 1963.

[3] Edwards, P. R. and Ryman, R. J., "Studies in Fretting Fatigue under Variable Amplitude Loading Conditions," *RAE-TR 75132*, 1975.

[4] Edwards, P. R. and Cook, R., "Frictional Force Measurements on Fretted Specimens under Variable Amplitude Loading," *RAE-TR 78059*, 1978.

[5] Edwards, P. R. and Cook, R., "Fracture Mechanics Predictions of Fretting Fatigue under Gaussian Random Loading," *RAE-TR 78086*, 1978.

[6] Mutoh, Y., Tanaka, K. and Kondoh, M., "Fretting Fatigue in SUP9 Spring Steel under Random Loading," *JSME. International Journal - Series I*, Vol.32, No.2, 1989, pp. 274–281.

[7] Van Leeuwan, H. P., Nederveen, A., and Ruiter, H., "Fretting Fatigue under Complex Fatigue Loads," *NLR-TR 73015*, 1973.

[8] Cortez, R., Mall, S. and Calcaterra, J. R., "Investigation of Variable Amplitude Loading on Fretting Fatigue Behaviour of Ti-6Al-4V," *International Journal of Fatigue*, Vol.21, 1999, pp. 709–717.

[9] Szolwinski, M. P., Matlik, J., and Farris, T. N., "Effects of HCF Loading on Fretting Fatigue Crack Nucleation," *International Journal of Fatigue*, Vol.21, 1999, pp. 671–677.

[10] Iyer, K. and Mall, S., "Effects of Cyclic Frequency and Contact Pressure on Fretting Fatigue under Two-Level Block Loading," *Fatigue and Fracture of Engineering Materials and Structures*, Vol.23, No.4, 2000, pp. 335–346.

[11] Mutoh, Y., Tanaka, K., and Kondoh, M., "Fretting Fatigue in JIS S45C Steel under Two-Step Block Loading, *JSME International Journal*," Vol. 30, No.261, 1987, pp. 386–393.

[12] Hooper, J. and Irving, P. E., "Fretting Fatigue under Conditions of Constant and Variable Amplitude Loading," *Proceedings of the International Tribology Conference – Volume I*, Japanese Society of Tribology, Nagasaki, 2000, pp. 419–424.

[13] Vincent, L., Berthier, Y., Dubourg, M. C., and Godet, M., "Mechanics and Materials in Fretting, *Wear*, Vol.153, 1992, pp. 135.

[14] Lindley, T. C., "Fretting Fatigue in Engineering Alloys," *International Journal of Fatigue*, Vol.19, 1997, S39-S49.

[15] Nowell, D. and Hills, D. A., "Crack Initiation Criteria in Fretting Fatigue," *Wear*, Vol.136, 1990, pp. 329–343.

[16] Zhou, Z. R., Fayeulle, S., and Vincent, L., "Cracking Behaviour of Various Aluminium Alloys During Fretting Wear," *Wear*, Vol.155, 1992, pp. 317–330.

[17] *Miscellaneous Papers by H.Hertz*, (Ed. Jones & Scott), 1896, McMillan, London.

[18] Mindlin, R. D., "Compliance of Elastic Bodies in Contact," *Trans.ASME – Series E – Journal of Applied Mechanics*, Vol.16, 1949, pp. 259–268.

John F. Matlik[1] and Thomas N. Farris[2]

High-Frequency Fretting Fatigue Experiments

REFERENCE: Matik, J. F., Farris, T. N., **"High-Frequency Fretting Fatigue Experiments,"** *Fretting Fatigue: Advances in Basic Understanding and Applications, STP 1425,* Y. Mutoh, S. E. Kinyon, D. W. Hoeppner, and, Eds., ASTM International, West Conshohocken, PA, 2003.

Abstract: The fretting problem is of particular interest to the damage tolerant design of turbine blades in today's gas turbine engines. The exotic environment, high-frequency, and variable amplitude load history associated with the dovetail blade/disk connection create a critical location for fretting induced crack nucleation. With little work having been done on investigating fretting contact behavior at high-frequencies and variable amplitude load spectra, sufficient impetus has been generated to better characterize these two currently ambiguous fretting factors. The threat of early crack nucleation and propagation due to these fretting conditions has led to several major research efforts aimed at explicating the high cycle fatigue (HCF) and low cycle fatigue (LCF) interaction and behavior of advanced materials used in modern aircraft turbomachinery. As a part of this effort, a well-characterized experimental setup has been constructed to aid the observation and analysis of the aforementioned frequency and loading factors in fretting. A detailed description of the designed high-frequency fretting rig is presented. Significant vibration and bending results observed during high frequency operation suggest further design modification for improved specimen and pad alignment. Preliminary experimental observations illustrate crack nucleation and failure in specimens subjected to operational frequencies between 100 and 350 Hz. A stress invariant equivalent stress life model is employed for comparison of predicted and observed experimental crack nucleations. The paper concludes with suggested future work aimed at experimentally explicating the frequency and variable amplitude factor effects in fretting fatigue.

Keywords: high frequency, fretting, fatigue, life prediction, friction

[1]Graduate Student, School of Aeronautics and Astronautics, Purdue University, 1282 Grissom Hall, West Lafayette, IN 47907-1282.
[2]Professor, School of Aeronautics and Astronautics, Purdue University, 1282 Grissom Hall, West Lafayette, IN 47907-1282.

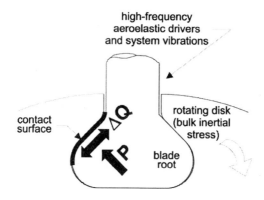

Figure 1 - *Representative two-dimensional geometry of dovetail notch at turbine blade root.*

Introduction

Fretting and fretting fatigue phenomena arise whenever there exist two contacting surfaces subjected to a vibration or repeated load. Though fretting fatigue plagues machinery of all kinds, aircraft components and materials that are continually driven to the limit of their load capacities for light weight design are of particular interest to the fretting research community. Specific components susceptible to fretting fatigue in aircraft include control surface hinge point bearings, spline connections, pin joints, clamps [1], riveted lap joints [2-4], and of particular interest to this work, dovetail turbine blade/disk connections (Figure 1) [5]. Fretting in such turbomachinery components has become a focus of current research [6-8].

Recent estimates hold that possibly one sixth of all damage events associated with United States Air Force (USAF) engine hardware are in relation to high-cycle fatigue damage via fretting and galling [9]. When considering the possible aftermath resulting from high speed turbomachinery and subsequent aircraft engine failure, the reality of such a threat becomes clear. The blade/disk contact surfaces see a complex variable amplitude, variable frequency load history during the course of any given mission. In particular, the interaction between HCF and LCF load spectra has the potential for significant reduction of aircraft component fatigue lives. Except for early work [10], research efforts in fretting fatigue have focused almost exclusively on low frequency (5–20 Hz) with constant amplitude loading waveforms [11]. However, high-frequency engine vibration and unsteady aerodynamic flow over the blade lead to complex, multi-axial, cyclic stress fields and sharp stress gradients at the fretting contact. To successfully characterize the effects of frequency and random load histories on fretting contact, a well-characterized, robust experimental setup must be developed.

Both frequency and variable amplitude (or random) loading are considered mechanical variables involved with the fretting and fretting fatigue processes. This fact suggests a mechanics based approach to modeling. Such a mechanics based approach has been proposed and experimentally verified for prediction of fretting fatigue crack nucleation for a specimen subjected to a constant amplitude, fully reversed loading history [12, 13]. Limited work has been done for both high frequency and variable amplitude loading schemes with regard to fretting fatigue, however.

Published literature has shown that the majority of all fretting fatigue tests have been carried out with constant amplitude stress cycling. Variable amplitude investigations have been greatly discouraged due to lack of success from current cumulative damage schemes particularly with regard multi-axial cumulative damage laws. Though there has been some success in showing Miner's rule to demonstrate qualitative agreement with random loading fretting fatigue experiments [14, 15], little confidence has been put in the varying quantitative results.

Waterhouse performed an excellent survey of the fretting fatigue factor effects [11] in which he observed that little has been done with the frequency factor in fretting fatigue since the work of Endo and Goto [10, 16]. This work as well as others [17, 18] showed that the fretting and fretting fatigue strength of the materials increased as the frequency increased. It was also observed that the coefficient of friction decreased with an increase in frequency of relative movement [16, 17] leading to a decrease in fretting wear rate at higher frequencies [18]. These results suggested that the frequency effect was related to the environment and corresponding oxidation rate. It was concluded that as the frequency increased less time was available for chemical oxidation to occur. By effecting oxidation, the frequency factor influenced the coefficient of friction which in turn effected the contact stresses, surface wear, and overall fretting fatigue life of the specimen.

Current research has shown a trend toward fretting fatigue investigation at relative slip frequencies of 200 Hz [15] and 300 Hz [19], more than an order of magnitude greater than previous research. This frequency shift was motivated by a growing interest in devising design methodologies capable of avoiding or minimizing HCF damage arising in gas turbine engine fan, compressor or turbine blade components.

No formal definition of high cycle fatigue (HCF) has been given except through its relation to its larger load amplitude, lower frequency counterpart – low cycle fatigue (LCF). HCF is best defined with regard to turbomachinery because blades see both LCF load histories resulting from engine spool-up and HCF load histories due to engine vibrations and unsteady flow over the blades. Historically, only the higher amplitude LCF loading component was considered without any inclusion of HCF loading or HCF/LCF interaction effects. However, it has been shown that HCF loading can in fact produce significant reductions in the LCF life of a component [8], and the effect becomes more pronounced as the amplitude of the HCF loading cycle increases.

With a growing need for aircraft safety while pushing both the performance and service life envelope, the necessary development of a robust damage tolerant design approach is evident. To achieve this goal, a clear experimental observation and modeling of currently ambiguous frequency and loading effects on fretting fatigue is required. It remains to design and implement a cost effective, well controlled, well characterized fretting rig capable of aiding a rigorous mechanics based investigation of frequency effects in the fretting process.

Designing the Contact

To produce the small amplitude, oscillatory, relative tangential motion associated with fretting contact, any fretting experimental rig must be designed to generate contact between two bodies via the application of a clamping or normal load, P, followed by an oscillatory tangential load, Q. The tangential load is associated with the development of global sliding if Q is greater than the average coefficient of friction times the normal load (μP) or partial slip if $Q < \mu P$. For meaningful analysis and determination of critical parameters associated with fretting contact, it is mandatory that a clear understanding of contact tractions and stresses be monitored continuously throughout the experiment. To allow for a rigorous analysis of contact pressure, shear traction distribution, and interior stress fields, the current design adopted a fretting contact geometry similar to that originally devised by Bramhall and O'Connor [20, 21] and implemented by Hills and Nowell [22]. This geometry was also chosen because of its successful use with mechanics based life predictions models [13]. The current design was built around a standard uniaxial test specimen which was subsequently clamped between two cylindrical pads to produce the classical Hertzian contact (Figure 2). The pad and specimen geometry choice allows for many variations in contact geometry to be investigated. Pad geometry can be machined to cylindrical profiles [22], spherical profiles [23], or any variation of a nominally flat pad with rounded edges [24]. The chosen specimen geometry conforms to currently accepted standards for constant amplitude uniaxial fatigue test specimens (American Society for Testing Materials E 466-96).

Fretting Test Fixture

A common fretting fixture scheme used in fretting fatigue experimental investigations can be seen in Figure 3a [13]. Both Figures 3a and 3b illustrate two contacting cylindrical pads of radius R subjected to a cyclic tangential load of small enough magnitude ($|Q| < \mu P$) to achieve a condition of partial slip. In Figure 3a after applying a normal load to introduce a contact, the oscillatory tangential load can be generated by applying a cyclic bulk load, ΔL, to the free end specimen. As pad moves with the specimen, the spring force resulting from the stiffness of the fretting chassis resists the motion of the pad and generates an in-phase tangential force, Q. The disadvantage in implementing this design scheme for high frequency applications arises from the large amplitude of cyclic bulk load required. One of the complications associated with high frequency operation is the apparent inverse relationship between load and frequency.

Previously successful low frequency (5–20 Hz) fretting fatigue test fixtures have relied on custom fretting fixtures mounted on standard uniaxial servohydraulic fatigue test frames [13, 22]. Higher frequencies put severe demands on the hydraulic actuators in these test setups and lead to significantly reduced tangential and bulk load histories. Significantly increased operational frequency can also compromise the

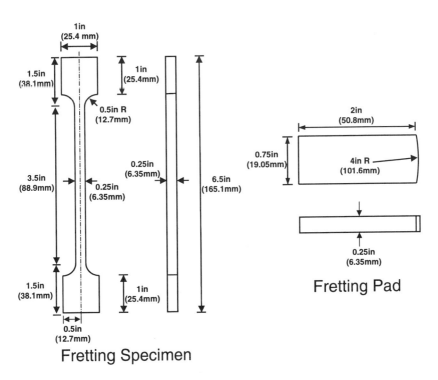

Figure 2 - *Designed fretting fatigue specimen and pad geometry and dimensions.*

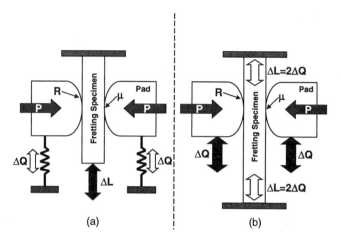

Figure 3 - *A schematic of single-actuator fretting fatigue fixture schemes where (a) the tangential load is indirectly generated via a cyclically applied bulk load [13], and (b) the tangential load is directly applied resulting in an indirectly generated cyclic bulk stress. Note that this is for a condition of partial slip ($|Q| < \mu P$).*

integrity of the hydraulic seals in the system. These two observations suggest two areas of design modification: (1) investigate the increased control and frequency range inherent to electric systems in efforts to realize the order of magnitude increase in testing frequency, and (2) devise a way to generate comparable tangential force magnitudes but circumvent the necessary cycling of the large amplitude bulk load. The second design modification follows necessarily from the first because electric actuator technology requires significantly lower operational loads than those characteristic of lower frequency servohydraulic systems.

Figure 3b proposes an alternate fretting fixture configuration which escapes the necessary cycling of the bulk load for indirect application of Q in Figure 3a by cyclically loading the fretting pads instead of the specimen. This represents the inverse configuration of Figure 3a in that the tangential load is directly applied and the bulk stress indirectly results from the cyclic tangential loading of the contact. This loading scheme exchanges an increase in magnitude of the cyclic tangential load with a reduction in magnitude of cyclic bulk load amplitude. By implementing this fixture style and state-of-the-art piezoelectric technology, a high frequency fretting fatigue load frame has been designed (Figure 4).

Superstructure

The designed high frequency fretting rig was constructed and mounted on a custom built machine base. Standard vibration dampening feet were used to minimize vibrational response of the rig during high frequency operation. The machine base also served as a house for the high voltage amplifier used to power the piezoelectric actuator, and as a place to mount the hydraulic hand pumps for the

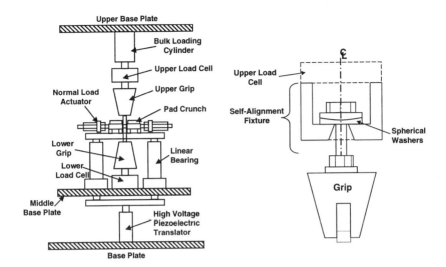

Figure 4 - *A schematic of the designed high-frequency fretting fatigue load frame and self-alignment fixture used for axial alignment of the fretting specimen.*

bulk stress hydraulic loading cylinder. The fretting test fixture and load train was built between three 25 mm (1 inch) thick hot rolled steel base plates each separated by four 25 mm diameter, solid, chrome plated, 440C Stainless Steel shafts (see Figure 4). The overall height of the rig is approximately 1.83 m (6 feet) from the ground to the top base plate. The 610x356x25 mm (24x14x1 inch) base plate of the rig was bolted to the 762x610x25 mm (30x24x1 inch) Mattison ground stainless steel top of the machine base. The middle base plate serves as an anchor for the bottom grip and a support plate to maintain alignment of the linear bearings used in tangential load transfer from the piezoelectric actuator to the contact surface. The top base plate is threaded to anchor the bulk loading hydraulic cylinder and allow for application of an initial tensile bulk stress to the fretting specimen.

Bulk Stress Application

Though the experimental rig generates an oscillating bulk stress via the cyclically applied tangential load, an initial static bulk load allowed for fretting fatigue tests at a variety of mean stress levels. The existence of a tensile bulk stress provided favorable conditions for crack growth and subsequent fracture of the specimen after crack initiation due to fretting contact. The static bulk stress was also necessary to provide sufficient gripping force for symmetric (R-ratio=Q_{min}/Q_{max}=-1.0) operation. The high frequency rig employed a pair of 22.3 kN (5000 lbs.) capacity wedge grips for specimen placement and bulk loading application. Due to the wedge-shaped grip faces, the gripping force was directly proportional to the applied tensile loading and required that the specimen always be loaded in tension before the tangential load could be applied.

To avoid the cost and control issues of a powered hydraulic pump and double-acting cylinder, the tensile static bulk stress was applied using a 107 kN (12 ton) capacity, single-acting, hollow plunger cylinder and 69 MPa (10,000 psi.) capacity hydraulic hand pump. Though most single-acting hydraulic cylinders have applications for only pushing applications, this cylinder was designed to allow the loading rod to pass through the center of the cylinder making a pulling or tensile application possible. Bulk load drift was arrested by incorporating a hydraulic check valve.

As with any axial fatigue load frame, axial alignment of the specimen must be maintained in order to avoid the development of unwanted stresses due to torsion or bending. The current rig addresses alignment through the use of a self-aligning fixture as seen in Figure 4. Recalling that the lower grip is *fixed*, the upper grip is threaded into a bolt which passes through a set of spherical washers and out of the alignment fixture via a tapered 10 degree exit hole. A light machine oil is applied to the spherical washers of diameter 38 mm (1.5 in.) to allow the mating surfaces of the washer and alignment fixture surface to slide and the bolt to rotate in all planes. As a result, the grip is free to rotate +/- 10 degrees and translate +/- 3 mm from the center of the alignment fixture's through hole. Axial translation of the grip occurs only with the travel of the bulk loading cylinder. The alignment fixture adjusts for non-axial load misalignments in the specimen by rotating or translating the spherical washer surfaces to attain an orientation which will minimize or eliminate these non-resisted forces.

Normal Load Application

In order to maintain alignment for the duration of a fretting test, the pads were mounted and clamped in place via the top and bottom pad crunch seen in Figure 5. The right side of the schematic shows the pad bolted into the crunch where the left side shows an exposed pad before the top crunch has been placed. The pad slots in the top and bottom pad crunch were machined to be several thousandths of an inch larger than the pad thickness. This allowed the pad orientation to be adjusted through the use of shims. After each pad had been shimmed into position, the uniformity of the contact was investigated using pressure sensitive paper. If a skewed contact was observed, the pads were adjusted accordingly and the pressure paper test was repeated.

Each side of the bottom pad crunch has a thin plate that extends to the load transfer base plate. These thin *diaphragm springs* provide significant stiffness in the tangential loading direction but give little stiffness in the normal load direction. As a result, the normal load is applied by tightening a set of load nuts on the normal load through rods which pass through the fretting test fixture. These through rods serve as a proving ring structure to maintain a symmetrically applied normal load. Calibration of the current design showed that 86% of the externally applied load was transferred to the contact surface. The normal load brings the pads into contact with the flat faces of the fretting specimen which is mounted in the upper and lower grips. Recall that the lower grip is connected to the lower half of the load train which is fixed in the middle base plate, and the upper grip is connected to the upper

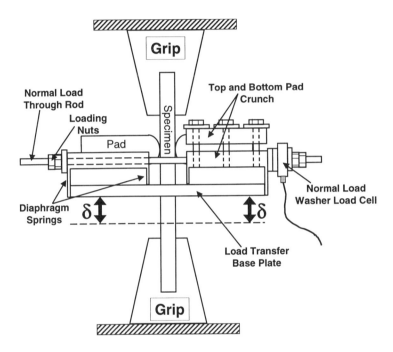

Figure 5 - *A schematic of the fretting fatigue fixture for normal and tangential load application.*

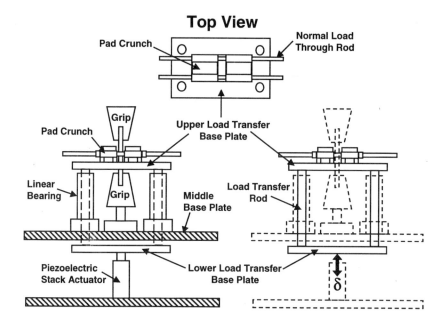

Figure 6 - *A schematic illustrating the tangential load application via the piezoelectric stack actuator. Note that in the right schematic the dashed lines represent portions of the rig that are* fixed *and solid lines represent structure that is being moved by the piezoelectric stack.*

half of the load train which is threaded into the bulk loading cylinder.

Tangential Load Application

As seen in Figure 5, the tangential load is applied by tangential displacement, δ, of the fretting pads after contact has been established via normal loading. The high frequency, oscillatory, tangential displacement is produced by a piezoelectric stackactuator. To avoid subjecting the piezoelectric stackactuator to the large static bulk stress present in the load train, a method had to be devised to translate the piezo displacement around the load train to the fretting fixture and pads while maintaining pad and specimen alignments during high frequency operation. As observed in Figure 5, the bottom pad crunch for each pad was fixed to a load transfer base plate. To circumvent the bulk stress and maintain clearance from the lower grip and load cell, four L-class rods were bolted into the four outside corners of the load transfer base plate and then passed through the middle base plate to a lower load transfer base plate to which the piezoelectric stack actuator was attached (Figure 6).

The four L-class load transfer rods connected the upper and lower load transfer base plates. Four rods were chosen to ensure plate alignments during high frequency operation. The rod alignment is achieved by passing each through a flange mounted Simplicity bearing from Pacific Bearing. The Simplicity bearing is characterized by a smooth Frelon surface lining on the inside of the linear bearing. The liner wears off during operation to serve as its own lubricant and aids in dampening vibration associated with high frequency operation. The bearings prevent rotation in all axes of motion and translation in all axes except for the direction of tangential loading.

Loading Sequence

With the outlined specifics of bulk, normal, and tangential load applications presented, it is now possible to outline the loading sequence of an actual fretting fatigue test in the following sequence.
1. After the specimen is mounted in the top and bottom grip, the bulk loading cylinder is used to introduce a static bulk or mean stress into the specimen. Note that at this point the pads have not been brought into contact and the bulk load at the top, L_T, equals the bulk load at the bottom, L_B.
2. The static normal load is applied by tightening the set of nuts on the normal load through rod to generate the contact.
3. The high frequency tangential load waveform is directly applied as the piezoelectric stack actuator displaces the pads tangentially along the contact surface. It is important to note that the oscillatory tangential load leads to different oscillatory, subsurface, bulk stresses in the specimen above and below the contact.

Data Acquisition System

Analog Measurement of Applied Loads

For successful quantification of the fretting damage process, accurate measurement of the normal load, P, tangential load, Q, and bulk load, L as applied by the experimental setup during the duration of the test is mandatory. The designed high frequency fretting rig monitors these fretting parameters using a series of analog sensors residing both in the fretting fixture and in the load train. The normal load is obtained by observing the calibrated output of a set of 9 kN (2000 lbs.) capacity compression load washers slipped over the ends of the normal load through rods as seen in Figure 5. The tangential load and upper and lower bulk stress measurements can be obtained from calibrated output from the 22 kN (5000 lbs.) capacity tension and compression load cells resident on the upper and lower half of the load train as seen in Figure 4. Assuming that the load applied by the left and right pad are equal, the tangential load, Q, could be obtained by taking half the difference between the lower load cell measurement, L_B, and the upper load cell measurement, L_T ($Q = (L_B - L_T)/2$). Each of the load measurement devices is constructed of 17-4 PH stainless steel externally with the actual load measurement occurring from foil strain gages bonded along an internal diaphragm to reduce off-center loading effects. The applied load could then be determined using the measured output load cell voltage change and the load cell calibration data as obtained from the manufacturer.

Analog Signal Conditioning and Storage

The excitation voltage and signal conditioning of the analog output from each load cell / load washer was achieved through use of a National Instruments manufactured general-purpose Signal Conditioning eXtensions for Instrumentation (SCXI) chassis (model SCXI-1000). Two of the four slots in the chassis each house a National Instruments manufactured four-channel isolation amplifier (model SCXI-1121) with excitation, isolation, and signal conditioning for the analog output from each load cell. Each module allows a user-defined excitation voltage of 3.333 or 10V, two-stage filters of 4 or 10000 Hz, and two-stage gains from 1 to 1000. A terminal block (model SCXI-1321) mounted to the front of each amplifier module allows for offset-null and shunt calibration of the strain gage signal. Each terminal block is equipped with four separate channels giving the current configuration a total of eight data acquisition channels with the potential for expansion to sixteen. The conditioned analog signals obtained from the load cell array are then sent from the chassis to a personal computer for data logging. The SCXI-1000 chassis is limited to data transfer rates of 333 thousand samples per second (kS/S).

To exploit the fast data transfer capabilities of the SCXI-1000 chassis, a 16-bit analog to digital (A/D) National Instruments data acquisition card (model PCI-6052E) was used to capture the strain gage data sent from the SCXI chassis. This card was capable of streaming the data to the PC hard drive at the maximum SCXI communication rate of 333 kS/S and allows for a total of 16 single-ended analog inputs. The card also features two 24-bit, 20 MHz counter/timers and 8 digital I/O lines. Though the current configuration does not allow for computer control of the piezoelectric actuator, the PCI-6052 does offer two 16-bit analog outputs for control of the piezo input voltage and therefore tangential loading waveform. Incorporating a closed loop control scheme will allow for fully automated, high frequency, variable amplitude fretting fatigue experiments in the future.

The continuous online monitoring of each fretting fatigue experiment is achieved though LabVIEW (Laboratory Virtual Instrument Engineering Workbench), a graphical programming language that allows the user to enjoy a user-friendly GUI environment by developing a series of Virtual Instruments (VIs) for interfacing and monitoring acquired data obtained from the SCXI chassis. The in-situ data collected can be presented in real-time via the graphical format of the devised VI or written to the hard drive in binary or ASCII format for post-processing of the data after the experiment. A graphical interpretation of the data acquisition system as used in the designed high frequency testing setup can be seen in Figure 7.

Control Issues

Control of the piezoelectric actuator is achieved by oscillating the input voltage via a Wavetek 2 MHz sweep / function generator. The Wavetek allows for variable frequency, variable R-ratio tangential loading of the contact. The magnitude of the applied tangential load can vary from zero to the maximum tangential load amplitude corresponding to full-scale piezo actuator displacement. However, the

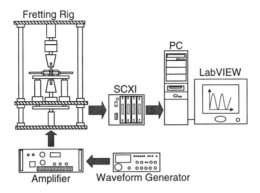

Figure 7 - *Hardware for data acquisition of high frequency fretting fatigue experiments.*

Wavetek cannot produce superposed load waveforms of variable amplitudes. Therefore, the implementation of a LabVIEW integrated waveform generator would not only allow for software generation and control of variable amplitude load waveforms but would also allow for fully automated, closed loop test control. For a final robust experimental design, it remains to integrate this proposed control scheme to meet the automated control demands associated with fretting fatigue tests.

Preliminary Experimental Investigations

High Frequency Operation

Before confidence can be placed in data obtained during high frequency operation, it is necessary to determine any critical frequency ranges in which resonant frequencies of the system interfere with accurate acquisition of data. A frequency scan between 0 and 400 Hz was performed on the current setup while comparing remotely attained strain data from the load cell with local strain data obtained from strain gages mounted above and below the contact on the fretting specimen. Significant differences between remotely and locally measured strain values reflected resonance frequencies at which tests could not be operated. Allowing a 5% variation between remote and local measurements results in acceptable operational frequencies within the ranges 0–165, 195–230, 255–270, and 335–400 Hz.

In dynamic operation of a real structure, modes of vibration can occur in all three dimensions. The frequency scan performed initially investigated variations of strains only in the axial direction. The dynamic bending behavior of the specimen was also investigated for the same frequency range (0 – 400 Hz) by employing an appropriate strain gaged specimen and Wheatstone bridge configuration. The average bending stress ranges were found to be 3-5% of the initial static bulk stress and 17-30% of the stress range associated with the applied tangential load.

During high frequency operation, a coupling between the normal and tangential load was observed. However, the normal load amplitude was found to be less than

5% of the load applied during operation at frequencies that did not cause resonances in the structure. This observed coupling between the normal and tangential load was assumed negligible in the subsequent analysis. Due to extended duration tests inherent to high cycle fatigue investigations, load drift also becomes an area of concern. For two day experiments operated at 100 Hz, a 10% load drift was observed in the normal load and a 5% drift in the mean bulk load was observed. Any analysis of the experimental results must address the effects of load drift in the setup.

Experimental Observations

Several preliminary tests were run as a break-in effort for the high-frequency fretting rig. The remaining efforts of this work are focused on the fretting crack nucleation observation and prediction in the similar contact of a common aircraft grade aluminum alloy, 2024-T4. The machined pad and specimen geometries chosen for this preliminary investigation are the same as those presented in Figure 2.

Before the experiment commences, the measurement of the as-received pad and specimen contact surface is achieved using a Talysurf surface profilometer. Due to the intimate relationship between contact geometry and resulting surface tractions and subsurface stresses, an accurate characterization of the contacting geometry is mandatory. This also allows for determination of average roughness values (R_a) and provides a way to compare the fretted and unfretted surfaces through subsequent surface profiling after the experiment has been performed. Average roughness values as obtained along the length of the AL 2024-T4 specimen in the contact region ranged from 0.16 to 0.5 μm.

In each of the preliminary tests presented, the similar material (Al 2024-T4) contact was subjected to an initial mean stress, σ_o, of 110 MPa (16 ksi), a constant normal load (P), an oscillatory tangential load of constant mean, Q_o, and amplitude, $\Delta Q/2$, and an oscillatory bulk stress resulting from the applied tangential load. Note that a different bulk stress history is experienced by the specimen above and below the contact. The specimen above the contact was subjected to a cyclic bulk stress, $\sigma_{To} \pm \Delta\sigma_T/2$, out-of-phase with the applied tangential load, Q. The specimen below the contact was subjected to a bulk stress, $\sigma_{Bo} \pm \Delta\sigma/2$, in-phase with the tangential load. As a result, the bulk stress means and amplitudes reported in Table 1 were determined from the maximum and minimum tangential loads and specimen cross sectional area, A, as calculated from the following static equilibrium equations.

$$Q_{max} = \frac{A}{2}(\sigma_{Bmax} - \sigma_{Tmin}) \qquad Q_{min} = \frac{A}{2}(\sigma_{Bmin} - \sigma_{Tmax}) \qquad (1)$$

$$Q_o = \frac{Q_{max} + Q_{min}}{2} \qquad \Delta Q/2 = \frac{Q_{max} - Q_{min}}{2} \qquad (2)$$

A summary of the five preliminary test loading parameters can be seen in Table 1 along with the associated operational frequency, ω, and the total number of cycles for which the test was run, N. The self-alignment fixture was used in all experiments except PH00. The absence of this fixture reduced the length of the upper load train leading to a greater amplitude of cyclic bulk stress as measured by the upper load cell in experiment PH00. Due to significant drift in the normal load during extended

Table 1 - *Summary of operational parameters and equivalent stress (σ_{eqv})*
calculations for preliminary high-frequency fretting fatigue experiments conducted on
Al 2024-T4 pads and specimens. Equivalent stress results were calculated using the
initial normal load, P_i.

Expt.	P_f (kN)	P_i (kN)	Q_o (N)	$\Delta Q/2$ (N)	ω (Hz)	N (cycles)	σ_{To} (MPa)	$\Delta\sigma_T/2$ (MPa)	σ_{eqvT} (MPa)	σ_{Bo} (MPa)	$\Delta\sigma_B/2$ (MPa)	σ_{eqvB} (MPa)
PH00[1]	4.194	4.882	169.4	663.3	350	2,600,000	96.2	18.9	82.60	104.6	14.0	98.78
PH04[2]	2.364	2.741	98.79	639.1	100	6,023,700	105.5	14.7	91.26	110.4	17.0	103.3
PH05	2.720	3.184	121.0	508.1	310	540,700	112.0	12.2	78.59	118.0	13.0	90.99
PH07[2]	4.563	5.026	223.8	687.5	100	9,864,000	106.4	15.4	83.88	117.5	18.7	103.1
PH19	4.925	5.024	110.9	741.9	225	895,000	104.6	13.1	95.20	110.1	23.7	112.0

Experiments PH04, PH05, and PH19 were stopped prematurely due to amplifier
overheat.
[1]Specimen failed
[2]Nucleated cracks observed

Figure 8 - *Photograph of a fretting wear scar on a pad surface as generated from a*
350 Hz fretting fatigue test.

duration tests at 100 Hz, the initial (P_i) and final (P_f) normal load measurements
have been reported.

The high-frequency, small-scale, oscillatory partial slip experienced by the
contact interface during these tests resulted in the generation of wear scars on each
contacting surface of the AL-2024 specimens and pads used. A representative wear
scar as generated during a preliminary 350 Hz fretting test can be seen in Figure 8.
Of the five preliminary tests run, one resulted in failure of a specimen at 350 Hz
operation. The specimen in experiment PH00 initially failed at the upper edge of
contact after $2.6x10^6$ cycles of loading. A second failure at the lower edge of contact
occurred after the initial fracture to produce the hourglass shaped specimen. A
photograph of the failed specimen can be seen in Figure 9. Note that the cracks
nucleated on all four edges of contact. A close up of the wear scar and twin crack
nucleation site can be seen in Figure 10. The right photograph of Figure 10 shows

Figure 9 - *Photograph of a failed Al 2024-T4 specimen after $2.6x10^6$ cycles of loading*
at 350 Hz.

Figure 10 - *Close-up picture of a failed Al 2024-T4 specimen crack nucleation site and wear scar, and hourglass shaped failure resulting from multiple edge of contact crack nucleations and failures.*

the resulting failure with the production of a small hourglass shaped piece bounded on each side by the fracture surface of each of the four nucleated fretting cracks.

Fretting Life Analysis

Subsurface Stress Calculation

Before a life parameter could be implemented, it is first necessary to obtain an accurate solution of the multi-axial, subsurface stress field associated with the fretting contact. This was achieved by assuming a two-dimensional, elastic, similar material contact of a cylindrical indenter (or Pad) of radius R and a semi-infinite half-space. With the assumptions, the application of a constant normal load (P) and oscillatory tangential load (Q) allows for straight forward development of the surface normal, $p(x)$, and shear tractions, $q(x)$, associated with the fretting contact. Classical Hertzian theory allows for the determination of the contact half-width, a, and classical elliptical or Hertzian pressure distribution, $p(x)$, arising from the two-dimensional line load P^{*} = P / contact thickness. By assuming like material contacts, the Hertzian contact pressure distribution and contact half-width are independent of the applied tangential load, Q^{*} = Q / contact thickness.

Unique to the fretting contact is a condition of *partial slip* which occurs when the globally applied tangential load is less than that required to produce sliding over the whole of the contact. The partial slip solution for monotonic loading was first solved by Cattaneo (1938) and Mindlin (1949) who considered the solution as the sum of two shear tractions: one which gave gross sliding over the whole of the contact, and one which restored a central region of stick of dimension c such that no relative motion occurs in $|x| < c$. Johnson provides an excellent summary of the interfacial mechanics associated with partial slip for monotonic tangential loading of a cylindrical contact [25].

Recent progress has identified a very efficient method for solving fretting contact problems. A set of singular integral equations (SIE) have been shown to relate the relative slip and the initial contact geometry to the contact tractions. These equations include the effect of the bulk stress on the surface shear stress distribution. An FFT based approach can then be used to obtain the subsurface stress distributions for similar material contacts [26]. A discrete Fourier transform technique has also been successfully implemented for solving subsurface stresses associated with dissimilar material contact [27]. This approach can also be used for any arbitrarily specified profile making it more flexible and computationally efficient than an equivalently fine set of finite element analyses [28]. This set of singular

Table 2 - *Summary of contact parameters used in multi-axial fretting fatigue life prediction. Material properties are representative of Al 2024-T4.*

Parameter	Symbol	Value
Pad radius	R	101.6 mm
Young's modulus	E	72.4 GPa ($10.5x10^6$ psi)
Poisson's ratio	ν	0.33
Coefficient of friction	μ	0.65

integral equations were used for solving the subsurface stresses in the following life analysis.

Fretting Fatigue Life Prediction

To analytically determine a critical location for the initiation and subsequent propagation of a fretting induced crack, it is necessary to choose a multi-axial crack initiation metric that best agrees with experimental data. By simplifying the three dimensional contact into the two dimensional geometry, a stress equivalent fatigue life parameter was chosen to capture the multi-axial nature of the subsurface stresses.

The contact configuration analyzed for the multi-axial life prediction was that of a cylindrical fretting pad of radius R=101.6 mm (4.0 inches) against a flat semi-infinite half-space. This model simplification allowed for complete characterization of the surface and subsurface stresses using the aforementioned singular integral equations. A uniform coefficient of friction, $\mu = 0.65$, was assumed across the whole of the contact. A summary of the literature obtained material constants can be seen in Table 2 [29, 30]. The contact was subjected to the constant normal load (P_i), oscillatory tangential load ($Q_o \pm \Delta Q/2$), out-of-phase bulk stress at the top, $\sigma_{To} \pm \Delta \sigma_T/2$, and in-phase bulk stress at the bottom, $\sigma_{Bo} \pm \Delta \sigma_B/2$, of the contact as measured in the preliminary high-frequency tests of Table 1. The applied bulk stress introduces an eccentricity in the distribution of stick and slip at the contact surface and therefore must be included in the analysis. Since the initial bulk stress is applied before contact generation in the high frequency setup, the bulk stress effect on the shear traction arises from the change in bulk stress term, $\Delta \sigma$, and not the initial bulk stress, σ_o. The effect of bulk stress on the contact tractions has been presented in other work [26].

Historically, critical plane models like the Smith-Topper-Watson approach have proven to successfully predict fretting crack formation [12], as well as the critical location and orientation for fretting crack nucleation [13, 31]. However, in addition to being computationally expensive, the critical plane approach fails to account for the gradients of stress inherent to fretting contact. As a result, the nucleation life was predicted using a stress invariant equivalent stress approach which has shown promise in predicting fretting fatigue lives in titanium contacts [24]. Stress invariant models have a distinct advantage in that they do not require the determination of critical plane location and orientation a priori. This leads to significant reduction in computational time as compared with critical plane life methods. The stress equivalent parameter, σ_{eq}, used in this life analysis is defined as:

$$\sigma_{eq} = 0.5(\Delta\sigma_{psu})^w(\sigma_{max})^{(1-w)}, \qquad (3)$$

where $\Delta\sigma_{psu}$ is the alternating pseudo-stress range, the Walker exponent, w, is a curve fit parameter, and σ_{max} is the maximum stress as obtained from the pseudo-stress and mean stress derived by Manson-McKnight.

To account for stress gradients associated with notches and fretting contacts, a weak link approach was then implemented by evaluating the following equation for the experimental subsurface stress history [24].

$$\Delta\sigma_{eq,1} = \left(\frac{Fs_2}{Fs_1}\right)^{\frac{1}{\alpha}}\Delta\sigma_{eq,2} \qquad (4)$$

$$Fs_j = \sum_{i=1}^{n}[-\left(\frac{\sigma_{i,j}}{\sigma_{max,j}}\right)^\alpha]\Delta A_{i,j} \qquad (5)$$

In the equations above, σ_i is the equivalent alternating stress, σ_{max} is the maximum calculated σ_i, Fs_1 is the stressed areas for the smooth bar, Fs_2 is the stressed area of the test specimen, and $\Delta\sigma_{eq,2}$ and $\Delta\sigma_{eq,1}$ are the equivalent stresses as obtained for the actual experiment and the smooth bar, respectively. This allows for the comparison of the equivalent stress obtained in the experiment to the equivalent stress in a smooth bar specimen having the same life as that in the experiment.

Szolwinski and Farris have shown some success predicting fretting fatigue nucleation life by relation to conventional strain-life fatigue data given that a clear knowledge of the near-surface contact stress field can be maintained [13]. Similarly, to investigate the ability of the stressed area approach to model fretting fatigue nucleation life of aluminum, this Al 2024-T351 data was analyzed. Smooth bar Al 2024-T3 fatigue life data from literature (Military Handbook 5G 1994) was used to select a Walker exponent, w = 0.56, and a smooth bar stressed area, Fs_1 = 161 mm^2. Due to lack of sufficient fretting data for aluminum, the value of the shape factor, α, was chosen by iterating until a minimum root mean square error between the smooth bar equivalent stresses for each experiment and the calculated equivalent stresses was found. The corresponding α was found to be 19. A plot of the estimated nucleation life against the calculated equivalent stresses for the Al 2024-T351 data can be seen in Figure 11. The predicted value of nucleation life is plotted as a solid line representing the smooth bar Al 2024-T3 equivalent stress.

Previous work on titanium showed that the same value of the shape factor, α, could be used for a range of notches and fretting contacts [24]. Therefore, the shape factor for the Al 2024-T4, high-frequency tests was chosen to be the same as used in the previous Al 2024-T351 life analysis (α = 19). The experimental loading conditions, subsurface stresses as found from the singular integral equations, and equivalent stress method were then used to predict the nucleation lives. The predicted and experimental high-frequency test results are shown in Figure 11 as squares. Calculated equivalent stress values can be seen explicitly in Table 1. Note

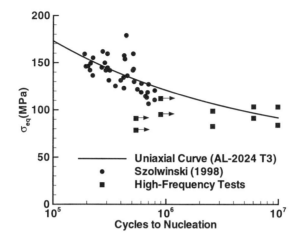

Figure 11 - *A plot of equivalent stress calculated from existing Al 2024-T351 fretting fatigue data [13] and equivalent stress calculated from the high-frequency Al 2024-T4 fretting fatigue data reported in Table 1 verses cycles to nucleation. The solid line represents the equivalent stress curve for Al 2024-T3 ($\sigma_{eq} = 2155N_n^{-0.2519} + 54$ MPa) as found from literature [30]. Calculations for the high-frequency tests were performed based on the initial normal load, P_i.*

that the high-frequency data is for Al 2024-T4 while the conventional fretting fatigue data is for Al 2024-T351.

Right arrows indicate tests where no cracks were visually observed upon termination of the experiment as in experiments PH05 and PH19. Experiments PH00, PH04, and PH07 were test runs in which cracks were observed after the experiment ended. Here failure is defined as the visual observation of a crack at the upper or lower edge of contact. Experiment PH00 fractured from a crack initiating at the top edge of contact followed by a second fracture from a crack initiating at bottom edge of contact. Note that all crack observations were performed at the termination of each test. Therefore, there is no observation of when or if the cracks arrested. The squares in Figure 11 are reflective of the total number of cycles.

The promise of the stress equivalent approach to capture fretting fatigue crack nucleation behavior as illustrated by Figure 11 and the high frequency experiments presented here build confidence in this approach as a method for fretting crack nucleation predictions. Recent successful fretting fatigue life predictions for titanium [24] and the computational benefits of the stress invariant life models provide sufficient impetus for further investigation of this approach as a method for fretting fatigue life modeling.

Conclusions and Future Work

A high-frequency fretting fatigue load frame capable of in-situ monitoring of critical fretting contact parameters has been designed. Preliminary experimental observations show successful fretting induced crack nucleation and failure at frequencies in excess of 300 Hz. Operation at or exceeding frequencies of 300 Hz allow for order of magnitude reduction in fretting fatigue test times. By pushing the frequency envelope in fretting, implications of contact dynamics on fretting and fatigue lives, increased wear due to highly localized multiple slip zones at the edge of contact, and evolution of the coefficient of friction during high frequency operation can be investigated. It remains to design a set of experiments aimed at explicating the contributions of frequency and variable amplitude loading effects to fretting fatigue crack nucleation and propagation. A computationally efficient stress invariant model showed promise in predicting nucleation behavior for a set of preliminary high-frequency fretting tests. Through identification of an appropriate HCF crack nucleation parameter, a set of crack initiation metrics can be developed for fail safe design. By integrating a fully automated, closed loop control scheme for waveform generation and control, a robust, high-frequency fretting fatigue load frame will be available for realizing these goals.

Acknowledgments

This research was supported in part by the National Science Foundation Fellowship for J. F. Matlik.

References

[1] Waterhouse, R. B., *Fretting Corrosion*, Pergamon Press, Oxford, 1972.

[2] Petiot, C., Vincent, L., Dang Van, K., Maouche, N., Foulquier, J., and Journet, B., "An Analysis of Fretting-Fatigue Failure Combined with Numerical Calculations to Predict Crack Nucleation," *Wear*, **181-183**, 1995, pp. 101–111.

[3] Muller, R. P. G., "An Experimental and Analytical Investigation on the Fatigue Behavior of Fuselage Riveted Lap Joints," 1995.

[4] Szolwinski, M. P., Harish, G., McVeigh, P. A., and Farris, T. N., "Experimental Study of Fretting Crack Nucleation in Aerospace Alloys with Emphasis on Life Prediction," In *Fretting Fatigue: Current Technology and Practices, ASTM STP 1367*, D. W. Hoeppner, V. Chandrasekaran, and C. B. Elliott, Eds., American Society for Testing and Materials, West Conshohocken, PA, 2000, pp. 267–281.

[5] Kalb, B. J., "Friction Stresses Between Blade and Disk Dovetail Possible Case of Numerous Dovetail Problems," In *USAF Structural Integrity Conference*, 1999.

[6] Ruiz, C., Boddington, P. H. B., and Chen, K. C., "An Investigation of Fatigue and Fretting in a Dovetail Joint," *Experimental Mechanics*, **24**, 1984, pp. 208–217.

[7] Ruiz, C., Wang, Z. P., and Webb, P. H., "Techniques for the Characterization of Fretting Fatigue Damage," In *Standardization of Fretting Fatigue: Test Methods and Equipment, ASTM STP 1159*, M. H. Attia and R. B. Waterhouse, Eds.,

American Society of Testing and Materials, Philadelphia, PA, 1992, pp. 170–177.

[8] Szolwinski, M. P., Matlik, J. F., and Farris, T. N., "Effects of HCF Loading on Fretting Fatigue Crack Nucleation," *International Journal of Fatigue*, 21, 1999, pp. 671–677.

[9] Thomson, D., "The National High Cycle Fatigue (HCF) Program," In *3rd National Turbine Engine High Cycle Fatigue (HCF) Conference*, CD-ROM proceedings, San Antonio, TX, Universal Technology Corp., 1998.

[10] Endo, K. and Goto, H., "Initiation and Propagation of Fretting Fatigue Cracks," *Wear*, 38, 1976, pp. 311–324.

[11] Waterhouse, R. B., "Fretting Fatigue," *International Materials Reviews*, 37, 1992, pp. 77–96.

[12] Szolwinski, M. P. and Farris, T. N., "Mechanics of Fretting Fatigue Crack Formation," *Wear*, 198, 1996, pp. 193–107.

[13] Szolwinski, M. P. and Farris, T. N., "Observation, Analysis, and Prediction of Fretting Fatigue in 2024-T351 Aluminum Alloy," *Wear*, 221, 1998, pp. 24–36.

[14] Mutoh, Y., Tanaka, K., and Kondoh, M., "Fretting Fatigue in SUP9 Spring Steel under Random Loading," *Japan Society of Mechanical Engineers International Journal*, 32(2), 1989, pp. 274–281.

[15] Cortez, R., Mall, S., and Calcaterra, J. R., "Investigation of Variable Amplitude Loading on Fretting Fatigue Behavior of Ti-6Al-4V," *International Journal of Fatigue*, 21, 1999, pp. 709–717.

[16] Endo, K., Goto, H., and Nakamura, T., "Effects of Cycle Frequency on Fretting Fatigue Life of Carbon Steel," *Bulletin of the Japan Society of Mechanical Engineers*, 12(54), 1969, pp. 1300–1308.

[17] Iwabuchi, A., Kayaba, T., and Kato, K., "Effect of Atmospheric Pressure on Friction and Wear of 0.45% C Steel in Fretting," *Wear*, 91(3), 1983, pp. 289–305.

[18] Feng, I. M. and Uhlig, H. M., "Fretting Corrosion of Mild Steel in Air and in Nitrogen," *Journal of Applied Mechanics*, 21(4), 1954, pp. 395–400.

[19] Hutson, A. L., Nicholas, T., and Goodman, R., "Fretting Fatigue of Ti-6Al-4V Under Flat-on-Flat Contact," *International Journal of Fatigue*, 21, 1999, pp. 663–669.

[20] Bramhall, R., *Studies in Fretting Fatigue*. PhD thesis, Oxford University, 1973.

[21] O'Connor, J. J., "The Role of Elastic Stress Analysis in the Interpretation of Fretting

Fatigue Failures," In *Fretting Fatigue*, R. B. Waterhouse, Ed., Applied Science Publishers, London, 1981, ch. 2, pp. 23–66.

[22] Hills, D. A. and Nowell, D., "The Development of a Fretting Fatigue Experiment with Well-Defined Characteristics," In *Standardization of Fretting Fatigue: Test Methods and Equipment, ASTM STP 1159*, M. H. Attia and R. B. Waterhouse, Eds., American Society for Testing and Materials, Philadelphia, PA, 1992, pp. 69–84.

[23] Fouvry, S., Kapsa, P., and Vincent, L., "A Multiaxial Fatigue Analysis of Fretting Contact Taking Into Account the Size Effect," In *Fretting Fatigue: Current Technology and Practices, ASTM STP 1367*, D. W. Hoeppner, V. Chandrasekaran, and C. B. Elliott, Eds., American Society for Testing and Materials, West Conshohocken, PA, 2000, pp. 167–182.

[24] Murthy, H., Farris, T., and Slavik, D., "Fretting Fatigue of Ti-6Al-4V Subjected to Blade/Disk Contact Loading," In *Developments of Fracture Mechanics for the New Century, 50*[th] *Anniversary of Japanese Society of Material Science* , 2001, pp. 41-48.

[25] Johnson, K. L., *Contact Mechanics*, Cambridge University Press, Cambridge, 1985.

[26] Murthy, H., Harish, G., and Farris, T. N., "Efficient Modeling of Fretting of Blade/Disk Contacts Including Load History Effects," *ASME Journal of Tribology* , In-Press.

[27] Rajeev, P. T. and Farris, T. N., "Analysis of Fretting Contacts of Dissimilar Isotropic and Anisotropic Materials," *Journal of Strain Analysis* , In-Press.

[28] McVeigh, P. A., Harish, G., Farris, T. N., and Szolwinski, M. P., "Modeling Interfacial Conditions in Nominally Flat Contacts for Application to Fretting Fatigue of Turbine Engine Components," *Int. J. of Fatigue*, **21**, 1999, pp. 5157–5165.

[29] Hertzberg, R., *Deformation and Fracture Mechanics of Engineering Materials*, Wiley, New York, 1976.

[30] MIL-HDBK-5G, *Metallic Material and Elements for Aerospace Vehichle Structures*, vol. 1, Defense Printing Service Detachment Office, Philadelphia, PA, 1994.

[31] Hills, D. A. and Nowell, D., "Crack Initiation Criteria in Fretting Fatigue," *Wear*, **136**, 1990, pp. 329–343.

H. Murthy,[1] P. T. Rajeev,[1] Masaki Okane,[2] and Thomas N. Farris[3]

Development of Test Methods for High Temperature Fretting of Turbine Materials Subjected to Engine-Type Loading

REFERENCE: Murthy, H., Rajeev, P.T., Okane, M., and Farris, T.M., "**Development of Test Methods for High Temperature Fretting of Turbine Materials Subjected to Engine-Type Loading,**" *Fretting Fatigue: Advances in Basic Understanding and Applications, ASTM STP 1425*, Y. Mutoh, S. E. Kinyon, and D. W. Hoeppner Eds., ASTM International, West Conshohocken, PA, 2003.

Abstract: Two bodies are said to be in fretting contact when they are clamped together under the action of a normal force and see an oscillatory motion of small amplitudes at the contact interface due to the effect of shear force and bulk stress. The contact stresses that drive crack nucleation are very sensitive to the shape of the contacting surfaces and the coefficient of friction. To have an understanding of fretting at the contacts in turbine engine components, it is important to simulate similar temperature, load, and contact conditions in the laboratory and develop tools to analyze the contact conditions. Efforts made to simulate the temperature and load conditions typical of engine hardware in a controlled laboratory setting, similar to that developed previously at room temperature, are described. It is shown that the temperature in the contact region can be held constant at nominally 600°C. Preliminary results are given in terms of loads and measured total lives along with thoughts on development of a total life model for single crystal materials.

Keywords: high temperature, fretting, fatigue, life prediction, turbo-machinery, friction, Ti-6Al-4V, single crystal nickel

Introduction

Fretting phenomenon is widely observed in aircraft structures [1]. In engine assemblies, fretting occurs at the blade/disk contact in the fan, compressor, and turbine stages (Figure 1). The influence of fretting on High Cycle Fatigue (HCF) of gas turbine engine components prompts interest in the contact mechanics, with

[1]Graduate Student, School of Aeronautics and Astronautics, Purdue University, West
   Lafayette, IN 47907-1282.
[2]Dept. of Mechanical and Intellectual Systems Engineering, Faculty of Engineering,
   Toyama University, Japan.
[3]Professor and Head, School of Aeronautics and Astronautics, Purdue University, West
   Lafayette, IN 47907-1282.

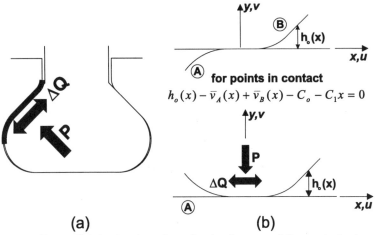

Figure 1 - *Dovetail joint in aircraft engine hardware and the equivalent profile.*

friction, of dissimilar isotropic as well as anisotropic materials. The loading is caused by the inertial forces due to engine spooling and by vibrations and aerodynamic forces acting on the blade airfoil. Severe stress gradients at the edge of contact may allow these high-frequency, small-amplitude loads to drive the crack growth leading to failure. The contact stresses are influenced by a large number of factors: the effective shape of the contact, coefficient of friction between the two bodies in contact, various loads acting on the components, and the operating temperature. This, in turn, affects the fretting-induced damage and the life of the component. Thus, a simulation of fretting contact phenomenon in blade/disk joints should take into account the effective contact shape, the various loads acting on the body, and the operating temperature.

As shown in Figure 1, the blade/disk contact is equivalent to a nominally flat profile in contact with a flat surface. Since the contact region is small compared to overall dimensions of the blade and disk, they may be approximated as half spaces. The forces acting on the contact can be resolved into a normal component and a tangential component over the contact region. The nominally flat geometry of contacting surfaces causes high contact stress levels near the edges of contact. An accurate estimate of the near surface stress-states is essential to characterize the performance and reliability of components.

One of the approaches used to calculate contact stresses is the Finite Element Method (FEM). However, the number of elements needed to resolve the stress peaks (and their locations), especially in nominally flat contacts, at the edges of the contact zone is very high resulting in a computer run time on the order of tens of hours [2]. This leads to very costly and slow design cycles. Alternatively, the theory of singular integral equations can be used to obtain the contact tractions very quickly. An FFT based method with the contact tractions as boundary conditions yields the sub-surface stresses in a matter of minutes [2–4]. In some instances, the blade and disk are made from dissimilar isotropic materials (Ti-6Al-4V blade and

Inconel 718 disk) or one of them is anisotropic (blade made from single crystal nickel). In such cases the contact problem is considerably more difficult to solve. The two-dimensional contact of dissimilar isotropic materials has been analyzed using a numerical (SIE) method, that solves the governing Singular Integral Equations (SIE) [5, 6]. The SIE method provides rapid calculation of the large edge of contact pressure and shear stress present in engine blade/disk type contact and is significantly faster than FEM while retaining the essential mechanics of fretting contacts. The fretting contact stress conditions relevant to blade/disk contacts described above are highly localized. Using these highly localized stresses in life prediction requires great care. Efforts aimed at developing such tools are described elsewhere[7].

The temperature at which some of the stages in a gas turbine engine operate is very high. This high temperature calls for the use of specialized materials like single crystal nickel (SCN) alloy and IN100 that have special charateristics at high temperature. Hence, an attempt has been made to characterize the fretting fatigue behavior of materials experimentally at high operating temperatures ($\approx 600^{\circ}$C). A new experimental setup has been designed to simulate the high temperature at the local region of contact that is subjected to fretting loads. A set of igniters are used to increase the local temperature to 600°C. The features of the experiment include a combination of an efficent cooling mechanism using water and an insulation mechanism using ceramic blocks and sheets. The characersitics of the rig, like the normal load transfer ratio, have been experimentally studied. The rig has been used to conduct an initial set of experiments using SCN specimens and IN100 pads, and was found to function well under the operating temperature and load conditions.

For motivation of the development of high temperature fretting fatigue capability, the paper begins with a brief description of existing room temperature capabilities. A brief description of life prediction is then given to support the relevance of laboratory fretting fatigue experiments to gas turbine engine hardware. A detailed description of the high temperature experiments is given next followed by results for a preliminary set of high temperature fretting fatigue experiments. The paper concludes with thoughts on extending existing life prediction capability to the single crystal materials used in the high temperature experiments.

Experimental Details

Nominally flat fretting pads, with rounded edges, and flat *dog-bone* specimens are used to duplicate the local geometry at the blade/disk attachments in engine hardware. To simulate the loading conditions, the nominally flat pad and the flat specimen are clamped together to generate a normal load and then subjected to an oscillatory tangential load. Such a configuration was achieved in the laboratory using a fretting chassis attached to a standard uniaxial servo-hydraulic testing machine. In addition to the contact loading, the laboratory experiments apply a bulk load to the specimen.

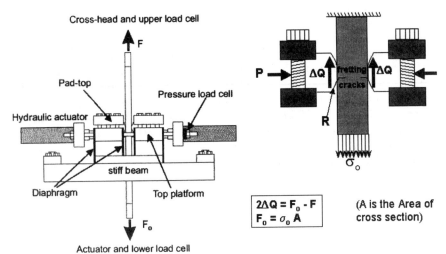

Figure 2 - *Components of the fretting chassis with a schematic detailing the definition of forces applied to the rig*

Fretting Chassis and Loads

The fretting chassis is a superstructure built on the fatigue machine that allows the generation of tangential loads that are in phase with the bulk load while applying a nominally constant normal load. Figure 2 shows the components of the fretting chassis schematically. The stiff beam provides the bulk of the stiffness of the chassis. The pads rest on the top platform which also provides the means of application of the normal load. The normal load is applied via a pair of hydraulic actuators that transmit the load onto the top platform. Note that the two pressure rods ensure that pressure is applied symmetrically to the pads. The thin steel diaphragms offer little resistance to the pressure loading, but offer a large resistance that carries the portion of the tangential load transmitted to the chassis. This ensures that almost all of the pressure is transmitted to the specimen through the pads, while maintaining the required stiffness to produce a large tangential force. A finite element analysis has shown that more than 98% of the pressure is transmitted to the specimen. The pad-tops fix the pads to the chassis. The tangential force produced was about 50% of the bulk load applied, subject to a maximum of the force required to produce gross sliding. This load is monitored throughout the fatigue experiments by recording the difference of the upper and lower load cell readings.

Figure 2 shows a schematic of the various loads. The crush load (normal load) per unit depth is P, Q is the tangential load per unit depth, F is the force applied at the bottom of the specimen (measured by the actuator load cell). The reaction force as measured by the cross-head load cell (top grip) is referred to as R. Note that the difference between F and R gives the tangential force, $2Q$. The bulk stress, σ_o, is F divided by the cross-sectional area of the specimen.

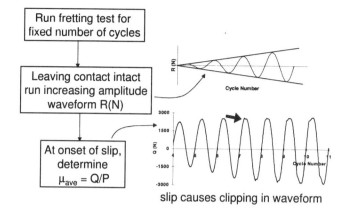

Figure 3 - *Estimation of the coefficient of friction.*

Evolution of Coefficient of Friction

Wear of the pads and specimens described in the previous section causes the coefficient of friction to increase during the fretting fatigue process. Many of the load-controlled fretting experiments are conducted in the regime of mixed-mode fretting. Due to low coefficient of friction in the beginning of the experiment, the applied tangential load causes sliding of the two contacting surfaces ($Q = \mu_o P$). After some time, the coefficient of friction increases leading to partial slip. The transition from sliding to partial slip is borne out by full-field infrared measurements of temperature [8]. This transition from sliding to partial slip is called mixed-mode fretting.

Careful experiments are required to characterize the coefficient of friction in the slip zones. Measurement of friction coefficient in this slip zone is required for accurate evaluation of contact stresses. Friction experiments aimed at characterizing the evolution of the coefficient of friction in the slip zones, μ_s, with number of cycles in a partial slip experiment were conducted. As the coefficient of friction increases with the fretting damage, the tangential force stabilizes at a value less than that required for gross sliding. Thus, the coefficient of friction cannot be evaluated as the magnitude of tangential force divided by the normal force. Hence, an alternative approach, that requires a special load history, is taken.

After running the test for a specified number of fretting cycles, the test is stopped. The average coefficient of friction is now determined in the following way. Without disturbing the pad/specimen contact, an increasing amplitude waveform is applied to the specimen. This causes the tangential force to increase during each cycle. Figure 3 shows this approach schematically. Ultimately, the tangential force reaches the value required for gross sliding, and the experiment is stopped. The maximum value of the tangential force before sliding commences divided by the normal force gives the average coefficient of friction. It must be noted that experiments conducted in the gross sliding regime may result in larger coefficients of friction.

Cylindrical pads with a pad radius of 178 mm were used in the friction experiments. When the pad and the specimen are made from similar isotropic materials an approach outlined in Hills & Nowell [9] is used to determine the slip zone coefficient of friction, μ_s, in terms of the average coefficient of friction, μ, and the initial coefficient of friction, μ_0 [4]. This formulation [9] is only suitable when the pressures distribution is given by the Hertzian pressure distribution. In the case of contact of dissimilar isotropic materials the pressure is not only not Hertzian but also the solutions for the contact pressure and the shear traction are coupled. Hence, a closed form relation between μ_s, μ, and μ_0 cannot be simply deduced.

A numerical method has been devised to obtain μ_s in terms of μ and μ_0. Essentially, μ_s is increased in an incremental manner, starting from μ_0, until the solution of the sliding contact problem yields the experimentally measured average coefficient of friction, μ. As with the Hills & Nowell approach μ_s is taken as constant in the slip zones, and the coefficient of friction in the stick zones does not change. Initially, the coefficient of friction is assumed to be constant ($= \mu_0$) through-out the contact zone. As the slip zone coefficient of friction increases so does the stick zone size. In each increment the ends of the stick zone are determined by solving the partial slip contact problem taking into account the remote tension applied on the specimen and the material dissimilarity of the pad and the specimen. Keeping track of the previous increments allows us to determine the distribution of the coefficient of friction within the contact zone in the current increment, $\mu_i(x)$. Hence, the sliding contact problem for a given increment can be written as,

$$\frac{\partial H}{\partial x} + C_1 = k_1 \mu_i(x)q(x) - \frac{k_2}{\pi}\int_{a_1}^{a_2}\frac{p(t)}{t-x}dt \tag{1}$$

where $H(x)$ is the gap function, $p(x)$ and $q(x)$ are the pressure and the shear traction respectively, a_1 and a_2 are the ends of the contact zone, and the constants k_1 and k_2 are obtained from the elastic material properties of the two contacting bodies (described in the next section). Once the sliding contact problem is solved the average coefficient of friction in a given increment is obtained as,

$$\mu = \frac{\int_{a_1}^{a_2}\mu_i(x)p(x)\,dx}{\int_{a_1}^{a_2}p(x)\,dx} \tag{2}$$

where a_1 and a_2 are the ends of the contact zone. The results of the friction experiments and the numerically calculated μ_s values are presented in Figure 4(a). The distribution of the coefficient of friction in the contact zone and the evolution of the stick zone as the fretting progresses can be seen from Figure 4.

Contact Stress Calculation

A schematic of two elastic bodies in partial slip contact is shown in Figure 5. Let the displacements of both the bodies due to the contact tractions be given by $_iu_x$ and $_iu_z$ respectively ($i = 1, 2$). The relative slip between the two bodies, $s(x)$, and the initial gap function, $H(x)$, in the contact zone can be related to the displacements as,

$$s(x) = {}_2u_x - {}_1u_x \tag{3}$$

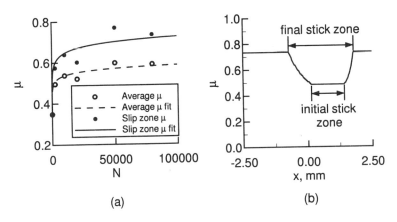

Figure 4 - (a)Evolution of the average coefficient of friction, μ, and the slip zone coefficient of friction, μ_s. (b)Distribution of the coefficient of friction. As μ_s increases from μ_0 to its final value, the stick zone size increases.

$$H(x) + C_1 x = {}_2 u_z - {}_1 u_z + H_0 \qquad (4)$$

where H_0 is a constant and C_1 is the rotation term.

It has to be noted that the in-plane shear traction is produced as a result of friction between the contacting bodies. For a general anisotropic body in-plane tractions might produce out of plane relative displacements resulting in an out of plane shear traction in addition to the in-plane shear traction. This effect is neglected in our analysis, i.e., the out of plane shear traction is assumed to be zero. By knowing the relationship between the surface displacements and the surface tractions for an elastic half-space one can express, as shown in Equation 5 the initial gap and relative slip functions in terms of the unknown contact pressure and shear traction. Stroh formalism is used to derive the Green's function for an anisotropic half-space resulting in the following system of coupled singular integral equations.

$$\frac{\partial}{\partial x} \left\{ \begin{array}{c} s(x) \\ H(x) \end{array} \right\} = -[\alpha] \left\{ \begin{array}{c} q(x) \\ p(x) \end{array} \right\} - [\beta] \int_{a_1}^{a_2} \left\{ \begin{array}{c} q(t) \\ p(t) \end{array} \right\} \frac{dt}{t - x} \qquad (5)$$

For isotropic bodies $\alpha_{11} = \alpha_{22} = \beta_{12} = \beta_{21} = 0$, $\alpha_{12} = -\alpha_{21} = k_1$, and $\beta_{11} = \beta_{22} = -k_2$ where

$$k_1 = \frac{(1 - 2\nu_1)(1 + \nu_1)}{E_1} - \frac{(1 - 2\nu_2)(1 + \nu_2)}{E_2} \qquad (6)$$

$$k_2 = \frac{2(1 - \nu_2^2)}{E_2} + \frac{2(1 - \nu_1^2)}{E_1} \qquad (7)$$

When both the bodies are made from similar isotropic materials, the fretting contact problem has been solved very efficiently by Murthy et al [4] using a Fast Fourier Transform (FFT) based approach. For the case of dissimilar isotropic bodies a different numerical approach, the details of which are given in Rajeev & Farris [5],

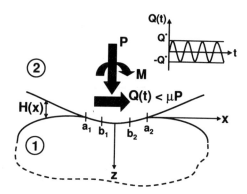

Figure 5 - *Schematic of two elastic bodies in partial slip contact. a_1 and a_2 denote the ends of the contact zone and b_1 and b_2 denote the ends of the stick zone.*

is used to obtain the contact tractions. The pressure and shear are decomposed into a sigular part and a non-singular part. The orders of singularity of the singular part are determined using an eigenvalue analysis of the system of Equations 5. Bounded contact tractions are obtained by iterating for the limits of the contact zone and the stick zone such that, the constant multipliers of the singular parts are zero, and the equilibrium conditions are satisfied. In both cases an FFT based approach [4] is used to obtain the sub-surface stress distributions.

Murthy *et.al.* [7] discuss the utilization of subsurface stresses obtained from the above approach to predict the life of specimens used in a well characterized set of experiments. A crack nucleation parameter, σ_{eq}, was used to predict the crack nucleation life [7] and conventional fracture mechanics based calculations were used to obtain the propagation lives [10]. The results are presented in the Figure 6. The consistency of the predictions with the experimental results provides confidence in the aforementioned approach as a tool for life predictions.

Design of Fretting Rig for High Temperature Testing

Motivation

The performance of a jet engine, measured in terms of thrust-to-weight ratio, can be increased by increasing turbine gas temperature and by reducing the weight of the engine using innovative designs and advanced materials. These materials have to be lighter and must have good structural integrity at high temperatures. The allowable metal temperature was very low in the earlier jet engines that had uncooled turbine blades cast from polycrystalline nickel-based alloys. In recent years, a significantly improved turbine airfoil cooling design and use of cast directionally solidified turbine airfoils have increased the allowable turbine inlet gas temperature. As a result higher thrust-to-weight ratio can be achieved in the engines. The anisotropic behavior and the elevated operating temperatures of these advanced materials pose significant problems for design and life prediction of advanced turbine blades. Fretting is one of the primary causes of HCF in turbine blades. Therefore, to

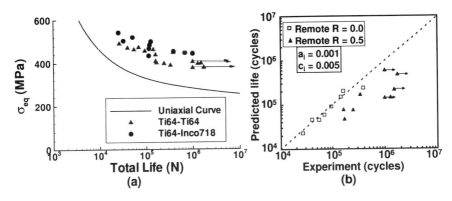

Figure 6 - *(a)Comparison of experimental lives with predicted nucleation lives using σ_{eq} as a critical parameter. (b)Comparison of measured failure lives with predicted total lives taken as sum of nucleation lives and propagation lives (estimated from fracture mechanics.)*

characterize the effect of fretting on the life of these components, experiments have to be conducted at the operating temperatures of the materials in consideration and the analysis has to include the effect of anisotropic behavior on the contact stresses.

The existing fretting rig at Purdue University had been designed to operate at room temperature. The rig was useful in testing the materials like Ti-6Al-4V, Ti17, and Inco718 that are used in components that operate at lower temperatures as described above. However, advanced alloys like Single Crystal Nickel (SCN), IN100, etc. that are used in high temperature applications have characteristic properties at elevated temperature. Therefore, to study the material fretting fatigue characteristics, the experiments must be conducted at elevated temperatures to which the material is exposed in the engine environment. To achieve an elevated temperature of 610°C at the contact region, a new rig was designed as is described next.

Features of the High Temperature Rig

Figure 7 shows a schematic of the rig designed for high temperature tests. Components of the rig closer to the zone of elevated temperature were designed using Ti-6Al-4V alloy, while 4140 steel was used to design the components that were far away from the region of high temperature.

The load transfer in the new rig was based on a principle similar to that of the room temperature fretting experiment. The webs connected to the platforms that hold the pads act as membranes. They transfer most of the normal load applied to it, but their stiffness in the tangential direction gives rise to tangential (shear) load at the contact. Normal load is applied using two hydraulic actuators. Two rods on either side of the contact ensure that the pressure is applied symmetrically to the pads. Since the temperatures that would be encountered during the experiments were high, assembly of different parts using welds or bolts were avoided. Hence, the

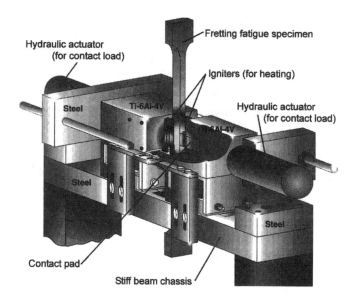

Figure 7 - *Schematic of the elevated temperature fretting fatigue experiments.*

webs and the platforms that hold the pads were machined from a single Ti-6Al-4V block. Further, the pad holder block was not made of two different parts that would be bolted together with pad in between them, as is done in the room temperature rig. As the normal load is applied, the pad gets wedged into a tapered groove (tapered at an angle of 5°) machined in the pad holder block.

The temperature of the local area of contact was increased using a pair of igniters on either side of the specimen. The igniters draw a current of 4.25 A to 4.75 A at 132 V, and the maximum temperature achieved by the igniter surface is 1550°C. The heat transfer from the igniter to the specimen and the pads is through radiation and convection with air as the medium. The temperature of the specimen was measured using a K type thermocouple. The voltage output of the thermocouple was used to control the temperature at the contact using on/off control. The controller was set up such that the igniter switches off if the temperature goes up by 1°C from the desired temperature and switches on if the temperature drops by 0.5°C. On/off control requires an igniter capable of reaching a very high temperature in a short period of time to minimize the fluctuations in temperature. Hence, an igniter which can achieve the maximum temperature from room temperature in 17secs was chosen. However, there was a fluctuation of ±5°C when the desired temperature was 610°C. The fluctuation may be due to air currents affecting the convective heat transfer, in addition to on/off type of control. The influence of air currents were minimized by forming a shield around the zone of elevated temperature, using ceramic blocks and sheets. Since the fluctuation was less than 1% of the desired value, it will be neglected in the subsequent modeling efforts.

The surface of the webs and the pad holder block facing the zone of elevated temperature was covered with ceramic sheets glued using high temperature adhesive,

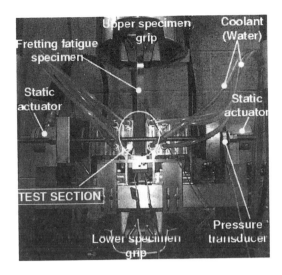

Figure 8 - *Photograph of the elevated temperature fretting fatigue experiment.*

hence preventing their exposure to high temperatures. It also prevents absorption of heat by the rig from the elevated temperature zone. Loss of heat to ambient air was further prevented using a shield made of a ceramic block placed behind the igniter. The ceramic shields along with the ceramic sheets covering the rig give rise to a furnace-like environment around the region of contact. Inspite of covering the surface of the webs with ceramic sheets, there is some heat transfer to the rig due to the air surrounding it which is at a very high temperature. In addition, there is a transfer of heat from the pad to the pad holder by conduction. To absorb the heat thus generated in the rig, the pad holders are cooled by passing water through channels machined in the block (Figure 8). Heat is also conducted to the wedges that hold the specimens. Hence, water cooled wedges were used for clamping the specimens.

A set of experiments were conducted to find the load transfer ratio when a

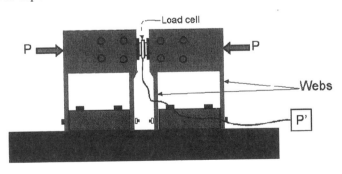

Figure 9 - *Schematic of the experiment conducted to obtain the load transfer ratio.*
P'/P was found to be > 92%.

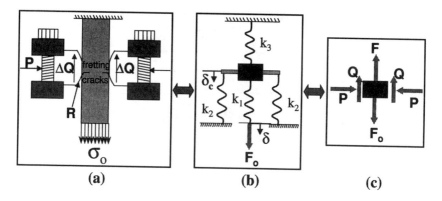

Figure 10 - *A simple linear model of the rig. (a) Schematic of the rig with specimen and pad (Figure 2). (b) A simple linear spring model of the setup. The part of the specimen below contact acts as a spring of stiffness k_1 and the part above acts as spring of stiffness k_3. The diaphragms act as springs of stiffness k_2.*

normal load was applied (Figure 9). The load transfer ratio was found consistently to be between 92% and 95%. The experiments conducted to obtain the load transfer ratio also provided insight into techniques for minimizing the moment and rotation at the contact.

Simple Model of Elevated Temperature Fretting Fatigue Experiment

A simple model was developed, assuming that the specimen and the diaphragms act as linear springs, to assess the effects of various parameters on the nature of mixed-mode fretting process (Figure 10). The portion of the specimen between the contact and the bottom grips acts as a spring of stiffness $k_1 = A_s E_s / l_1$, where A_s is the area of cross-section of the specimen, E_s is the Young's modulus of the specimen material and l_1 is the length of the specimen below the contact. Similarly, the portion of the specimen above the contact acts as a spring of stiffness $k_3 = A_s E_s / l_3$, where l_3 is the length of the specimen between the contact and the top grips. The two diaphragms act as springs of stiffness $k_2 = A_d E_d / l_d$, where A_d is the area of cross-section of the diaphragm, E_d is the Young's modulus of the diaphragm material and l_d is the length of the diaphragm. Though the assumption of linear spring behavior is not entirely accurate, it provides a qualitative assessment of the effect of the dimensions and the material properties of the rig. The displacement at the bottom grip, where load F_o is applied, is taken as δ and the displacement at the contact is taken as δ_c. The forces can be related to the displacements as,

$$F_o = k_1(\delta - \delta_c), \quad F = k_3 \delta_c, \quad Q = k_2 \delta_c. \tag{8}$$

From the above relations and from the equilibrium condition $F_o = F + 2Q$, the following relations can be deduced,

$$F = \frac{F_o}{1 + 2\frac{k_2}{k_3}}, \quad Q = \frac{F_o}{2 + \frac{k_3}{k_2}}. \tag{9}$$

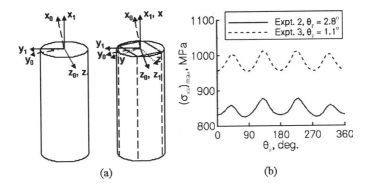

Figure 11 - *(a) Schematic showing the material principal axes with respect to the specimen. (b) Effect of the change in orientation of principal axes with respect to contact surface on the subsurface stresses.*

Equation 9 gives a relation between the applied load, F_o, and the recorded loads, F and Q, in terms of the stiffnesses of the specimen and the diaphragms, which in turn are affected by the geometry and the material properties. This provided a valuable tool to modify the design as described in the subsequent section.

Experimental Procedure

   A procedure for setting up the experiment was arrived at based on the experience with the experiments conducted to estimate the load transfer ratio. This procedure was arrived at to minimize the moment and the rotation at the contact. First the pads are pushed into contact with the specimen by pushing the pad holder and held in place using screw-jacks. The pad holder moves towards the specimen in a groove in the stiff beam made of steel. The bolts holding the block on to the stiff beam chassis are tightened and the screw-jacks are removed. The small load applied at the bottom, using the screw-jacks, creates a moment about the contact. To relieve this moment, once a small normal load is applied using actuators, the bolts are loosened and tightened again. Finally, the normal load is applied using the hydraulic actuators. This causes the pad to get wedged into the groove in the pad holder. In the process of wedging, the pad moves into the groove. Hence, the webs now have to bend more while applying the normal load. This leads to a drop in load transfer ratio. To avoid the problem, once the total normal load is applied and the pad gets wedged into the grove, the rig is unloaded. The entire process of setting up the contact is repeated again.

Preliminary Experiments

   Cylindrical rods (0.2m long, 25.4mm in diameter) cast from SCN were used to machine the fretting fatigue specimens. The primary principal axis (x_0) is possibly tilted from the cylinder axis (x_1) by a small angle ($< 10^o$). A rectangular cross section (15mm×9.6mm) oriented at an arbitrary angle about the x_1-axis was

Table 1 - *Experimental conditions for High Temperature fretting fatigue experiments with SCN on IN100. (σ^t=stresses measured by top load cell, σ^b=stresses measured by bottom load cell). ^{s} gross sliding oberved throughout the experiment, ^{p} partial slip or mixed fretting regime during the experiments.*

Expt. No.	Controllable Loads			Measured Loads				Failure Life	Comments
	P (kN/m)	σ^t_{max} (MPa)	σ^t_{min} (MPa)	σ^b_{max} (MPa)	σ^b_{min} (MPa)	Q_{max} (kN/m)	Q_{min} (kN/m)	(cycles)	
1	1873	212	52	323.5	3.95	853	-365	415,000	Stopped^s
2	1852	400	53	441.8	14.07	315	-297	944,495	Failed^s
3	2193	290	78	357.8	22.71	519	-422	2,000,000	Run-out^s
4	3345	401	89	532.9	14.56	1004	-569	174,973	Failed^p
5	3394	407	84	501.6	0.43	723	-634	169,815	Failed^p
6	3431	337	71	444.8	-0.67	821	-543	61,900	Failed^p
7	3466	369	100	441.3	1.78	550	-746	491,292	Failed^p

machined. Tabs made from WASP alloy were inertia welded to the ends to produce the specimens. The nominally flat fretting pads were made from IN100. A schematic of the specimen orientation with respect to the material principal axes is shown in Figure 11(a). Three preliminary experiments were conducted to understand the working of the rig and to study the nature of contact between the materials used at high temperatures. The details of the experiments are listed in Table 1. The first experiment provided valuable insight into the mechanics of the rig and the temperature variation at the contact. Based on the behavior of the rig in the first experiment, changes were made in the alignment procedure. The next two experiments served the purpose of verifying the working of the rig. Coefficient of friction was found to be $0.17 - 0.23$ for the given contacting materials, due to which gross sliding (implying $Q = \mu P$) was observed during the course of experiments. It is possible that the relatively large displacements associated with sliding wear introduced non-linearity into the load transfer relationship. Thus, P could be smaller than the measured value.

In order to achieve partial slip, representative of engine hardware, the tangential load has to be decreased such that $Q < \mu P$. From Equation 9, Q can be reduced either by increasing k_3 or by reducing k_2. Therefore, either the length of the specimen between the contact and the top grips could be reduced or the area of cross-section (i.e, the thickness) of the diaphragms could be reduced. The length of the specimen was reduced due to the ease of modification. This was achieved by raising the fretting chassis, relative to the specimen, using spacers. By this process, partial slip conditions were achieved in the subsequent experiments. Q/P ratios of up to 0.27 were achieved implying that the coefficient of friction could be higher than what was anticipated from the first few sliding experiments.

The sliding contact between the two materials was analyzed using the SIE approach [6]. The orientation of the primary axis of the material with respect to the axis of the specimen was obtained from the manufacturer. However, since the orientation of the secondary axis with respect to the contacting surface was not known, SIE analysis was performed for all possible orientations. As can be seen from Figure 11(b) the effect of orientation of the material principal axes on sub-surface stresses is significant. Lauè x-ray diffraction method will be used to determine the orientation of the material principal axes to facilitate an accurate stress analysis for each experiment.

Conclusions

An experimental setup for elevated temperature tests (610°C) has been designed and built based on the principle of load transfer similar to that of room temperature fretting fatigue experiments. The normal load transfer ratio of the rig was experimentally measured to be > 92%. An experimental procedure has been designed to minimize the pad rotation and the moment at the contact location during elevated temperature testing. The elevated temperature experiment is capable of generating mixed-mode or partial slip fretting conditions representative of engine hardware. Q/P ratios of up to 0.27 were achieved during the partial slip tests, implying that the slip zone coefficient of friction is higher than 0.27. The orientation of the material principal axes, which was found to affect the sub-surface stresses significantly will be determined by Lauè x-ray diffraction. A preliminary set of data is given that can be used to begin assessing life prediction models. Further work will include additional experiments as well as the extension of isotropic life prediction models to the case of single crystal materials.

Acknowledgment

This work was supported in part under University of Dayton Research Institute (UDRI) subcontracts to Purdue University as part of the United States Air Force (USAF) High-Cycle Fatigue (HCF) program. M. Okane would also like to acknowledge the Ministry of Education, Science, and Culture of Japan for support during his stay at Purdue University.

References

[1] Farris, T. N., Szolwinski, M. P., and Harish, G., "Fretting in Aerospace Structures and Materials," In *Fretting Fatigue: Current Technology and Practices, ASTM STP 1367*, D. W. Hoeppner, V. Chandrasekaran, and C. B. Elliott, Eds., American Society for Testing and Materials, West Conshohocken, PA, 2000, pp. 523–537.

[2] McVeigh, P. A., Harish, G., Farris, T. N., and Szolwinski, M. P., "Modeling Interfacial Conditions in Nominally Flat Contacts for Application to Fretting of Turbine Engine Components," *International Journal of Fatigue,* **21**, 1999, pp. 157–165.

[3] Murthy, H., Harish, G., and Farris, T. N., "Efficient Modeling of Fretting of Blade/Disk Contacts Including Load History Effects," *ASME Journal of Tribology* , In-Press.

[4] Murthy, H., Harish, G., and Farris, T. N., "Influence of Contact Profile on Fretting Crack Nucleation in a Titanium Alloy," In *Collection of Technical Papers - 41st AIAA/ASME/ASCE/AHS/ASC Structures, Structural Dynamics and Materials Conference*, vol. 1, AIAA, 2000, pp. 1326–1333.

[5] Rajeev, P. T. and Farris, T. N., "Two Dimensional Contact of Dissimilar/Anisotropic Materials," In *Collection of Technical Papers - 42nd AIAA/ASME/ASCE/AHS/ASC Structures, Structural Dynamics and Materials Conference*, vol. 1, AIAA, 2001, pp. 515–522.

[6] Rajeev, P. T. and Farris, T. N., "Analysis of Fretting Contacts of Dissimilar Isotropic and Anisotropic Materials," *Journal of Strain Analysis* , In-Press.

[7] Murthy, H., Farris, T. N., and Slavik, D. C., "Fretting Fatigue of Ti-6Al-4V Subjected to Blade/Disk Contact Loading," *Developments in Fracture Mechanics for the New Century, 50th Anniversary of Japan Society of Materials Science* , 2001, pp. 41–48.

[8] Harish, G., Szolwinski, M. P., Farris, T. N., and Sakagami, T., "Evaluation of Fretting Stresses Through Full-Field Temperature Measurements," In *Fretting Fatigue: Current Technology and Practices, STP 1367*, D. W. Hoeppner, V. Chandrasekaran, and C. B. Elliott, Eds., American Society for Testing and Materials, West Conshohocken, PA, 2000, pp. 423–435.

[9] Hills, D. A. and Nowell, D., *Mechanics of Fretting Fatigue*, Kluwer, 1994.

[10] Golden, P. J., *High Cycle Fatigue of Fretting Induced Cracks*. PhD thesis, Purdue University, West Lafayette, 2001.

TITANIUM ALLOYS

David W. Hoeppner,[1] Amy M. H. Taylor,[1] and Venkatesan Chandrasekaran[1]

Fretting Fatigue Behavior of Titanium Alloys

Reference: Hoeppner, D. W., Taylor, A. M. H., and Chandrasekaran, V., **"Fretting Fatigue Behavior of Titanium Alloys,"** *Fretting Fatigue: Advances in Basic Understanding and Applications, STP 1425*, Y. Mutoh, S. E. Kinyon, and D. W. Hoeppner, Eds., ASTM International, West Conshohocken, PA, 2003.

Abstract: A discussion of experimental studies to characterize the fretting fatigue behavior of titanium alloys is presented. This includes studies to evaluate the effects of normal pressure, contact material conditions, surface treatments, and environment. In addition, concepts of normal pressure threshold and damage threshold as related to fretting fatigue of titanium alloys are discussed. Finally, this paper is concluded with suggestions for future research efforts that are essential to improve understanding of the fretting fatigue behavior of titanium alloys.

Keywords: fretting fatigue, titanium alloys, fatigue, fretting

Introduction

Fretting fatigue can be described as a combination of wear and fatigue processes. For fretting fatigue to occur two components should be in contact and in addition one or both components may be subjected to cyclic stresses resulting in "small" relative displacement between them. All materials are affected by fretting. Titanium alloys are more susceptible to fretting initiated fatigue degradation mechanisms when compared to other metal alloys. This may be because titanium alloys exhibit a greater tendency for material transfer when it slides over other materials or over itself. Moreover, fretting produces different forms of damages such as pits, scratches including fretting and/or wear tracks, oxide, debris, material transfer, surface plasticity, subsurface cracking, and fretting craters. It is believed that cracks nucleate from these damages after only a fewer number of fretting fatigue cycles when compared to cycles needed to form cracks under fatigue conditions without fretting. Therefore, the basic effect of fretting is to accelerate the crack nucleation process. More importantly, the effect of fretting is to eliminate the apparent "fatigue limit" of a material. Although fretting produces damage on the contact surfaces after they are subjected to a single fretting fatigue cycle a damage threshold exists for crack(s) to form from the damage. This phenomenon indicates that fretting may not be harmful if fretting damages are continuously monitored, controlled and repaired before they reach the threshold stage. Also, it should be noted that the fretting fatigue damage threshold is very specific to the applied normal pressure as well as the maximum fatigue stress. In addition, titanium alloys are very sensitive to normal pressure changes although there also

---

[1] The authors are, respectively, Professor, research engineer, and research assistant professor, Department of Mechanical Engineering, University of Utah, 50 S. Central Campus Dr., Room 2202, Salt Lake City, UT 84112-9208.

is an existence of normal pressure threshold. These aspects will be discussed in the latter part of this paper. However, under conditions of contact and applied cyclic stresses cracks form from fretting damages after certain a damage threshold is reached. After this, elimination of fretting may not have any favorable influence as nucleated cracks eventually propagate to fatigue failure by cyclic stress alone. As a result fretting shortens the fatigue life of a structural component. Thus it is imperative to inspect structural components for fretting damages to eliminate the possibility of these damages becoming crack nucleation sites.

It is well known that the influence of fretting on the fatigue life of a metal alloy is dependent on numerous parameters. This paper will discuss some important fretting fatigue variables such as normal load, surface conditions, environment, and contact conditions. With the support of experimental studies these effects will be discussed as related to aerospace components with emphasis on titanium alloys. First, experimental data with regard to effects of normal pressure on fretting fatigue life of titanium alloys are given in the following section.

Normal Load Effects on Fretting Fatigue Life of Titanium Alloys

As mentioned before, the basic effect of fretting is to decrease the fatigue life of a material. One of the parameters that significantly contributes to the reduction in fretting fatigue life of any metal alloy is the magnitude of the normal pressure that one component exerts on the other component. This section summarizes experimental studies conducted to illustrate the effects of normal pressure on the fatigue life of Ti-6Al-4V [1-16]. All experiments discussed in this section were performed in the laboratories of the principal author of this paper during the past four decades. Both fatigue specimens and fretting pads were made using Mill Annealed Ti-6Al-4V. The fretting fatigue experimental methods are described elsewhere in references 3, 5, and 10. It is important to note that the results in all fretting fatigue experiments are influenced by experimental conditions a great deal.

Figure 1 [1,2] provides some unpublished Ti-6Al-4V fatigue and fretting fatigue experimental results generated in the early 1970s by the principal author of this paper. The reduction in fatigue life of Ti-6Al-4V specimens is evident from this study when they were subjected to the normal pressure of 20.7 MPa (3 ksi) at different maximum fatigue stresses. The results from another study also showed that fretting fatigue life was reduced when tested under the normal pressure of 20.7 MPa (3 ksi) (see Figure 2) [4]. It also was observed from this study that at higher fatigue stresses the difference between fretting fatigue life and fatigue life of Ti-6Al-4V is small whereas the difference is large at lower fatigue stresses. This undoubtedly is due to the relatively high applied cyclic stress at the shorter lives thus causing the fretting to provide less effect in life reduction. It was stated that this might be because at lower fatigue stresses the fretting acting in conjunction with fatigue produces crack(s) in a significantly smaller number of cycles whereas "pure" fatigue conditions need a larger number of cycles to produce cracks. In another study [10], the effect of fretting on the fatigue life of Ti-6Al-4V under different normal pressures was evaluated. As shown in Figure 3, the fatigue life of Ti-6Al-4V was reduced when the normal pressure was increased and vice versa. The scatter in the data presented in the

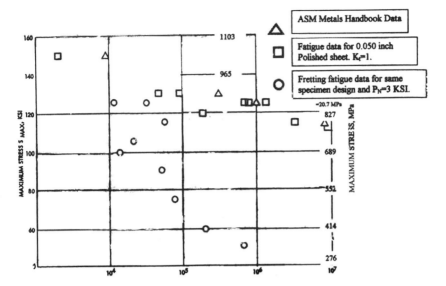

Figure 1: Fatigue and fretting fatigue data for Ti-6Al-4V annealed R = +0.1 (1970-75)
[1, 2].

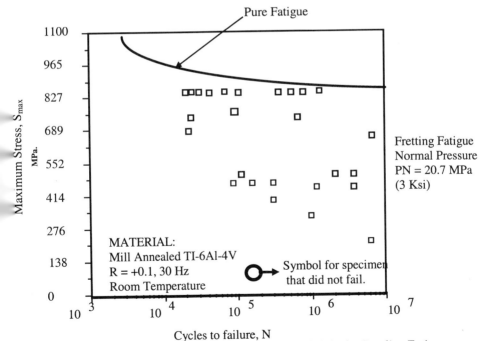

Cycles to failure, N
Figure 2: Maximum Stress vs. Cycles to Failure for Ti-6Al-4V for Baseline Fatigue
and Fretting Fatigue Conditions [4].

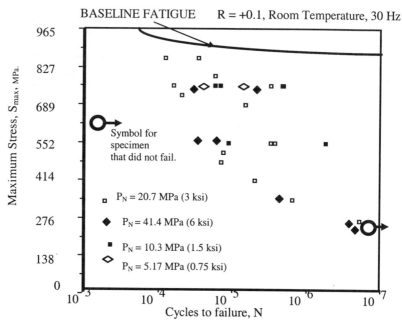

Figure 3: Maximum Stress vs. Cycles to Failure for Ti-6Al-4V (Ma) Material for Normal Pressure Conditions as Indicated. From: [4]

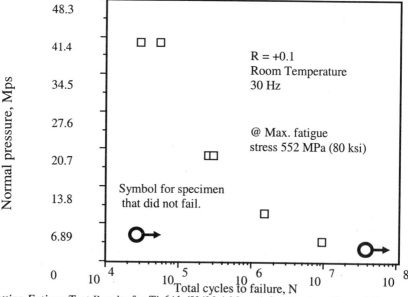

Figure 4: Fretting Fatigue Test Results for Ti-6Al-4V (Ma) Material at Various Normal Pressure

figure was related to the local microstructural variations in Ti-6Al-4V. Also, it is important to note that normal pressure effects on fretting fatigue life of Ti-6Al-4V are dependent on the maximum fatigue stress. As shown in Figures 4,5 and 6 [7], increasing the maximum fatigue stress from 552 MPa to 758 MPa (80 Ksi to 110 ksi) then the fatigue life of Ti-6Al-4V was decreased. This behavior was observed when Ti-6Al-4V was tested under all the different normal pressures. Metallographic and fractographic analysis of fretted Ti-6Al-4V specimens revealed that fretting resulted in damage states which were related to mechanical action between the contacting surfaces. In addition, it was observed that these damage states were nucleation sites from which numerous secondary cracks were formed. Figure 6 compares the normal pressure effect in 7075-T6 and Ti-6Al-4V.

When tested simulating hip prosthesis loading conditions, Ti-6Al-4V also exhibited fretting fatigue life reduction at different normal pressures (see Figure 7) [13]. It was clearly evident from the test results that fretting reduced the fatigue life of Ti-6Al-4V significantly from ten million fretting-fatigue cycles at 41.4 MPa (6 ksi) normal pressure, 1.65/0.165 kN (371/37 lbf.) cyclic fatigue load, to 724 040 fretting-fatigue cycles at 82.7 MPa (12 Ksi) normal pressure, 1.65/0.165 KN (371/37 lbf.) cyclic fatigue load. Furthermore, when the cyclic fatigue load was increased to 3.30/0.33 KN (742/74 lbf.) R= 0.1 the fatigue life of Ti-6Al-4V was reduced to a mere 69 660 fretting-fatigue cycles at 41.4 MPa (6 ksi) normal pressure.

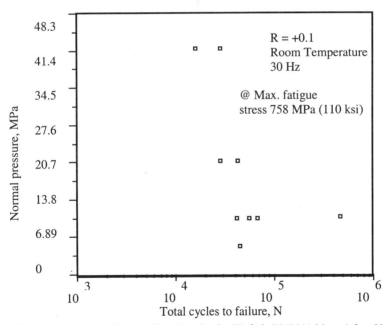

Figure 5: Fretting Fatigue Test Results for Ti-6Al-4V (MA) Material at Various Normal Pressures, [7].

These studies show that the variable normal pressure is a major factor in reducing the fatigue life of Ti-6Al-4V. However, there exists a damage threshold beyond which eliminating the cause of fretting does not have any effect on the remaining fatigue life of Ti-6Al-4V [3, 4, 10]. The concept of damage threshold as related to fretting fatigue behavior of Ti-6Al-4V is discussed in the next section.

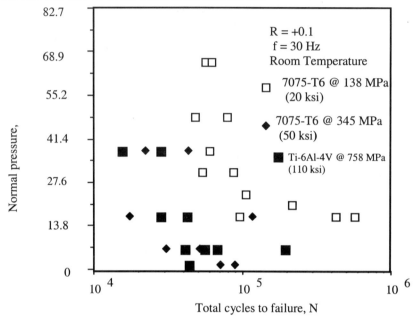

Figure 6: Fretting Fatigue Normal Pressure Threshold for 7075-T6 Aluminum Alloy and Ti-6Al-4V Titanium Alloy at Various Maximum Fatigue Stress. [10]

Damage Threshold Concept

As mentioned before, fretting produces various damage states on the contact surfaces. Cracks nucleate from these damage states after only a smaller number of applied cycles. Experimental studies indicated that at all loading conditions a minimum amount of fretting damage is needed to cause any reduction in fatigue life whatsoever. Interrupted fretting fatigue experiments were conducted on Ti-6Al-4V specimens at two different fatigue stresses; viz., 862 MPa and 483 MPa (125 ksi and 70 ksi) at 20.7 MPa (3 ksi) normal pressure to observe the existence of the damage threshold. The results from this study are plotted in Figure 8. As shown, at 100% fretting fatigue life, that is with the fretting pad in
contact until the test specimen fractures, the cycles to failure was approximately 1.5×10^4. Fretting damage was assessed by observing the remaining (residual) life after the fretting pad was removed. Thus, at 60% fretting fatigue cycles, that is when the fretting pad was removed, the specimen fractured after only 4×10^4 cycles. This is in fact little more than the 100% value. In addition, at 20% fretting fatigue cycles it was observed that there was no reduction in the fatigue life of Ti-6Al-4V as the cycles to failure are the same. This

means that the damage produced under fretting condition does not contribute any fatigue life reduction. In other words these results suggested that no fretting damage was produced to cause a reduction in fatigue life at 60 to 0% fretting pre-cycles. Therefore, the existence of a damage threshold in Ti-6Al-4V helps to alleviate fretting related challenges only if these damages are removed before they become potential crack nucleation sites. Once the crack nucleation has occurred from the fretting further exposure to the fretting has virtually no continual effect on the life.

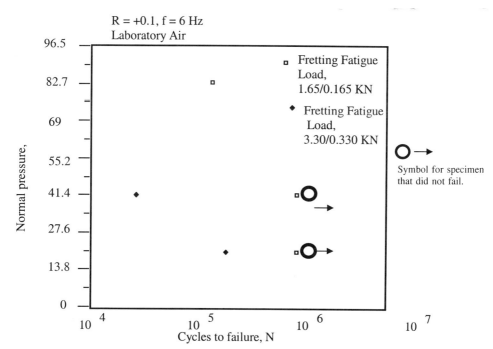

Figure 7: Fretting Fatigue Test Results for Ti-6 Al-4V Simulating Hip Prosthesis Loading Conditions, [13]

Effect of Dissimilar Materials in Contact under Fretting Fatigue Conditions

Titanium alloys are known for their susceptibility to fretting degradation mechanisms whether they are in contact with themselves or with other metal alloys. To evaluate the fretting fatigue behavior of Ti-6Al-4V when it is in contact with itself or with other materials, studies were conducted. In this study fretting pads were made from 7075-T6, Al-1100, Copper, 0.40/0.50 C Steel, and Ti-6Al-4V as well. Ti-6Al-4V fatigue specimens were fretted with pads of different materials as mentioned above. Figure 9 shows the fretting fatigue test results resulted from this study [14]. All tests were conducted at room temperature under 20.7 MPa (3 ksi) normal pressure at different fatigue stresses. As can be seen from the plot, Ti-6Al-4V in contact with Ti-6Al-4V (4, 7) exhibited comparatively more reduction in fretting fatigue life. Moreover, as shown in Figure 10, in general, the harder the fretting pad the larger the reduction in fatigue life of Ti-6Al-4V under these conditions. This behavior was observed to be same for two different fatigue stresses as shown.

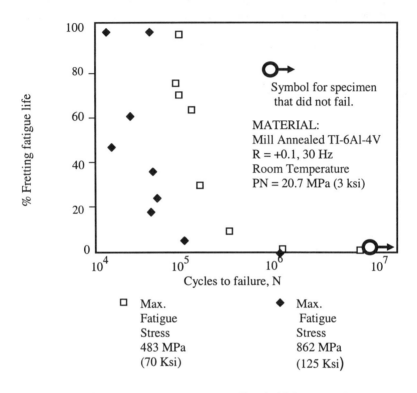

Figure 8: Fretting Fatigue Damage Threshold Concept, [4]

Environmental Effects to Characterize Fretting Fatigue Behavior of Ti-6Al-4V

Chemical environment is one of the most important parameters that affect the fretting fatigue behavior of a metal alloy including Ti-6Al-4V. Yet very little work has been performed to study this issue especially on Ti-6Al-4V. However, to provide some additional insight to environmental effects on fretting on Ti-6Al-4V this section discusses some experimental results of the primary author's unpublished work. This study was performed in different environmental conditions such as laboratory air, distilled water, 3.5% NaCl under different normal pressure and fatigue stress levels. At both lower and higher fatigue stress levels fretting fatigue tests in 3.5% NaCl environment resulted in greater reduction in fatigue life when compared to other environments (see Figure 11).

Effect of Surface Treatment on Fretting Fatigue of Titanium Alloys

It is now well known that in applications which involve wear and fretting conditions titanium alloys cannot be used without any surface treatments as they are known to exhibit more material transfer. This may be the result of the breakdown in the oxide protective film and the relatively rapid nucleation of cracks from the fretting damage. Therefore, some studies were conducted to evaluate surface treatment as a means to alleviate the reductions of life from fretting induced damage of titanium alloys [11, 16]. In aircraft structural components, especially in the aircraft fuselage and wing structures, fretting mechanisms are observed to be one of the potential causes which nucleate cracks from rivet or other fastener holes. In recent years, the solution-treated and aged (STA), and the beta processed Ti-6Al-4V are extensively used to manufacture critical aircraft components. In addition, Boeing Commercial Airplane Company is using extensive amounts of beta-processed Ti-6Al-4V on the B777.

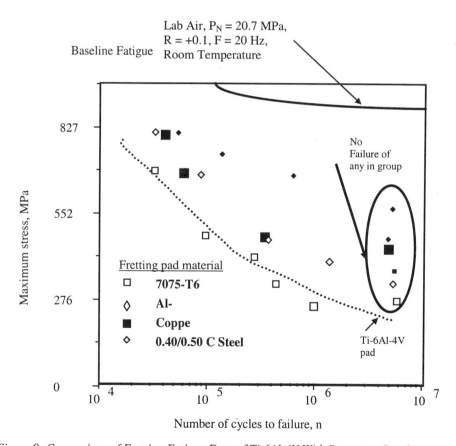

Figure 9: Comparison of Fretting Fatigue Data of Ti-6Al-4V With Respect to Baseline Data for all Fretting Pad Materials [14].

Fretting corrosion and fretting fatigue are potential problems in maintaining the structural integrity of aircraft components made using Ti-6Al-4V with aforementioned processing conditions. It also will be of serious consequence when the fretting fatigue behavior of Ti-6Al-4V alloy systems under different environments and temperatures is not characterized. This section discusses some experimental studies that were conducted to determine the effect of airframe design parameters and coatings on fretting fatigue life of titanium alloys [16].

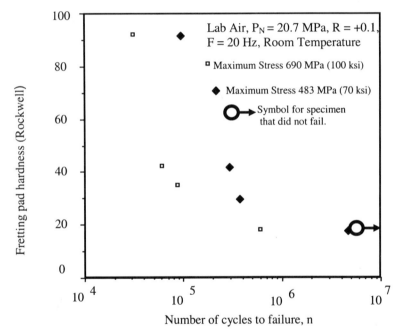

Figure 10: Harder the Fretting Pad The Larger the Reduction in Fatigue Life of Ti-6Al-4V [14].

Simulating aircraft structural joint design parameters fretting fatigue tests were conducted first on uncoated Ti-6Al-6V-2Sn fatigue specimen with fretting pad made of same material. The fretting pad was fastened to the fatigue specimen using different types of fasteners, viz. Ti-6Al-4V screw clearance fit, Ti-6Al-4V taperlok interference fit, and Ti-6Al-4V Hilok clearance fit. The objective of this study is to evaluate parameters such as material, fastener type, interference, installation torque, stress level, cyclic loading frequency, and coatings. Different types of fasteners were selected because they induced varying amounts of normal pressures at the interface. All experiments were conducted at laboratory air. Fretting fatigue test results at different fatigue stresses for uncoated Ti-6Al-6V-2Sn with different fastener types are summarized in Figure 12. From this study it was concluded that Ti-6Al-4V taper-lok fasteners exhibited better fretting fatigue life than Ti-6Al-6V-2Sn specimens. Subsequently, titanium alloy specimens made of Ti-6Al-6V-2Sn, Ti-6Al-2Sn-4Zr-6Mo, and Ti-6Al-4V were coated with Molykote 106 and were fretting fatigue tested with Ti-6Al-4V taper-lok fasteners.

The work showed that coated Ti-6Al-4V specimens exhibited greater reduction in fretting fatigue life when compared other titanium alloys. However, additional work showed that a significant improvement in the fretting fatigue life of coated (with Molykote 106) Ti-6Al-6V-2Sn occurred when compared to uncoated Ti-6Al-6V-2Sn.

Recent Studies

In recent years other studies have shown similar effects to those discussed [17-26]. The studies reported by Antoniou and Radke [17] are an excellent summary of the current state of the understanding of fretting fatigue of Ti alloys and validate many of the ideas discussed earlier in this paper.

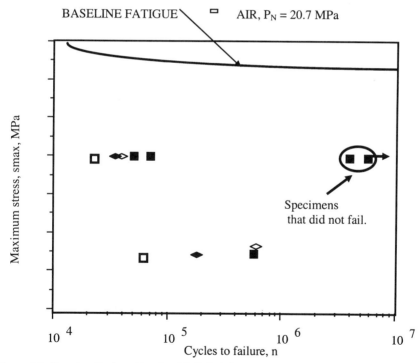

Figure 11: Fretting Fatigue Results Of Ti-6Al-4V under Different Environment Tested At Room Temperature, R = +0.1, f=30Hz. [15]

Distilled
water,

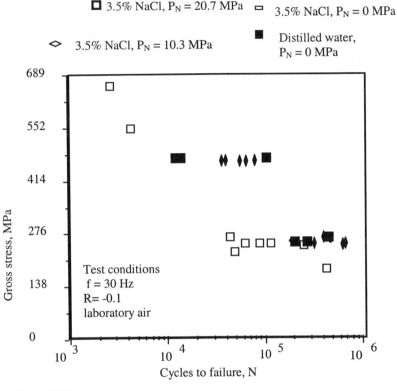

Figure 12: *Fretting Fatigue Test Data for Uncoated Ti-6Al-6V-2Sn Simulating Aircraft Structural Joint Design Parameters. HiLok and TaperLok are two Commercial Fastener Types used in Industry [16].*

Conclusions

A review of some fretting fatigue experimental studies on titanium alloys were presented in this paper. The following conclusions can be made from the aforementioned studies.

- Fretting can reduce the fatigue allowable of Ti-6Al-4V by various amounts depending on parameters related to the application or study.
- Fretting damage accumulates during the process of fretting fatigue and reduces time/cycles to form cracks under the relevant conditions.

- Fretting may have an influence on crack propagation rates in the microstructurally dependent crack propagation regime - especially in titanium alloys.
- A fretting damage threshold exists for titanium alloys.
- The fretting damage threshold is dependent on normal pressure and potentially on slip amplitude. This threshold also may be dependent on chemical environment.
- Chemical environment affects fretting-fatigue behavior of titanium alloys.
- Fretting appears to have its greatest effects as the stress/strain decreases, thus, fretting has its greatest effect in the "long" life regime.
- Surface treatments can significantly improve the effect of fretting in titanium alloys and deserve much study as a means to alleviate fretting fatigue.
- Spectrum loading has a significant effect on the fatigue of titanium alloys as with other materials.

Future Needs

Although some studies were conducted on fretting fatigue behavior of titanium alloys during the past several years, much more work is to be done in order to understand more about this complex mechanism. Following are some recommendations to be pursued to enhance our understanding of fretting fatigue of titanium alloys.

- Some studies have been reported on fretting-fatigue of titanium alloys under variable amplitude loading. This important issue must be evaluated for many variable amplitude-loading conditions.
- Analytical or empirical models for fretting-fatigue life estimation need extensive development. [friction, slip amplitude, pressure, temperature, environment, material]. Some work has been done on this aspect but more needs to be done. The time dependency of fretting development needs further study in this regard.
- The role of microstructure of titanium alloys on fretting-fatigue needs much study. In view of the significant effect of microstructure on fatigue (S-N) and crack propagation of titanium alloys this could bear important results for the future.
- Surface palliatives (treatments) need study to evaluate the effects and formulate general principles for application.
- Studies to establish transfer functions to assure information developed on specimen test elements for fretting fatigue can be accurately transferred to components.
- Extensive study of mechanics and fracture mechanics applied to fretting fatigue must be carried out.
- The role of temperature in fretting fatigue of titanium alloys needs evaluation.
- The role of multiaxial stresses in fretting fatigue needs study-especially for aircraft applications.
- Since the confocal microscope is another useful tool to understand the metrics of fretting damage development from initial damage states to fretting damage to crack transition it will be important to use this tool and others to continue the mapping aspects and to develop the metrics to be used for the models. Clearly,

this could be one of the most significant aspects of research to be done if a quantitative method of fretting fatigue life prediction is to be developed.

- The effect of chemical environment related to fretting fatigue effects of Titanium alloys needs study since many applications of Ti alloys occur in environments that cause corrosion of one type or another.
- A compilation of fretting fatigue terminology needs development.
- A design guide (simplistic yet thorough) for fretting fatigue needs development.

References

[1] *ASM Handbook*, Vol. 18, Friction, Lubrication, and Wear Technology, 1994, 778.

[2] Hoeppner, D.W. "Mechanisms of Fretting Fatigue", *Fretting Fatigue, ESIS 18,* R.B. Waterhouse and T.C. Lindley, Eds., Mechanical Engineering Publications, London, (1994) 3-19.

[3] Hoeppner, D.W. and Goss, G.L., "Research on the Mechanism of Fretting Fatigue, *Corrosion Fatigue: Chemistry, Mechanics, and Microstructure, NACE-2*, pp. 617-626, 1971.

[4] Hoeppner, D.W. and Goss, G.L., "A Fretting Fatigue Damage Threshold Concept", *Wear,* 27, pp. 61-70, 1974.

[5] Hoeppner, D.W. and Goss, G.L., A New Apparatus for Studying Fretting Fatigue, *The Review of Scientific Instruments*, 42, pp. 817-821, 1971.

[6] Hoeppner, D.W., Unpublished work.

[7] Hoeppner, D.W. and Goss, G.L., "Normal Load Effects in Fretting Fatigue of Titanium and Aluminum Alloys," *Wear*, 27, pp. 153-159, 1974.

[8] Hoeppner, D.W. and Goss, G.L., "Metallographic Analysis of Fretting Fatigue Damage in Ti-6Al-4V MA and 7075-T6 aluminum, *Wear*, 27, pp. 175-187, 1974.

[9] Hoeppner, D.W. and Goss, G.L., "Characterization of Fretting Fatigue Damage by SEM analysis", *Wear*, 24, pp. 77-95, 1973.

[10] Adibnazari, S., Hoeppner, D.W., "The Role of Normal pressure in Modeling Fretting Fatigue," *Fretting Fatigue, ESIS 18,* R.B. Waterhouse and T.C. Lindley, Eds., Mechanical Engineering Publications, London, (1994) pp. 125-134.

[11] Betts, R.E., "Development and Evaluation of Coatings for Alleviating Fretting on Fatigue Life of Titanium Alloys", General Electric Co., Contract AF33615-70-C-1537, 1971.

[12] Hoeppner, D.W. and Chandrasekaran, V., "Fretting in Orthopaedic Implants – a Review," *Wear*, 173, pp. 189-197, 1994.

[13] Hoeppner, D.W. and Chandrasekaran, V., "Characterizing the Fretting Fatigue Behavior of Ti-6Al-4V in Modular Joints," *Medical Applications of Titanium and Its Alloys: The Material and Biological Issues, ASTM STP 1272*, S.A. Brown and J.E. Lemons, Eds., pp. 252-265, 1996.

[14] Hoeppner, D.W., Unpublished work, 1975-78.

[15] Hoeppner, D.W., Unpublished work, 2000-current.

[16] Padberg, D.J., "Fretting Resistant Coatings for Titanium alloys", McDonnell Aircraft Co., *AFML TR-71-184*, 1971.

[17] Antoniou, R.A., Radke, T.C., "The Mechanisms of Fretting Fatigue of Titanium Alloys", *Materials Science and Engineering*, A237 (1997)pp229-240.

[18] Kinyon, S.E., Hoeppner, D.W., "Spectrum Load Effects on the Fretting Behavior of Ti-6Al-4V", Fretting Fatigue: Current Technology and Practices, ASTM STP 1367, Hoeppner, D.W., Chandrasekaran, V., Elliott, C.B., Editors, American Society for Testing and Materials, West Conshohocken, PA, 2000, pp 100-115.

[19] Hutson, A.L., Nicholes, T., "Fretting Behavior of Ti-6Al-4V Against Ti-6Al-4V Under Flat-on Flat Contact With Blending Radii, *Fretting Fatigue: Current Technology and Practices, ASTM STP 1367*, Hoeppner, D.W., Chandrasekaran, V., Elliott, C.B., Editors, American Society for Testing and Materials, West Conshohocken, PA, 2000, pp 308-321.

[20] Anton, D.L., Lutian, M.J., Favrow, L.H., Logan, D., Annigeri, B., "The Effects of Contact Stress and Slip Distance on Fretting Fatigue Damage in Ti-6Al-4V/17-4PH Contacts", *Fretting Fatigue: Current Technology and Practices, ASTM STP 1367*, Hoeppner, D.W., Chandrasekaran, V., Elliott, C.B., Editors, American Society for Testing and Materials, West Conshohocken, PA, 2000, pp119-140.

[21] Satoh, T., "The Influence of Microstructure on Fretting Fatigue Behavior of Near Alpha Titanium", *Fretting Fatigue: Current Technology and Practices, ASTM STP 1367*, Hoeppner, D.W., Chandrasekaran, V., Elliott, C.B., Editors, American Society for Testing and Materials, West Conshohocken, PA, 2000, pp295-307.

22] Hutson, A.L., Nicholes, T., "Fretting Behavior of Ti-6Al-4V Against Ti-6Al-4V Under Flat-on Flat Contact With Blending Radii, *Fretting Fatigue: Current Technology and Practices, ASTM STP 1367*, Hoeppner, D.W., Chandrasekaran, V., Elliott, C.B., Editors, American Society for Testing and Materials, West Conshohocken, PA, 2000, pp 308-321.

23] Anton, D.L., Lutian, M.J., Favrow, L.H., Logan, D., Annigeri, B., "The Effects of Contact Stress and Slip Distance on Fretting Fatigue Damage in Ti-6Al-4V/17-4PH Contacts", *Fretting Fatigue: Current Technology and Practices, ASTM STP*

1367, Hoeppner, D.W., Chandrasekaran, V., Elliott, C.B., Editors, American Society for Testing and Materials, West Conshohocken, PA, 2000, pp119-140.

[24] Satoh, T., "The Influence of Microstructure on Fretting Fatigue Behavior of Near Alpha Titanium", *Fretting Fatigue: Current Technology and Practices, ASTM STP 1367*, Hoeppner, D.W., Chandrasekaran, V., Elliott, C.B., Editors, American Society for Testing and Materials, West Conshohocken, PA, 2000, pp295-307.

[25] Cortez, R., Mall, S., Calcaterra, J., "Investigation of Variable Amplitude Loading on Fretting Fatigue Behavior of Ti-6Al-4V", *International Journal of Fatigue*, 21 (1999) 709-717.

[26] Wallace, J., Neu, R., "Fretting Fatigue Crack Nucleation in Ti-6Al-4V", submitted to *Fatigue and Fracture of Engineering Materials and Structures*, 2001.

Alisha L. Hutson,[1] Noel E. Ashbaugh,[2] and Ted Nicholas[3]

An Investigation of Fretting Fatigue Crack Nucleation Life of Ti-6Al-4V under Flat-on-Flat Contact

---

Reference: Hutson, A. L., Ashbaugh, N. E., and Nicholas, T., **"An Investigation of Fretting Fatigue Crack Nucleation Life of Ti-6Al-4V under Flat-on-Flat Contact,"** *Fretting Fatigue: Advances in Basic Understanding and Applications, STP 1425,* Y. Mutoh, S. E. Kinyon, and D. W. Hoeppner, Eds., ASTM International, West Conshohocken, PA, 2003.

Abstract: A study was conducted to investigate fretting fatigue damage of Ti-6Al-4V against Ti-6Al-4V under flat-on-flat contact with blending radii at room temperature. Both the location of and the time required to nucleate fretting fatigue cracks were investigated for two static clamping stress values representative of those estimated for turbine engine blade attachments. Fatigue limit stresses for a 10^7 cycle fatigue life were evaluated for the selected conditions. Fretting fatigue nucleated cracks were identified *in situ* using shear wave ultrasonic NDI. The effect of fretting fatigue on uniaxial fatigue life was quantified by interrupting fretting fatigue tests at 10% of life, and conducting uniaxial residual fatigue life tests. Metallography, scanning electron microscopy and spectral analysis were used to characterize the fretting damage.

Keywords: fretting fatigue, Ti-6Al-4V, crack nucleation

Introduction

Fretting fatigue is the phenomenon caused by localized relative motion between contacting components under vibratory load. The damage due to fretting fatigue can lead to premature crack initiation and failure. Such damage has been indicated as the cause of many unanticipated disk and blade failures in turbine engines, and as a result, has been the focus of numerous studies [1-5] over the years. Fretting fatigue often takes place in the presence of other types of damage, leading to confusion regarding the source of and micro-mechanisms responsible for premature failures. With the numerous conditions influencing the fretting fatigue phenomenon, a comprehensive understanding of how

---

[1] Assistant Research Engineer, University of Dayton Research Institute, 300 College Park, Dayton, OH 45469-0128.
[2] Senior Research Engineer, University of Dayton Research Institute, 300 College Park, Dayton, OH 45469-0128.
[3] Senior Scientist, Air Force Research Laboratory, Materials and Manufacturing Directorate, Metals, Ceramics & NDE Division (AFRL/MLLMN), Wright-Patterson AFB, OH 45433-7817.

fretting fatigue occurs and subsequent accurate life prediction modeling has been elusive.

Many different test systems have been developed to aid in the study of the fretting fatigue phenomenon. Most research has been conducted on Hertzian [3, 5], or square cornered punch-on-flat [3, 5] geometries because of the availability of closed form analytical solutions for the resulting stress distributions. However, the applicability of these geometries to real components is limited. Some work has been done to directly test real component geometries [1], but specimens are expensive, determination of the stress distribution is tedious and costly, and the results are difficult to model. A geometry that provides information at an intermediate level of complexity is required as the next step toward the development of an accurate life prediction model that bridges the gap between laboratory and service conditions.

The authors have published work on a novel test geometry that simulates the essential features of a turbine engine blade root attachment while maintaining the simplicity of a more generalized geometry. Previously published work established the viability of the test apparatus in investigating fretting fatigue behavior and documented preliminary results on the effect of several parameters critical to fretting fatigue [6, 7].

The purpose of the present investigation is to extend the scope of prior work and answer questions regarding the mechanisms of fretting fatigue crack nucleation in the current test geometry. Of interest to the current work are the location of crack nucleation sites, the portion of life spent in fretting fatigue crack nucleation, and quantification of the effects of fretting fatigue damage on residual fatigue properties. Identification of crack nucleation sites using post-mortem fractography has produced less than conclusive results [6, 7]. In this study, crack nucleation life and location will be investigated by conducting fretting fatigue tests that are interrupted based on two independent criteria. The resultant damage will be characterized using scanning electron microscopy, metallographic analysis and semi-quantitative elemental analysis.

Experimental Approach

Fretting Fatigue Apparatus

The fretting experiments in this study were designed to simulate the primary loading conditions under which fretting fatigue damage occurs in turbine engine blade attachments. The dovetail geometry of a blade attachment is, in part, composed of flat pressure faces that blend into the mating component via rather large radii. The pressure faces are subjected to a bending moment produced by the dovetail flank angle, and to cyclic normal and shear forces. Either all of the shear load is transferred across the contact or the threshold of gross sliding is achieved and one component moves with respect to the mating component.

The fretting fatigue test apparatus used in this and related studies incorporates a flat specimen tested against a flat pad with blending radii at the edges of contact to simulate the dovetail pressure face geometry. Static clamping loads are imposed using instrumented bolts to measure the clamping load. A cyclic bulk load applied to the specimen produces a negligible bending moment and a cyclic shear stress at the contact that results in fretting fatigue, as indicated in the magnified view in Figure 1. The contribution of the bending moment in the dovetail geometry is assumed to be negligible.

Valid tests performed in this apparatus always result in full transfer of the shear load to the fixture, thus producing a mixed slip/stick condition that results in fretting fatigue crack nucleation. The test assembly produces two nominally identical fretting fatigue tests for each specimen, when similar clamping conditions are imposed in both grips (see Figure 2) [8]. Additional details on the fixture can be found in reference [6]. For this study, fretting fatigue cycles were applied at 300 Hz.

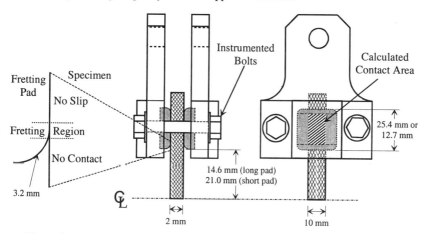

Figure 1 — *Test fixture schematic indicating pad lengths, distance from center of specimen to lower edge of pad, and blending radius at edge of contact.*

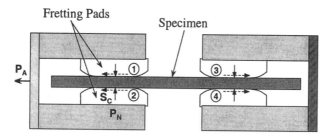

Figure 2 — *Test load train schematic. Numbers indicate edge of contact locations.*

Uniaxial Fatigue Apparatus

For part of the tests conducted in this study, fretting fatigue conditions were discontinued after a small portion of total fretting fatigue life, and the residual fatigue lives were evaluated in uniaxial fatigue. A servo-hydraulic test system was used to conduct the uniaxial fatigue tests. The system was equipped with a 5000 lbf (22.24 kN) capacity load cell, an analog controller operated in the load control mode, and hydraulic wedge grips with a 25 kN grip capacity and anti-fret inserts made of surfalloy. Tests were conducted at 3 Hz. All tests were conducted under ambient laboratory conditions.

Material and Machining Parameters

Fretting fatigue specimens and pads were machined from forged Ti-6Al-4V plates with a α+β duplex microstructure. Since no cracks were expected in the pads, orientation with respect to loading for the pads was not specified or identified. Axial specimens were machined with the fatigue axis oriented in the longitudinal direction. The material yield strength is 930 MPa; the tensile strength is 980 MPa, and the modulus is 116 GPa. Microstructure details can be found in reference [9]. Specimens and pads were low stress ground to an RMS 8 μinch surface finish.

Test Condition Selection

Three objectives were defined for the current investigation: identify the location of fretting fatigue crack nucleation, identify the portion of life spent nucleating fretting fatigue cracks, and quantify the reduction of fatigue properties caused by fretting fatigue damage accumulated by 10% of fretting fatigue life. Attempts to identify primary crack nucleation sites and crack nucleation lives in prior work [6, 7] have been inconclusive.

Three series of experiments were designed for this investigation, as a continuation of the previous work. First, the fretting fatigue limit stresses for a 10^7 cycle fatigue life were identified using a step loading technique, designated as 100%-of-life tests. Then, shear wave ultrasonic crack detection was used *in situ* to try to identify fretting fatigue cracks early in the propagation portion of life, designated as N_{DET} tests. Finally, some of the fretting fatigue experiments were interrupted at 10%-of-life, and the effect of cycling was quantified as residual fatigue life.

Geometry and loading conditions were the same as those used in prior work. Fretting fatigue sample dimensions were 10 mm x 2 mm x 150 mm. Nominal fretting pad lengths were 12.7 mm and 25.4 mm. A contact radius of 3.2 mm was used, resulting in contact lengths of 6.35 mm and 19.05 mm. Clamping stresses, σ_N, of 200 MPa to 620 MPa were selected to approximate the stress range identified for the blade attachment region. All of the fretting fatigue experiments were conducted at R = 0.5.

For the 100%-of-life series of tests each specimen was cycled to failure using a step loading procedure to obtain the fretting fatigue limit for a 10^7 cycle fatigue life. The procedure involved cycling the specimen at a constant stress, which was below the expected fatigue limit stress, for 10^7 cycles. Then the stress level was increased by a small amount, usually five percent, and the cycling was continued for an additional 10^7 cycles. The stress increments were continued until the specimen fractured. Then, the fatigue limit was calculated from the final stress, number of cycles at the final stress, and the stress from the prior block, using linear interpolation [10]. Six tests were conducted at the nominal 620 MPa clamping stress condition and three were conducted at the 200 MPa clamping stress condition. The purposes of these tests were to establish an average fatigue strength for a 10^7 cycle fatigue life including data on the repeatability of each condition, and to provide fretted surfaces at 100 percent of life. Fretting scars were characterized in the SEM (Scanning Electron Microscope) for comparison to partial life damage characterization from later tests.

For the N_{DET} series of tests, specimens were cycled at a far field bulk stress of 350 MPa for the lower clamping stress and 260 MPa for the higher clamping stress. These stress levels are roughly equal to the fatigue limit stresses identified for a 10^7 cycle fatigue life in the 100%-of-life tests. During each test, two transducer/coupling wedge assemblies were attached to the specimen, using a fast curing epoxy, to monitor cracking in each grip independently. When applied to uniaxial fatigue specimens, the technique is capable of detecting surface cracks with depths of ~200 μm. In the configuration employed in this study, the transducers both transmitted the 10 MHz ultrasonic wave and received reflections from cracks present in the structure. Identification of cracks was achieved by comparing successive signals acquired at 500,000 mechanical fatigue cycle intervals, and identifying changes in shape and amplitude of various peaks occurring in the vicinity of the edge of contact. If a crack was detected, the specimen was removed for inspection in the SEM, heat tinted at 420°C for four hours to mark the location of the crack front at the time of inspection, and then reinstalled in the test frame for cycling to fracture at the same conditions used to nucleate the crack. Three fretting fatigue specimens for each condition were tested using this technique.

For the 10%-of-life series of tests each specimen was cycled for 1 million cycles at a far field bulk stress of 350 MPa for the lower clamping stress and 260 MPa for the higher clamping stress. Then the fretting condition was discontinued and residual fatigue life tests were conducted. These experiments are referred to as 10%-of-life tests, since the fretting fatigue damage is removed at 10% of the expected fretting fatigue life. The 10%-of-life damage level was selected, in part, to allow comparison to work in the literature [3, 5, 11, 12], where significant damage progression up to 10% of life has been reported. Damage characterization was conducted through SEM (Scanning Electron Microscope) imaging and uniaxial residual fatigue life testing.

In previous work interrupted test specimens were re-machined into dogbone samples in the vicinity of the fretting damage. The residual fatigue life test results obtained from those samples did not indicate a reduction in fatigue life compared to undamaged samples, except in the case of features that could not be attributed to fretting damage [7]. SEM inspection of these fretting scars prior to re-machining indicated that the highest levels of wear damage occurred at the sample edge. Thus, the results of these tests provided information only on the location of the highest level of damage, which was outside of the region of the fretting scars after dogbone specimen machining.

In the current work, the region of highest wear damage was isolated by cutting straight sided (SS) specimens from the fretted regions, as shown in Figure 3. The resulting straight sided sample dimensions were 2 mm X 2.5 mm X 100 mm. To minimize grip failures, the ends of the specimen were covered with 0.127 mm (5 mil) thick copper tabs. Samples were tested at 950 MPa maximum stress at R = 0.5. Several baseline tests were also conducted on unfretted specimens to compare against the fretted specimen data.

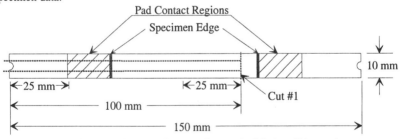

Figure 3 — *Machining schematic for residual fatigue life specimens.*

Characterization

Specimens were characterized in the SEM after the application of fretting damage to provide qualitative information on the level of wear generated in each test. Uniaxial fatigue fracture surfaces were also inspected in the SEM after fracture. Crack nucleation sites and causes were documented for all samples that fractured in the gage section. For specimens that failed in the gage, but away from the fretting damage, the distance between the fracture surface and the fretting scar was measured.

The fretting fatigue damage was also characterized with a metallographic cross-section of a step load sample and with representative spectral analysis of the fretting debris. The section was taken from the unfailed end of a failed sample bearing non-failure cracks. Similar sections taken from samples subjected to fretting fatigue under Hertzian or punch-on-flat contact conditions have been reported in the literature [2, 4, 5]. Those authors reported the presence of numerous non-propagating cracks and primary crack nucleation angles of approximately 45°. A similar cross-section from the current test geometry was cut to acquire information on the crack front location at the point at which the section was taken and to allow future comparison to other geometries. Spectral analysis of the fretting debris was conducted using Energy Dispersive Spectrometry (EDS) to identify elemental differences between fretting debris and the base material.

Experimental Results

Step Load Test Results

The results of the step load tests are presented in Figure 4, which shows the estimated fatigue limit for 10^7 cycles versus average applied clamping stress. Some data are included from previous work conducted at R = 0.5 on specimens having the same contact radius, 3.2 mm, as that used here [6]. At 200 MPa average clamping stress the average fatigue strength was ~ 330 MPa; at 620 MPa the average fatigue strength was ~ 250 MPa. Error bars representing two standard deviations are included. A subtle increase in fatigue limit was observed as a function of decreasing clamping stress in these results, although relatively large amounts of scatter tend to obscure the trend.

N_{DET} Results

Three tests were conducted at the 200 MPa clamping stress, in which the intent was to interrupt the tests based on crack detection using shear wave Non-Destructive Evaluation (NDE) technique. None of the attempts to detect cracks on specimens subjected to this clamping stress were successful. Table 1 lists cycles to failure as well as the cycle count of the last shear wave data acquisition. One of the tests at the 200 MPa clamping stress (#99-B57) was stopped for inspection based on changes in the waveform that were thought to be indications of crack nucleation at ~11 million cycles. However, careful inspection in the SEM did not reveal any cracks. If the shear wave NDE results for the other two tests are accurate - that no cracks were present at the last cycle count for which shear wave data were acquired - it can be concluded that less than 4 percent of life was spent in crack propagation. This conclusion is a direct contradiction of the findings from earlier researchers [3, 5, 13] that indicate the presence of cracks very early in life.

Table 1 – *N_{DET} test results for 200 MPa clamping stress*

SPECIMEN	N (LAST SW SIGNAL)	CYCLES TO FAILURE
99-B56	14 110 000	14 628 924
99-B57	10 900 000	>11 000 000
99-B59	3 000 000	3 099 251

Cracks were successfully detected in two of the three tests conducted at the 620 MPa clamping stress (#s 99-B60 and 99-B61). Results for these tests are given in Table 2. The detected cracks were fairly large, so that the time to propagate the detected cracks to failure was very small compared to the total specimen life. It should be noted that application of this inspection technique has not previously been documented for fretting fatigue. The viability of detecting cracks under these conditions was unknown, so the detection of any cracks was considered a success. Ultimately, the observation of new peaks in the waveforms lead to the identification of cracks in 99-B60 and 99-B61. While these results do not provide adequate information on the crack nucleation life for the 620 MPa clamping stress condition, they do support the feasibility of using the shear wave technique *in situ* in the high frequency contact laboratory environment. Feasibility of using the shear wave ultrasonic technique was demonstrated despite the significant amount of acoustic and electro-magnetic noise generated by the high frequency mechanical fatigue apparatus.

Table 2 – N_{DET} test results for 620 MPa clamping stress

SPECIMEN	CYCLES TO N_{DET}	CYCLES TO FAILURE	CRACK LENGTH ,mm
99-B60	7 019 649	7 349 240	2.3*, 0.90, 0.35
99-B61	6 125 123	6 227 169	2.6*, 0.50, 0.34, .025
99-B62	NA	4 126 650	NA

* indicates crack that ultimately produced failure

For the tests subjected to the lower clamping stress, total lives ranged from 3.1 million to 14.6 million cycles, thus bracketing the estimated 10^7 cycle fatigue life. At the higher clamping stress, the total lives ranged from 4.1 million to 7.3 million. For both conditions, the cycles-to-failure results recorded were within the expected level of scatter for such tests. Thus, these results may be used to substantiate the validity of the step loading technique for fretting fatigue conditions, a point not previously demonstrated for such complex loading.

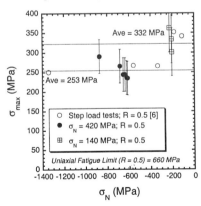

Figure 4 — Fatigue limit stress at R = 0.5 versus average clamping stress for a 10^7 cycle life.

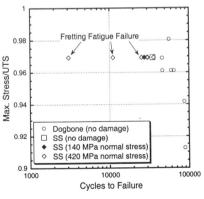

Figure 5 — S-N results for residual fatigue life specimens with 10%-of-life damage.

Residual Fatigue Life Results

Results of the residual fatigue life tests on straight sided specimens are shown in Table 3 and Figure 5. Table 3 lists cycles to failure and the clamping stress under which each fretting specimen was tested. In addition, the "failure location" column indicates failure in the grip, at the fretting damage, or the distance from the fretting damage that failure occurred. Figure 5 presents the residual fatigue life test results in S-N form along with baseline data described below.

Of the six residual fatigue life specimens whose damage was generated with 200 MPa clamping stress and 350 MPa axial fatigue stress, only one failed in the gage section when subjected to uniaxial fatigue. The resulting fracture occurred ~ 2.6 mm away from the line of fretting damage and, so was not a fretting fatigue related fracture. No appreciable reduction in fatigue life was observed (note "◆"in Figure 5). The remaining samples failed in the grip, and thus indicate that the fretting damage on those samples was less severe than the stress concentration produced by contact with the grip.

Table 3 – *Residual axial fatigue life test results at 950 MPa, R = 0.5.*

SPECIMEN	AVE. NORMAL STRESS, MPa	CYCLES TO FAILURE	FAILURE LOCATION
00-071	620	11 042	@ fretting
00-072	620	36 097	2.27 mm
00-073	620	30 343	86 μm
00-074	620	29 886	2.46 mm
00-075	620	3055	@ fretting
00-076	620	25 357	@ fretting
00-077	200	32 680	Grip
00-078	200	27 475	2.63 mm
00-081	200	21 794	Grip
00-083	200	12 563	Grip
00-084	200	16 282	Grip
00-262	0	35 037	-

Of the six specimens subjected to fretting fatigue damage at 620 MPa clamping stress and 260 MPa axial fatigue stress, all six failed in the gage section when subjected to uniaxial fatigue. (note "◇"symbol in Figure 5). One failed at approximately 3000 cycles at a sizeable fretting fatigue nucleated crack. Two more specimens failed in less than 25 000 cycles at fretting damage where no obvious cracks were present. The remaining three specimens failed away from the fretting damage in the gage section, and produced residual lives of 30 000 cycles or greater. The wear scars on specimens subjected to this clamping stress condition appeared nominally identical from the SEM imaging in the absence of fretting fatigue nucleated cracks.

Additional baseline data are provided in Figure 5 for comparison. These tests were conducted on as-received dogbone samples machined to the same surface condition as the fretting fatigue samples, and are represented by "O". One unfretted straight sided specimen was tested and fractured in the gage section, represented by "□". Only specimens that failed at the line of fretting damage showed a marked decrease in residual fatigue life compared to the baseline tests.

Characterization Results

Fretted surface characterization

Characteristic fretting scars from each clamping stress condition for both 100% and 10%-of-life tests are shown in Figures 6 through 9. All of the images were created using backscatter emission imaging in the SEM to enhance wear features. Wear was minimal in the 200 MPa samples (Figures 6 and 8) compared to the levels of wear observed in other test geometries [2, 3]. By comparison, almost no wear was observed in the 620 MPa samples (Figures 7 and 9), except where fretting fatigue nucleated cracks resulted in increased wear due to interaction of the crack and the edge of contact (Figure 10). Debris was spread over a larger area in the 200 MPa specimens than in the 620 MPa specimens. However, cracks of 1 mm or more in length were observed on all 620 MPa specimens at 10^6 cycles, indicating that cracks nucleate in less than 10% of life for this clamping stress condition. The crack nucleation life of specimens subjected to the lower clamping stress was indicated to be greater than 95% from the N_{DET} test results.

The presence of cracks at 10% of life in the specimens subjected to the higher clamping stress condition provides an explanation for the residual fatigue life results discussed above. If crack nucleation required most of the fretting fatigue life for the tests

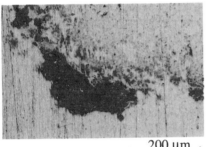

| 200 μm |

Figure 6 — *Wear debris for 100%-of-life at 200 MPa clamping stress. (Nf = 144 000 000 cycles)*

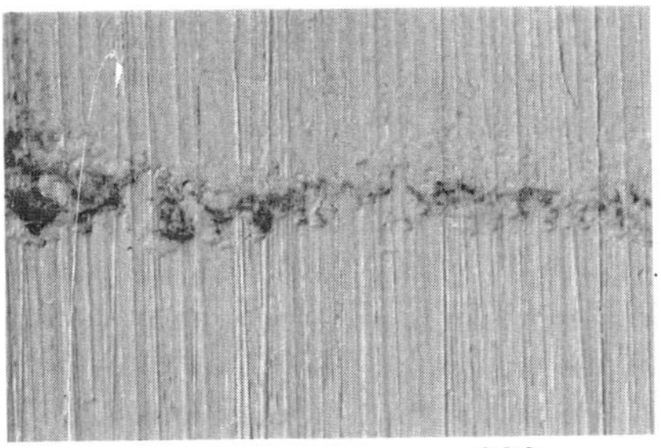

| 200 μm |

Figure 7 — *Non-crack wear debris for 100%-of-life at 620 MPa clamping stress. (Nf = 22 600 00 cycles)*

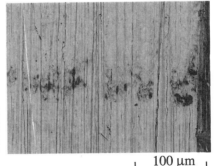

| 100 μm |

Figure 8 — *Wear debris for 10%-of-life at 200 MPa clamping stress.*

| 100 μm |

Figure 9 — *Non-crack wear debris for 10%-of-life at 620 MPa clamping stress.*

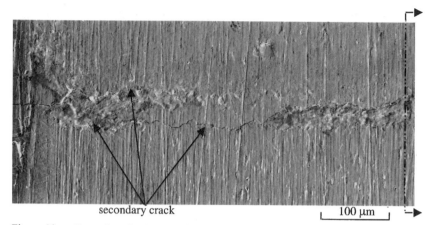

secondary crack   ∟ 100 μm ∣

Figure 10 — *Secondary fretting crack nucleated at 420 MPa clamping stress. (Nf ~ 4.1 million cycles)*

conducted at the lower clamping stress, then reduction in residual fatigue life after only 10% of the estimated fretting fatigue life would not be expected. However, if crack nucleation occurred early in life, as with the specimens tested at the higher clamping stress, the reduction in residual fatigue life is likely.

Results from characterization of the N_{DET} specimens, conducted prior to fracture, support these observations. On a given specimen, as many as eight crack nucleation sites are present (two for each pad/specimen contact region indicated by the numbers in Figure 2), and so simultaneous nucleation of cracks at two or more of these sites may occur. Many sizeable cracks were observed on the specimens subjected to the higher clamping stress, ranging in length from 200 μm to 1.6 mm as noted in Table 1. All of the cracks appeared to have nucleated near the specimen edges. By comparison, no cracks larger than 20 μm were observed on the specimens subjected to the lower clamping stress, and those cracks were contained entirely within the fretting debris. This observation is consistent with the results of the ultrasonic waveform data, which indicated no crack presence during testing. Previous work [14] has shown that cracks such as those observed on the specimens subjected to lower clamping stress were either non-propagating, or propagated in a manner that led to spalling of the wear products. Inspection of the specimens subjected to the lower clamping stress was performed after fracture, when interruption of the test prior to fracture was not achieved.

Fracture surface characterization

Representative fracture surfaces from 100%-of-life tests for both clamping stresses are shown in Figures 11 and 12. The micrographs show the nucleation regions of the fracture surfaces for 200 MPa and 620 MPa clamping stresses, respectively, and include the entire thickness of the specimen. Figures 13 and 14 give closer views of the estimated nucleation sites, indicated by the boxes in Figures 11 and 12. All four images were taken in backscatter mode, where darker regions represent different alloy phases, different levels in the fracture topography, or cracking parallel to the fatigue axis.

Inspection of both fracture surfaces reveals that the cracks nucleated from roughly the same region with respect to the edge of the specimen, and propagated into the specimen in the direction of the arrows (see Figures 11 and 12). The 200 MPa specimen fracture surface has a more planar appearance with no evidence that any cracks nucleated other than the dominant one that fractured the specimen. The 620 MPa specimen shows evidence of multi-plane cracking and chipping (note damage to the right of the box in Figure 12) resulting from the propagation of multiple cracks. The nucleation sites in Figures 11 and 12 are within 1 mm of the specimen edge.

Closer inspection of the primary nucleation site of the 200 MPa specimen also reveals a very planar fracture with no cracks parallel to the fatigue loading axis, pointing to the dominance of the bulk axial load on crack propagation for this contact configuration and loading condition. The magnified view of the 620 MPa specimen fracture surface shows evidence of a higher stress state and the multi-axial nature of that stress state in the form of cracking parallel to the fatigue axis.

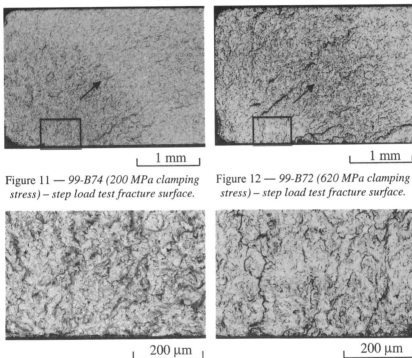

Figure 11 — *99-B74 (200 MPa clamping stress) – step load test fracture surface.*

Figure 12 — *99-B72 (620 MPa clamping stress) – step load test fracture surface.*

Figure 13 — *Magnified view of estimated primary crack nucleation site in Figure 11. (note region indicated by black box.)*

Figure 14 — *Magnified view of estimated primary crack nucleation site in Figure 12. (note region indicated by black box.)*

After residual fatigue life tests were completed on the 10%-of-life specimens, the resulting fracture surfaces were inspected in the SEM to determine if the fractures were fretting fatigue related. A representative fracture surface of a non-fretting fatigue gage section failure is shown in Figure 15. This image was taken with the sample tilted at a 45° angle and includes both the fretting scar and the fracture surface. In each non-fretting gage section failure an unintentional notch was identified as the cause of crack nucleation. Fretting damage was a sufficient distance from the crack nucleation site, as noted in Table 3, that fretting could not have played a role in the specimen fracture, except in one case. One specimen fractured under the pad contact area so that a conclusive cause of crack nucleation could not be determined. The fracture surface for this specimen is shown at a 45° angle in Figure 16. A notch was present at the crack nucleation site, but it occurred under the contact region. Since the source of the damage was unknown, the role of the contact in crack nucleation could not be determined, although the fatigue life of this specimen was not reduced appreciably.

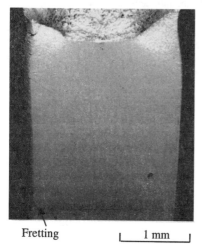

Fretting   1 mm

Figure 15 — *Residual fatigue life sample with fracture surface and fretting scar shown simultaneously at 45° to the fatigue loading axis. (Nf = 36,097 cycles)*

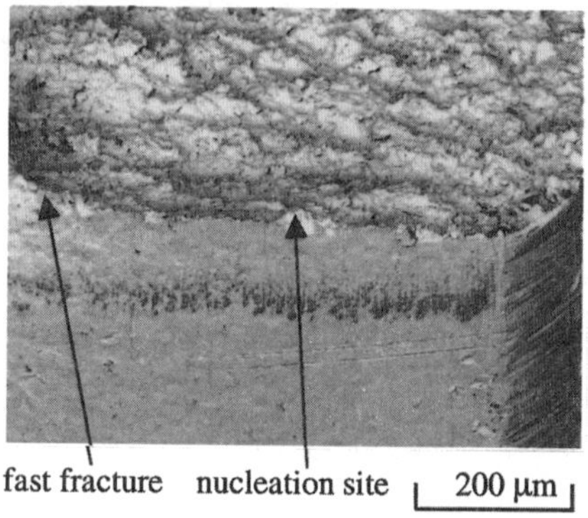

fast fracture   nucleation site   200 μm

Figure 16 — *Residual fatigue life sample with fracture surface shown at 45° to the fatigue loading axis. (Nf = 30,343 cycles – indeterminate failure)*

Two of the specimens failed at the fretting edge of contact without the presence of identifiable fretting nucleated cracking. A representative photo of one of the fracture surfaces is shown in Figure 17. In both cases crack nucleation occurred within the line of fretting fatigue damage, and without the presence of unintentional damage sites from

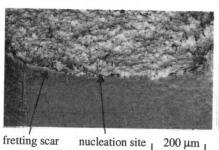

fretting scar   nucleation site   200 μm

Figure 17 — *Residual fatigue life sample with fracture surface shown at 45° to the fatigue loading axis. (Nf = 11,042 cycles – fretting nucleated failure)*

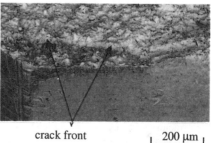

crack front   200 μm

Figure 18 — *Residual fatigue life sample with fracture surface shown at 45° to the fatigue loading axis. (Nf = 3,055 cycles – failure from fretting nucleated crack)*

other sources. The fretting "damage" that resulted in these failures was not noticeably different from the "damage" observed on specimens where fracture occurred away from the fretting site.

The last specimen failed at the fretting damage in the presence of a crack 400 μm long by 160 μm deep. The 45° view of the fracture surface is shown in Figure 18. Since the presence of the crack was known prior to uniaxial fatigue testing, the specimen was heat tinted to mark the location of the crack front. The crack was nearly elliptical, and nucleated within 100 μm of the specimen edge. This fracture surface supports the hypothesis of crack nucleation location near, but not at the edge of the specimen.

One unfretted specimen was also tested for comparison purposes. Crack nucleation was caused by surface damage perpendicular to the fatigue loading axis. While the fatigue lives of the straight samples did not differ in a statistically significant manner from the baseline tests on dogbone samples, some sort of stress riser was required to achieve a gage section failure. This point suggests that damage at 10% of fretting fatigue life obtained under 200 MPa clamping stress is less severe than a small damage site, such as might be incurred during machining or installation of the specimen into the test frame. Fretting damage obtained at 620 MPa clamping stress may be more severe than the types of incidental damage described here, but is not necessarily so.

Fretting debris characterization

The fretting debris from an un-cracked fretting scar was characterized for elemental content. The 200 MPa clamping stress condition was selected, since a greater volume of wear debris was available and the expected anomalous elements were small. Figure 19 shows a portion of a spectral analysis chart for the unfretted material. Only titanium, aluminum, and vanadium are present in significant quantities. Carbon and oxygen are marked on the graph, but the amount present with respect to the primary alloy constituents is negligible. A representative scan of the fretting debris is shown in Figure 20. Titanium, aluminum and vanadium are still the primary constituents, but the oxygen and carbon peaks are more pronounced, and can no longer be described as noise. Note that the primary titanium peak that occurs at ~4.5 keV and the vanadium peak that occurs at ~4.9 keV are not shown in Figures 19 and 20, but were present in the original scans at counts of ~500 and 100, respectively.

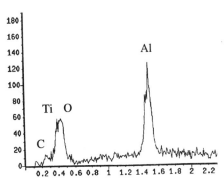 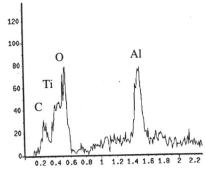

Figure 19 — *As-received material element analysis results.*   Figure 20 — *Fretted material analysis results.*

Oxidation of the original material elements has been linked to premature crack nucleation in both fretting fatigue and fretting wear in other materials [13, 15]. So the presence of significant levels of oxygen, as found here, presented a possible mechanism for crack nucleation in the specimens tested at the 200 MPa clamping stress. Brittle reaction products such as TiO_2 or Al_2O_3 could present ideal crack nucleation sites that would allow crack propagation into the bulk material, if the oxide were fixed firmly to the substrate. For each of the 200 MPa samples inspected in the SEM, the oxide was fixed firmly enough to the base material that it was left undisturbed by rigorous ultrasonic cleaning. This nucleation mechanism is not applicable to the higher clamping stress condition, based on the observed volume of wear debris in the absence of cracks.

A metallographic cross-section was cut in the region of a secondary crack on the unfractured end of a fractured specimen tested under the 620 MPa clamping stress condition. The crack extended ~ 1 mm from the specimen edge. Two nucleation sites were identified by the presence of relatively large amounts of fretting debris, and the crack appeared to propagate outward from both sites along a complex propagation path (Figure 10). The crack was sectioned to approximately 600 μm from the specimen edge near the second nucleation site (Figure 10 – note section line at right of photo).

The depth of the crack at the section location was ~ 500 μm. As shown in Figure 21, the dominant crack nucleated at ~ 45° to the loading axis, as reported by other researchers [4], and within 5-10 μm, turned perpendicular to the loading axis. Both propagating and non-propagating cracks are observed. In this case, very little material was removed from this part of the specimen face during the crack nucleation process. Also, the crack propagated across the α+β lath before turning perpendicular to the fatigue loading axis, indicating that the early stages of crack propagation under this contact condition are not heavily influence by environmental conditions such as oxidation or corrosion, as has been reported [13, 15]. Since no secondary cracking was observed on specimens subjected to 200 MPa clamping stress, the role of environment on fretting fatigue nucleated cracks could not be verified.

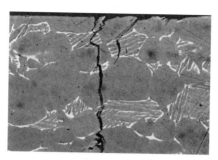

Figure 21 — *Cross-section of secondary fretting crack ~600 μm from sample edge.*

Summary and Conclusions

Taken together, the results from the different aspects of this investigation provide new insight into the fretting fatigue behavior for the current contact geometry. Using two clamping stress conditions representing the approximate range of normal stresses to which a turbine engine blade attachment region is subject, two types of crack nucleation behavior were identified and determined to be dependent on the applied clamping stress.

The number of cracks present on a single specimen was found to be dependent on the clamping stress. The higher clamping stress produced multiple cracks, of which one would eventually dominate. This trend in behavior was also reflected in the time required for nucleation of cracks for the two clamping stress conditions. The 200 MPa condition spent over 95% of life nucleating cracks, while the 620 MPa condition nucleated cracks within 10% of fretting fatigue life.

From this difference in crack nucleation behavior, the residual fatigue life results may be explained. If damage in the form of cracks can occur within 10% of life, as in the case of the high clamping stress condition, then a debit in fatigue life is expected for samples subjected to 10% of fretting fatigue life. However, the fretting fatigue limit stresses for the two clamping stress conditions were similar to produce a fatigue life of 10^7 cycles.

Some behavior was not dependent on the clamping stress. Cracks nucleated in the region near specimen edges, between 100 μm and 1 mm of the edge, regardless of the applied normal stress. This trend is likely a function of the contact geometry and the manner of load application. While such a trend may appear to be a deficiency in the apparatus design, it gives some indication of what mechanisms are responsible for the observed behavior. A general observation from the experiments performed here and in prior work is that all contact related behavior is dependent on pad, specimen and fixturing geometry.

In addition to the observations made regarding the objectives of this study, the step loading technique, which was developed to rapidly determine fatigue limit stresses for individual specimens, was verified for fretting fatigue conditions.

References

[1]   Ruiz, C., Boddington, P. H. B., and Chen, K. C., "An Investigation of Fatigue and Fretting in a Dovetail Joint," *Experimental Mechanics*, Vol. 24, 1984, pp. 208-217.

[2]   Bryggman, U. and Söderberg, S., "Contact Conditions and Surface Degradation Mechanisms in Low Amplitude Fretting", *Wear*, Vol. 125, 1988, pp. 39-52.

[3]   *Fretting Fatigue, ESIS 18*, R.B. Waterhouse and T.C. Lindley, Eds., Mechanical Engineering Publications, London, 1994, pp. 219-238.

[4]   Zhou, Z. R. and Vincent, L., "Mixed Fretting Regime," *Wear*, Vol. 181-183, 1995, pp. 551-536.

[5]   *Standardization of Fretting Fatigue Test Methods and Equipment, ASTM STP 1159*, M. Helmi Attia, and R. B. Waterhouse, Eds., American Society for Testing and Materials, Philadelphia, PA, 1992, pp. 153-169.

[6]   Hutson, A., Nicholas, T., and Goodman, R., "Fretting Fatigue of Ti-6Al-4V Under Flat-on-Flat Contact," *International Journal of Fatigue*, Vol. 21, No 7, 1999, pp. 663 – 670.

[7]   Hutson, A. and Nicholas, T., "Fretting Fatigue Behavior of Ti-6Al-4V against Ti-

6Al-4V under Flat-on-Flat Contact with Blending Radii," *Fretting Fatigue: Current Technologies and Practices, ASTM STP 1367,* D.W. Hoeppner, V. Chandrasekaran, and C.B. Elliot, Eds., American Society for Testing and Materials, West Conshohocken, PA, 1999, pp. 308 - 321.

[8]   Hutson, A.L., "Fretting Fatigue of Ti-6Al-4V Under Flat-on-Flat Contact with Blending Radii," M.S. Thesis, School of Engineering, University of Dayton, Dayton, OH, August, 2000.

[9]   Haritos, G.K., Nicholas, T. and Lanning, D., "Notch Size Effects in HCF Behavior of Ti-6Al-4V," *International Journal of Fatigue,* Vol. 21, No 7, 1999, pp. 643 – 652.

[10]   Maxwell, D.C. and Nicholas, T., "A Rapid Method for Generation of a Haigh Diagram for High Cycle Fatigue," *Fatigue and Fracture Mechanics: 29th Volume, ASTM STP 1321,* T. L. Panontin and S.D. Sheppard, Eds., American Society for Testing and Materials, 1999, pp. 626-641.

[11]   Endo, K. and Goto, H., "Initiation and Propagation of Fretting Fatigue Cracks," *Wear,* Vol. 38, 1976, pp. 311-324.

[12]   Nix, K.J. and Lindley, T.C., "The Application of Fracture Mechanics to Fretting Fatigue," *Fatigue and Fracture of Engineering Materials Structures,* Vol. 8, No. 2, 1985, pp. 143-160.

[13]   Blanchard, P., Colombie, C., Pellerin, V., Fayeulle, S. and Vincent, L. "Material Effects in Fretting Wear - Application to Iron, Titanium, and Aluminum Alloys," *Metallurgical Transactions,* Vol. 22A, 1991, pp. 1535-1544.

[14]   Fayeulle, S., Blanchard, P., and Vincent, L., "Fretting Behavior of Titanium Alloys," *Tribology Transactions,* Vol. 36, No. 2, 1993, pp. 267-275.

[15]   Waterhouse, R.B., "Fretting Wear," *Wear,* Vol. 100, 1984, pp. 107-118.

Kazuhisa Miyoshi,[1] Bradley A. Lerch,[1] Susan L. Draper,[1] and Sai V. Raj[1]

Evaluation of Ti-48Al-2Cr-2Nb Under Fretting Conditions

Reference: Miyoshi, K., Lerch, B. A., Draper, S. L., and Raj, S. V., **"Evaluation of Ti-48Al-2Cr-2Nb Under Fretting Conditions,"** *Fretting Fatigue: Advances in Basic Understanding and Applications, ASTM STP 1425,* Y. Mutoh, S. E. Kinyon, and D. W. Hoeppner Eds., ASTM International, West Conshohocken, PA, 2003.

Abstract: The fretting behavior of Ti-48Al-2Cr-2Nb[2] (γ-TiAl) in contact with a typical nickel-base superalloy was examined in air at temperatures from 296 to 823 K (23 to 550 °C). The interfacial adhesive bonds between Ti-48Al-2Cr-2Nb and superalloy were generally stronger than the cohesive bonds within Ti-48Al-2Cr-2Nb. The failed Ti-48Al-2Cr-2Nb debris subsequently transferred to the superalloy. In reference experiments conducted with Ti-6Al-4V against superalloy under identical fretting conditions, the degree of transfer was greater for Ti-6Al-4V than for Ti-48Al-2Cr-2Nb. Wear of Ti-48Al-2Cr-2Nb generally decreased with increasing fretting frequency. The increasing rate of oxidation at elevated temperatures led to a drop in wear at 473 K. However, fretting wear increased as the temperature was increased from 473 to 823 K. At 723 and 823 K, oxide film disruption generated cracks, loose wear debris, and pits on the Ti-48Al-2Cr-2Nb wear surface. Both increasing slip amplitude and increasing load tended to produce more metallic wear debris, causing severe abrasive wear in the contacting metals.

Keywords: Ti-48Al-2Cr-2Nb, fretting wear, fatigue, oxidation, high temperature, Ni-base superalloy

[1]National Aeronautics and Space Administration, Glenn Research Center, Cleveland, Ohio 44135-3191.
[2]Composition, at.%: titanium, 47.9; aluminum, 48.0; niobium, 1.96; chromium, 1.94; carbon, 0.013; nitrogen, 0.014; and oxygen, 0.167.

323

Introduction

Adhesion, a manifestation of mechanical strength over an appreciable area, has many causes, including chemical bonding, deformation, and the fracture processes involved in interface failure. A clean metal in contact with another clean metal will fail either in tension or in shear because some of the interfacial bonds are generally stronger than the cohesive bonds within the cohesively weaker metal [1]. The failed metal subsequently transfers material to the other contacting metal. Adhesion undoubtedly depends on the surface cleanliness; the area of real contact; the chemical, physical, and mechanical properties of the interface; and the modes of junction rupture. The environment influences the adhesion, deformation, and fracture behaviors of contacting materials in relative motion.

Clean surfaces can be created by repeated sliding in vacuum, making direct contact of the fresh, clean surfaces unavoidable in practical cases [2]. This situation also applies in some degree to sliding contact in air, where fresh surfaces are continuously produced on interacting surfaces in relative motion. Microscopically small, surface-parallel relative motion, which can be vibratory (in fretting or false brinnelling) or creeping (in fretting), produces fresh, clean interacting surfaces and causes junction (contact area) growth in the contact zone [3–5].

Fretting wear produced between contacting elements is adhesive wear taking place in a nominally static contact under normal load and repeated microscopic vibratory motion [6–10]. The most damaging effect of fretting is the possibly significant reduction in the fatigue capability of the fretted component, even though the wear produced by fretting appears to be quite mild [10]. For example, Hansson, et al. reported that the reduction in fatigue strength by fretting of Ti-47Al-2Nb-2Mn containing 0.8 vol.% TiB_2 was approximately 20 percent.

Fretting fatigue is a complex problem of significant interest to aircraft engine manufacturers [11–14]. Fretting failure can occur in a variety of engine components. Numerous approaches, depending on the component and the operating conditions, have been taken to address the fretting problem. The components of interest in this investigation were the low-pressure turbine blades and disks. The blades in this case were titanium aluminide and the disk was a nickel-base superalloy. A concern for these airfoils is the fretting in fitted interfaces at the dovetail where the blade and disk are connected. Careful design can reduce fretting in most cases, but not completely eliminate it, because the airfoils frequently have a skewed (angled) blade-disk dovetail attachment, which leads to a complex stress state. Further, the local stress state becomes more complex when the influence of the metal-metal contact and the edge of contact is evaluated.

Because titanium and titanium-base alloys in the clean state will exhibit strong adhesive bonds [2, 15] when in contact with themselves and other materials, this adhesion causes heavy surface damage and high friction in practical cases. Therefore, it is possible that fretting will be a serious concern in this application.

The objective of this investigation was to evaluate the extent of fretting damage on Ti-48Al-2Cr-2Nb (γ-TiAl) in contact with a typical nickel-base superalloy at temperatures from 296 to 823 K. Selected reference experiments were also conducted with Ti-6Al-4V. There is a large experience base with Ti-6Al-4V, which has been used extensively as a compressor blade material. The parameters of microscopic, surface-parallel motion, such as fretting frequency, slip amplitude, and load, were systematically examined in this study. Scanning interference microscopy (noncontact optical profilometry) was used to evaluate surface characteristics, such as topography, roughness, material transfer, and wear volume loss. Scanning electron microscopy with energy-dispersive spectroscopy was used to determine the morphology and elemental composition of fretted surfaces, transferred material, and wear debris.

Materials

The Ti-48Al-2Cr-2Nb specimens were determined to be of the following composition (in atomic percent): titanium, 47.9; aluminum, 48.0; niobium, 1.96; chromium, 1.94; carbon, 0.013; nitrogen, 0.014; and oxygen, 0.167. The tensile properties are shown in Table 1.

Table 1—*Tensile Properties of Ti-48Al-2Cr-2Nb*

Temperature, K	Modulus, GPa	Ultimate tensile strength, MPa
293	170	410
923	140	460

The nickel-base superalloy was solutioned and aged according to Aerospace Material Specification AMS 5596G, SAE, Warrendale, PA, 1987, yielding Rockwell C-scale hardness H_{RC} of 36. The tensile properties [16] are shown in Table 2. The ultimate tensile strength of superalloy is greater than that of Ti-48Al-2Cr-2Nb by a factor of ~3.5 at room temperature and ~2 at high temperature (~1000 K).

Table 2—*Tensile Properties of Nickel-Base Superalloy*

Temperature, K	Modulus, GPa	Ultimate tensile strength, MPa
293	200	1434
811	171	1276
1033	154	758

The reference Ti-6Al-4V specimens were of the following nominal composition (in weight percent): titanium, balance; aluminum, 5.5–6.75; vanadium, 3.5–4.5; iron, ≤0.50; carbon, ≤0.08; nitrogen, ≤0.05; oxygen, ≤0.20; and hydrogen, ≤0.015 [17].

Experiments

Figure 1 presents the fretting wear apparatus used in this investigation. Fretting wear experiments were conducted with 9.4-mm-diameter, hemispherical nickel-base superalloy pins in contact with Ti-48Al-2Cr-2Nb flats or with 6-mm-diameter, hemispherical Ti-48Al-2Cr-2Nb pins in contact with nickel-base superalloy flats in air at temperatures from 296 to 823 K. All the flat and pin specimens used were polished with 3-μm-diameter diamond powder. Both pin and flat surfaces were relatively smooth, having centerline-average roughness R_a in the range 18 to 83 nm (Table 3). The Vickers hardness, measured at a load of 1 N, for the polished flat and pin specimens is also shown in Table 3.

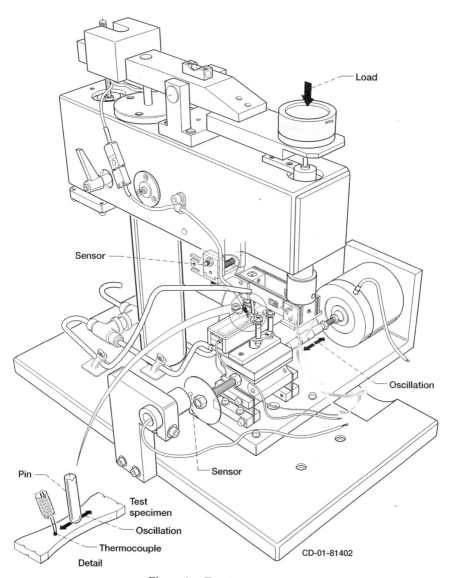

Figure 1—*Fretting apparatus.*

Table 3—*Surface Roughness and Vickers Hardness of Specimens*

Specimen	Centerline-average roughness, R_a, nm		Vickers hardness[a], H_V, GPa	
	Mean	Standard deviation	Mean	Standard deviation
9.4-mm-diameter, hemispherical, nickel-base superalloy pin	40	8.9	5.52	0.44
6-mm-diameter, hemispherical Ti-48Al-2Cr-2Nb pin	42	7.1	4.12	0.42
9.4-mm-diameter, hemispherical Ti-6Al-4V pin	83	2.0	3.85	0.092
Ti-48Al-2Cr-2Nb flat	35	3.3	3.78	0.57
Nickel-base superalloy flat	18	7.2	4.78	0.21

[a]Load, 1 N.

All fretting wear experiments were conducted at loads from 1 to 40 N, frequencies of 50, 80, 120, and 160 Hz, and slip amplitudes between 50 and 200 µm for 1 million to 20 million cycles. Both pin and flat surfaces were rinsed with 200-proof ethyl alcohol before installation in the fretting apparatus.

Two or three fretting experiments were conducted with each material couple at each fretting condition. The data were averaged to obtain the wear volume losses of Ti-48Al-2Cr-2Nb and Ti-6Al-4V. The wear volume loss was determined by using an optical profiler (noncontact, vertical scanning, white-light interferometer). It characterizes and quantifies surface roughness, height distribution, and critical dimensions (such as area and volume of damage, wear scars, and topographical features). It has three-dimensional profiling capability with excellent precision and accuracy (e.g., profile heights ranging from ≤1 nm up to 5000 µm with 0.1-nm height resolution). The shape of a surface can be displayed by a computer-generated map developed from digital data derived from a three-dimensional interferogram of the surface. A computer directly processes the quantitative volume and depth of a fretted wear scar. Reference fretting wear experiments were conducted with 9.4-mm-diameter hemispherical Ti-6Al-4V pins in contact with nickel-base superalloy flats.

Results and Discussion

Observations

Surface and subsurface damage always occurred on the interacting surfaces of the Ti-48Al-2Cr-2Nb fretted in air. The surface damage consisted of material transfer, pits, oxides and debris, scratches, fretting craters and/or wear scars, plastic deformation, and cracks.

Adhesion and Material Transfer—Figure 2 presents a backscattered electron image and an energy-dispersive x-ray spectrum (EDS) taken from the fretted surface of the nickel-base superalloy pin after contact with the Ti-48Al-2Cr-2Nb flat. Clearly, Ti-48Al-2Cr-2Nb transferred to superalloy. The Ti-48Al-2Cr-2Nb failed either in tension or in shear because some of the interfacial adhesive bonds (solid state or cold welding) were stronger than the cohesive bonds within the cohesively weaker Ti-48Al-2Cr-2Nb.

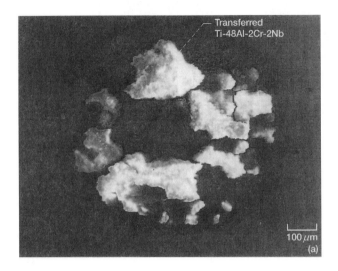

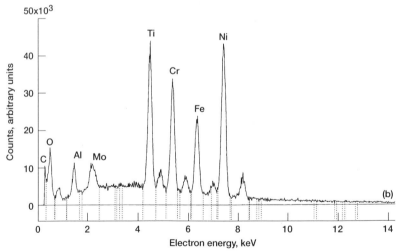

Figure 2—*Wear scar on superalloy pin fretted against Ti-48Al-2Cr-2Nb flat. (a) SEM backscattered electron image. (b) X-ray energy spectrum with EDS. Fretting conditions: load, 1.5 N; frequency, 80 Hz; slip amplitude, 50 μm; total number of cycles, 1 million; environment, air; and temperature, 823 K.*

The ultimate tensile strength of superalloy is greater than that of Ti-48Al-2Cr-2Nb by a factor of ~3.5 at room temperature and ~2 at high temperature (~1000 K). The failed Ti-48Al-2Cr-2Nb subsequently transferred to the superalloy surface in amounts ranging from 10 to 60 percent of the superalloy contact area at all fretting conditions in this study. The thickness of the transferred Ti-48Al-2Cr-2Nb ranged up to ~20 µm.

As with the materials pair of Ti-48Al-2Cr-2Nb and superalloy, material transfer was observed on the superalloy flat surface after fretting against the Ti-6Al-4V pin at 696 and 823 K in air. However, the degree of material transfer was remarkably different and greater, ranging from 30 to 100 percent of the superalloy contact area for identical fretting conditions. The thickness ranged up to 50 µm.

Fretting Wear—Figure 3 shows typical wear scars produced on the Ti-48Al-2Cr-2Nb pin and the superalloy flat with fretting. Because of the specimen geometry a large amount of wear debris was deposited just outside the circular contact area. Pieces of the metals (both Ti-48Al-2Cr-2Nb and superalloy) and their oxides were torn out during fretting. It appears that the cohesive bonds in some of the contact area of both metals fractured. Scanning electron microscopy (SEM) and EDS studies of wear debris produced under fretting verified the presence of metallic particles of both Ti-48Al-2Cr-2Nb and superalloy. In the central region of wear scars produced on Ti-48Al-2Cr-2Nb there was generally a large, shallow pit, where Ti-48Al-2Cr-2Nb had torn out or sheared off and subsequently transferred to superalloy. The central regions of wear scars produced on Ti-48Al-2Cr-2Nb and on superalloy were morphologically similar (Fig. 3), generally having wear debris, scratches, plastically deformed asperities, and cracks.

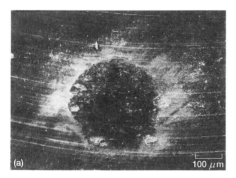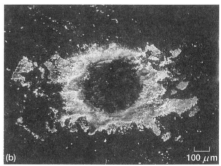

Figure 3—*Wear scars (a) on Ti-48Al-2Cr-2Nb pin and (b) on superalloy flat. Fretting conditions: load, 1 N; frequency, 80 Hz; slip amplitude, 50 µm; total number of cycles, 1 million; environment, air; and temperature, 823 K.*

Figure 4 shows examples of surface damage: metallic wear debris of Ti-48Al-2Cr-2Nb and superalloy, oxides and their debris, scratches (grooves), small craters, plastically deformed asperities, and cracks. The scratches (Fig. 4(a)) can be caused by hard protuberances (asperities) on the superalloy surface (two-body conditions) or by wear particles between the surfaces (three-body conditions). Abrasion is a severe form of wear. The hard asperities and trapped wear particles plow or cut the Ti-48Al-2Cr-2Nb surface. The trapped wear particles have a scratching effect on both surfaces; and because they carry

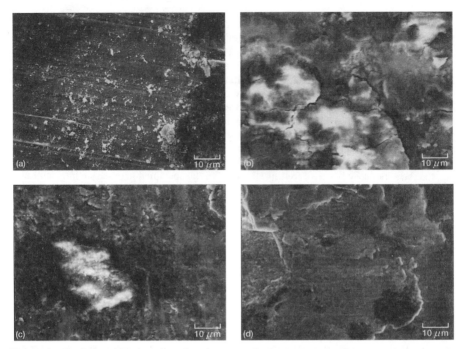

Figure 4—*Surface and subsurface damage in Ti-48Al-2Cr-2Nb flat in contact with super-alloy pin. (a) Scratches. (b) Cracks in oxide layers. (c) Cracks in metal. (d) Fracture pits and plastic deformation. Fretting conditions: load, 1 N; frequency, 80 Hz; slip amplitude and total number of cycles: (a) 50 µm and 1 million, (b) 60 µm and 10 million, (c) 50 µm and 1 million, (d) 70 µm and 20 million; environment, air; and temperature, 823 K.*

part of the load, they cause concentrated pressure peaks on both surfaces. The pressure peaks may well be the origin of crack nucleation in the oxide layers and the bulk alloys. Two types of crack were observed on the wear surface of Ti-48Al-2Cr-2Nb: cracks in the oxide layers, and cracks in the bulk Ti-48Al-2Cr-2Nb.

Oxide layers readily form on the Ti-48Al-2Cr-2Nb surface at 823 K and are often a favorable solution to wear problems. However, if the bulk Ti-48Al-2Cr-2Nb is not hard enough to carry the load, it will deform plastically or elastically under fretting contact. With Ti-48Al-2Cr-2Nb, cracks occurred in the oxide layers both within and around the contact areas (Fig. 4(b)).

Fractures in the protective oxide layers produced cracks in the bulk Ti-48Al-2Cr-2Nb (Fig. 4(c)) and also produced wear debris; chemically active, fresh surfaces; plastic deformation; and craters or fracture pits (Fig. 4(d)). The wear debris caused third-body abrasive wear (Fig. 4(a)). Local, direct contacts between the fresh surfaces of Ti-48Al-2Cr-2Nb and superalloy resulted in increased adhesion and local stresses, which may cause plastic deformation, flake-like wear debris, and craters (e.g., the fracture pits in the Ti-48Al-2Cr-2Nb shown in Fig. 4(d)).

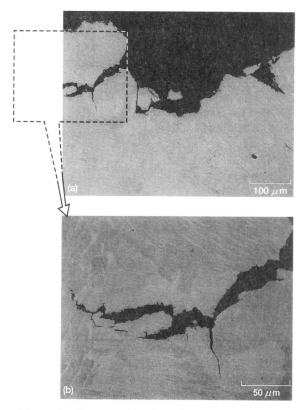

Figure 5—*Cross-section view of wear scar on
Ti-48Al-2Cr-2Nb flat in contact with superalloy pin.
(a) Overview. (b) Crack growth. Fretting conditions:
load, 30 N; frequency, 80 Hz; slip amplitude, 70 μm;
total number of cycles, 20 million; environment, air;
and temperature, 823 K.*

Cross sections of a wear scar on Ti-48Al-2Cr-2Nb revealed subsurface cracking and craters. For example, Fig. 5 shows propagation of subsurface cracking, nucleation of small cracks, formation of a large crater, and generation of debris. Cracks are transgranular and have no preference to the microstructure.

Parameters Influencing Wear Loss of Ti-48Al-2Cr-2Nb

Figure 6 shows the wear volume loss measured by the optical interferometer as a function of fretting frequency for Ti-48Al-2Cr-2Nb in contact with superalloy. Although there were some exceptions, the wear volume loss generally decreased with increasing

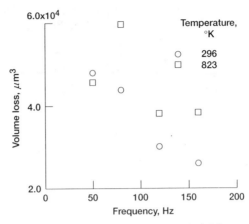

Figure 6—*Wear volume loss of Ti-48Al-2Cr-2Nb flat in contact with superalloy pin in air as function of fretting frequency. Fretting conditions: load, 30 N; slip amplitude, 50 μm; total number of cycles, 1 million; environment, air; and temperatures, 296 and 823 K.*

fretting frequency. A reasonable amount of material transfer from the Ti-48Al-2Cr-2Nb specimen to the superalloy specimen was observed at all frequencies. At the lowest frequency of 50 Hz remarkable plastic deformation (grooving) and surface roughening in the Ti-48Al-2Cr-2Nb wear scar were observed. At high frequencies wear scars were noticeably smooth with bulk cracks in the Ti-48Al-2Cr-2Nb surface.

Temperature influences the adhesion, deformation, and fracture behaviors of contacting materials in relative motion. It is known that temperature interacts with the fretting process in two ways: first, the rate of oxidation or corrosion increases with temperature; and second, the mechanical properties, such as hardness, of the materials are also temperature dependent [9]. Figure 7 presents the wear volume loss measured by optical interferometry as a function of temperature for Ti-48Al-2Cr-2Nb in contact with superalloy. Also, SEM images and EDS spectra were taken from the fretted Ti-48Al-2Cr-2Nb surfaces. The wear volume loss dropped to a low value at 473 K. The worn surface at 473 K was predominantly oxide and relatively smooth. A protective oxide film prevented direct metal-to-metal contact and ensured, in effect, that a mild oxidative wear regime prevailed. However, fretting wear increased as the temperature was increased from 473 to 823 K. The highest temperatures of 723 and 823 K resulted in oxide film disruption with crack generation, loose wear debris, and pitting of the Ti-48Al-2Cr-2Nb wear surface.

Figure 8 shows the volume loss measured by optical interferometry as a function of slip amplitude for Ti-48Al-2Cr-2Nb in contact with superalloy. The fretting wear volume loss increased as the slip amplitude increased. Increases in amplitude tended to produce

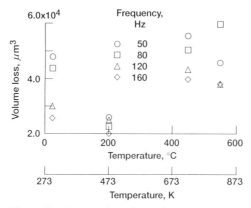

Figure 7—*Wear volume loss of Ti-48Al-2Cr-2Nb flat in contact with superalloy pin in air as function of fretting temperature. Fretting conditions: load, 30 N; slip amplitude, 50 μm; total number of cycles, 1 million; environment, air; and fretting frequencies, 50, 80, 120, and 160 Hz.*

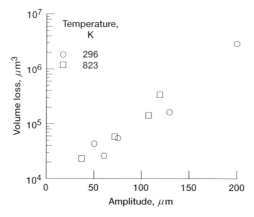

Figure 8—*Wear volume loss of Ti-48Al-2Cr-2Nb flat in contact with superalloy pin in air as function of slip amplitude. Fretting conditions: load, 30 N; frequency, 50 Hz; total number of cycles, 1 million; environment, air; and temperatures, 296 and 823 K.*

more metallic wear debris, causing severe abrasive wear in the contacting metals. Figure 9 presents a three-dimensional, optical interferometry image of the Ti-48Al-2Cr-2Nb wear scar at a slip amplitude of 200 μm and a temperature of 296 K. In the wear scar are large, deep grooves where the wear debris particles have scratched the Ti-48Al-2Cr-2Nb surface in the slip direction.

Figure 10 shows the measured wear volume loss as a function of load for Ti-48Al-2Cr-2Nb in contact with superalloy at a temperature of 823 K, a fretting frequency of 80 Hz, and a slip amplitude of 50 μm for 1 million cycles. The fretting wear volume loss generally increased as the load increased, generating more metallic wear debris in the contact area, the primary cause of abrasive wear in both Ti-48Al-2Cr-2Nb and superalloy.

Figure 9—*Wear scar on Ti-48Al-2Cr-2Nb flat in contact with superalloy pin, showing scratches. Fretting conditions: load, 30 N; frequency, 50 Hz; slip amplitude, 200 μm; total number of cycles, 1 million; environment, air; and temperature, 296 K.*

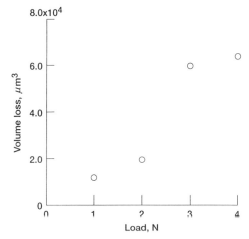

Figure 10—*Wear volume loss of Ti-48Al-2Cr-2Nb flat in contact with superalloy pin as function of load. Fretting conditions: frequency, 80 Hz; slip amplitude, 50 μm; total number of cycles, 1 million; environment, air; and temperature, 823 K.*

Concluding Remarks

The fretting behavior of γ-TiAl (Ti-48Al-2Cr-2Nb) in contact with nickel-base superalloy in air at temperatures of 296 to 823 K was examined with the following results:
1. The Ti-48Al-2Cr-2Nb transferred to the superalloy at all fretting conditions, such that from 10 to 50 percent of the superalloy contacting surface area became coated with the Ti-48Al-2Cr-2Nb. The maximum thickness of the transferred Ti-48Al-2Cr-2Nb was approximately 20 μm. In reference experiments Ti-6Al-4V transferred to superalloy under identical fretting conditions. Compared with Ti-48Al-2Cr-2Nb transfer, the degree of Ti-6Al-4V transfer was greater, such that from 30 to 100 percent of the superalloy contacting surface area became coated with the Ti-6Al-4V. The thickness of the transferred Ti-6Al-4V ranged up to 50 μm.
2. The wear scars produced on Ti-48Al-2Cr-2Nb contained metallic and oxide wear debris, scratches, plastically deformed asperities, cracks, and fracture pits.
3. Although oxide layers readily formed on the Ti-48Al-2Cr-2Nb surface at 823 K, cracking readily occurred in the oxide layers both within and around the contact areas.
4. The wear volume loss of Ti-48Al-2Cr-2Nb generally decreased with increasing fretting frequency, increased with increasing temperature, and increased with increasing slip amplitude.
5. Mild oxidative wear and low wear volume were observed at 473 K.

Acknowledgment

The authors are grateful to Dr. Gary R. Halford and Dr. Michael V. Nathal for very helpful discussions.

References

[1] Miyoshi, K., "Foreword: Considerations in Vacuum Tribology (Adhesion, Friction, Wear, and Solid Lubrication in Vacuum)," *Tribology International*, Vol. 32, 1999, pp. 605–616.

[2] Miyoshi, K., "Aerospace Mechanisms and Tribology Technology: Case Study," *Tribology International*, Vol. 32, 1999, pp. 673–685.

[3] Tallian, T. E., "Failure Atlas for Hertz Contact Machine Elements," *Fretting Wear*, ASME Press, New York, 1992, pp. 141–154.

[4] Waterhouse, R. B., "Plastic Deformation in Fretting Processes—a Review," *Fretting Fatigue: Current Technology and Practices, ASTM STP 1367*, D. W. Hoeppner, V. Chandrasekaran, and C. B. Elliot III, Eds., American Society for Testing and Materials, West Conshohocken, PA, 2000, pp. 3–18.

[5] Waterhouse, R. B., "Occurrence of Fretting in Practice and Its Simulation in the Laboratory," *Materials Evaluation Under Fretting Conditions, ASTM STP 780*, American Society for Testing and Materials, Philadelphia, PA, 1982, pp. 3–16.

[6] Kusner, D., Poon, C., and Hoeppner, D. W., "A New Machine for Studying Surface Damage due to Wear and Fretting," *Materials Evaluation Under Fretting Conditions, ASTM STP 780*, American Society for Testing and Materials, West Conshohocken, PA, 1982, pp. 17–29.

[7] Satoh, T., "Influence of Microstructure on Fretting Fatigue Behavior of a Near-Alpha Titanium," *Fretting Fatigue: Current Technology and Practices, ASTM STP 1367*, D. W. Hoeppner, V. Chandrasekaran, and C. B. Elliot III, Eds., American Society for Testing and Materials, West Conshohocken, PA, 2000, pp. 295–307.

[8] Lutynski, C., Simansky, G., and McEvily, A. J., "Fretting Fatigue of Ti-6Al-4V Alloy," *Materials Evaluation Under Fretting Conditions, ASTM STP 780*, American Society for Testing and Materials, West Conshohocken, PA, 1982, pp. 150–164.

[9] Bill, R. C., "Review of Factors That Influence Fretting Wear," *Materials Evaluation Under Fretting Conditions, ASTM STP 780*, American Society for Testing and Materials, West Conshohocken, PA, 1982, pp. 165–182.

[10] Hansson, T., et al., "High Temperature Fretting Fatigue Behavior in an XD γ-Base TiAl," *Fretting Fatigue: Current Technology and Practices, ASTM STP 1367*, D.W. Hoeppner, V. Chandrasekaran, and C. B. Elliot III, Eds., American Society for Testing and Materials, West Conshohocken, PA, 2000, pp. 65–79.

[11] VanStone, R. H., Lawless, B. H., and Hartle, M., "Fretting in Ti-6Al-4V at Room Temperature," *Proceedings of 5th National Turbine Engine High Cycle Fatigue (HCF) Conference*, 2000, Session 4, paper 1.

[12] Chakravarty, S., et al., "The Effect of Surface Modification on Fretting Fatigue in Ti Alloy Turbine Components," *JOM*, Vol. 47, No. 4, 1995, pp. 31–35.

[13] Hoeppner, D., Adibnazari, S., and Moesser, M. W., "Literature Review and Preliminary Studies of Fretting and Fretting Fatigue Including Special Applications to Aircraft Joints," DOT/FAA/CT–93/2, Defense Technical Information Center, Ft. Belvoir, VA, 1994.

[14] Hutson A. L. and Nicholas, T., "Fretting Fatigue Behavior of Ti-6Al-4V Against Ti-6Al-4V Under Flat-on-Flat Contact With Blending Radii," *Fretting Fatigue: Current Technology and Practices, ASTM STP 1367*, D. W. Hoeppner, V. Chandrasekaran, and C. B. Elliot III, Eds., American Society for Testing and Materials, West Conshohocken, PA, 2000, pp. 308–321.

[15] Miyoshi, K., et al., "Sliding Wear and Fretting Wear of Diamondlike Carbon-Based, Functionally Graded Nanocomposite Coatings," *Wear*, Vol. 225–229, 1999, pp. 65–73.

[16] Hunt, M. W. (Ed.), Materials Selector 1993, Materials Engineering, Penton Publishing, Cleveland, OH, 1992.

[17] Donachie, M. J. (Ed.), *Titanium and Titanium Alloys*, Source Book, American Society for Metals, Metals Park, OH, 1982.

Shankar Mall,[1]* Vinod K. Jain,[2] Shantanu A. Namjoshi,[3] Christopher D. Lykins[4]

Fretting Fatigue Crack Initiation Behavior of Ti-6Al-4V

Reference: Mall S., Jain V. K., Namjoshi S. A., and Lykins C. D., **"Fretting Fatigue Crack Initiation Behavior of Ti-6Al-4V,"** *Fretting Fatigue: Advances in Basic Understanding and Applications, STP 1425,* Y. Mutoh, S. E. Kinyon, and D. W. Hoeppner, Eds., ASTM International, West Conshohocken, PA, 2003.

Abstract: Fretting fatigue crack initiation behavior of titanium alloy, Ti-6Al-4V was investigated by using different geometries of the fretting pad. Specimens were examined to determine the crack initiation location and the crack angle orientation along the contact surface. Fretting fatigue experiments were analyzed with the finite element analysis. Several critical plane based parameters were used to predict the number of cycles to fretting fatigue crack initiation, crack initiation location and crack orientation angle along the contact surface. These predictions were compared with the experimental counterparts to study the role of normal and shear stresses on the critical plane in order to characterize fretting fatigue crack initiation behavior. From these comparisons, the fretting fatigue crack initiation mechanism in the tested titanium alloy appears to be governed by the shear stress on the critical plane. However, the role of normal stress on the critical plane at present appears to be unclear.

Keywords: fretting fatigue, titanium alloy, crack initiation, critical plane parameters

Introduction

Aircraft engine components are currently less prone to the failures associated with

---

[1] Air Force Research Laboratory, Materials and Manufacturing Directorate, Wright Patterson AFB, OH 45433-7817, *To whom correspondence be addressed at AFIT/ENY, Bldg. 640, 2950 P. St., Air Force Institute of Technology, Wright-Patterson AFB, OH, 45433-7765, email: Shankar.Mall@afit.edu.
[2] Mechanical and Aerospace Engineering Department, University of Dayton, Dayton, OH 45469-0210.
[3] Department of Aeronautics and Astronautics, Air Force Institute of Technology, Wright-Patterson Air Force Base, OH 45433-7765.
[4] Formerly with Air Force Research Laboratory, Aero Propulsion and Rockets Directorate, Wright Patterson AFB, OH 45433-7251.

low cycle fatigue (LCF) as a great deal of attention has been paid to the LCF area over the last two decades [1,2]. On the other hand, high cycle fatigue (HCF) has received much less attention and has resulted recently in higher incidence of failures related to HCF in many military aircraft engines. Under HCF conditions, engine components experience high frequency, vibrational type loads, often superimposed on a high mean stress. There is a lack of understanding of materials response and capability in the presence of in-service damage under HCF loading conditions. In the LCF arena, the "damage tolerant" approach has worked well where the remaining life is predicted from the crack propagation of an inspectable flaw size to a critical size. The direct extension of such an approach to HCF is a challenging task since a large fraction of life is spent during the crack nucleation and growth to a detectable size, and a very small fraction of life is spent in crack propagation from the detectable or inspectable flaw size to a critical size.

Fretting fatigue related failures in disk/blade attachments are one of the most insidious problems related to HCF in gas turbine engines. It is further exacerbated by the difficulty in detecting fretting fatigue crack initiation and propagation. Fretting is the process that occurs from the relative surface displacements, on the micro-scale (i.e. of the order of microns), at the interface between two mating components subjected to fatigue in the presence of normal pressure. Fretting acts as a damage generator that leads to crack nucleation, growth to a detectable size and propagation on the surface or subsurface level between two mating components, generally at a fraction of fatigue lives under the plain fatigue condition. In HCF failures, crack propagation does not constitute a large part of life and hence its analysis using a damage tolerant approach may not be applicable. In HCF failures a major part of the life is spent in 'crack initiation', which is the sum total of crack nucleation and growth to a detectable size. Thus, the long-term goal of this study is to develop an understanding of the high cycle fretting fatigue related failures in dovetail joints.

There have been several phenomenological life prediction models proposed [3-8], where fretting fatigue life predictions are made based on the life debit due to fretting (correlated using one or more test variables) as compared to plain fatigue. The presence of applied normal, tangential and axial loads during fretting causes a stress magnification at the edge of contact. Researchers have proposed fracture mechanics methods, where this stress intensity factor can be used to determine loading conditions that could lead to crack growth and arrest behavior [9,10,11] or the onset of fretting fatigue crack propagation [12]. However, in these methods, either an initial crack length and/or the crack orientation are assumed or an estimate of threshold (or maximum allowable value) of the flaw size that can be tolerated for infinite fretting fatigue life is provided. Currently, these approaches do not provide/predict the remaining life or life consumed to reach a certain initial crack length.

Two parameters, specifically developed for the fretting fatigue condition, are based on the work done by the frictional force between contacting bodies [13]. The first parameter, a measure of the frictional energy expenditure density, is defined as the maximum value of the product of the shear stress (τ) and the slip amplitude (δ) at the contact surface between two bodies. The second parameter; a modification to the first parameter recognizing the fact that crack nucleation in fretting fatigue can also depend on the maximum tangential stress, σ_T, on the interface; is defined as the maximum value of

$\sigma_T \tau \delta$. There has been mixed success in predicting the location, orientation, and initiation time for the fretting fatigue crack using this approach [14,15].

Researchers have also used characteristic descriptions of the local conditions (such as extrusion of material along persistent slip bands, dislocation glides along the crystallographic slip planes etc) to model crack nucleation mechanisms [16,17]. These micro level approaches are desirable from a fundamental point of view but are not sufficiently developed for use in industry. Crack initiation in fretting fatigue can be modeled by correlating the number of cycles to develop a crack of a certain size with continuum field variables or some parameters dependent on the state of cyclic stress or strain, or any combination of these. This approach though empirical, has potential to bring in the physics of crack formation into the analysis.

Many models/criteria/parameters have been proposed for fatigue crack initiation under the multiaxial stress state, which is the case in fretting fatigue [18]. A majority of the current multiaxial fatigue approaches can be grouped into two main categories: a) equivalent (invariant) stress models (such as Goodman, Gerber, Soderberg relations, von Mises equivalent stress amplitude, Modified Manson-Mcknight etc.), and b) critical plane models [19, 20]. The equivalent stress models require the determination of an equivalent cyclic scalar parameter often in terms of mean stress to be used in conjunction with uniaxial stress versus fatigue life data, while the critical plane models have been developed based on observations that fatigue cracks often nucleate on a particular plane. In the critical plane models, a combination of normal and shear stresses (strains) on the critical plane can be used to model crack initiation. It is hypothesized that shear stresses induce dislocation movement along slip lines, causing nucleation and growth of cracks while normal stresses open the crack and reduce the friction between the crack surfaces. The critical plane methods have the distinct advantage that they predict the orientation of crack, and they have a potential for providing an estimate of the crack size. Thus, the combination of critical plane methods and fracture mechanics approaches could eventually provide the total (i.e. the sum of nucleation, initiation and propagation) fretting fatigue crack behavior.

In the past, Szolwinski and Farris [21] proposed a critical plane approach for fretting fatigue based on a plain fatigue approach developed by Smith et al [22] which was formulated to account for the stress ratio (or the mean stress) effects. Their modified parameter assumed that crack initiation occurs on the plane where the product of normal strain amplitude, ε_a, and maximum normal stress, σ_{max}, is the maximum. Neu et al [23] found that for PH-13-8 Mo steel, the normal stress dominated Smith-Watson-Topper (SWT) [23] critical plane parameter predicted the crack location well but not the initial crack orientation while the shear stress dominated Fatemi-Socie (FS) parameter [24] predicted not only the location but also the orientation of the primary fretting fatigue crack. The present authors observed similar features with the SWT critical plane parameter in titanium alloy, Ti-6Al-4V [25], hence proposed a shear stress based critical plane parameter [26]. The fretting crack location as well as orientation was effectively characterized by this parameter involving the maximum shear stress range along with the local shear stress ratio effect on a critical plane. However, it has been observed that both the shear and normal stresses may play a role in the crack initiation under multiaxial fatigue conditions, e.g. Van and Maitournam have shown that fretting fatigue may be characterized by the shear amplitude and hydrostatic pressure on a critical plane [27].

Therefore, this study is an effort in this direction. The specific objective is to identify/develop a multiaxial fatigue model/parameter/criterion involving both shear and normal stresses to characterize fretting fatigue crack initiation under high cycle fatigue load conditions. It should be clarified here that the determination of the critical plane depends on all components of stress state (i.e. both normal and shear stresses), however the definition of a parameter may or may not include both the shear and normal stresses on the critical plane.

Experiments

Test Setup

The fretting setup used in this study incorporated a rigidly mounted fretting fixture on a servo-hydraulic fatigue test machine, as shown in Figure 1. This test-set was capable to apply cyclic axial stress, σ_{axial} and cyclic shear load, Q on the test specimen, while the normal load, P was held constant. This setup is capable of testing up to a frequency of 200 Hz, and it has one load cells on each side of the specimen so that all of the applied (σ) and reacted (Q) forces can be recorded at any point of interest during the test. A detailed explanation of the experimental setup is provided in a previous study [28].

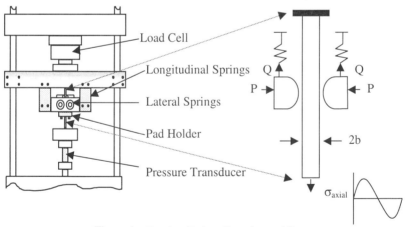

Figure 1 - *Fretting Fatigue Experimental Setup*

Specimen and Fretting Pad Details

Three types of pads (two cylindrical and one flat with edge radii) were used. Figure 2 shows a schematic of the dogbone specimen used in this work The cylindrical pads had radii of 50.8 mm and 101.6 mm, while the flat pad had an edge radius of 5.08 mm with a flat center section of 5.08 mm. The latter geometry of the pad was selected to simulate contact geometry in the dovetail joint of gas turbine engines. The specimens and pads were machined by the wire electrical discharge method from Ti-6Al-4V forged plates which had been preheated and solution treated at 935°C for 105 minutes, cooled under flowing air, vacuum annealed at 705°C for 2 hours, and then cooled under flowing argon. This resulted in a duplex microstructure consisting of 60% (volume) of primary-α (hcp) and 40% (volume) of transformed-β (α platelets in a β matrix - bcc) phases with a grain size of about 10 μm. The longitudinal tensile properties (along the loading axis) were determined to be elastic modulus = 127 GPa and yield strength = 930 MPa. The fretting fatigue specimen had thickness, 2b = 1.93 mm and width, w = 6.35 mm.

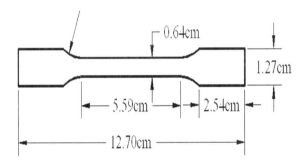

Figure 2 - *Fretting Fatigue Specimen*

Test Details

Constant amplitude fretting fatigue tests were conducted at 200 Hz, over a wide range of axial stress levels, $\sigma_{axial,max}$ = 400 MPa to 700 MPa with stress ratios, R, ranging from a nominal value of 0.0 to 0.7. The normal load for both the cylindrical pad with 50.8 mm radius and the flat with edge radii pad was 1.33 kN which resulted in the peak normal pressure of 304 MPa and 961 MPa respectively. The normal load for the cylindrical pad with 101.6 mm radius was 2.22 kN which resulted in the peak normal pressure of 267 MPa. The coefficient of friction was determined to be 0.5 for this experimental configuration by Iyer and Mall [29].

Finite Element Analysis

The finite element method was used to determine stress, strain and displacement fields at the contact interface. The analysis used was similar to that in a previous study [30]. Four node, plane strain quadrilateral elements were used with the "master-slave" interfacial algorithm developed for contact modeling in the finite element code, ABAQUS [31].   The measured loads were applied to the model in three steps.   The normal load, P was applied in the first step. In the second step, the maximum tangential load, Q_{max} and the maximum axial stress, $\sigma_{axial,max}$ were applied to match the experimental maximum cyclic load condition. Finally in the third step, the minimum tangential load, Q_{min} and minimum axial stress, $\sigma_{axial,min}$ were applied to match the experimental minimum cyclic load condition.

Damage Mechanisms

The main or primary crack, which caused failure in all specimens, initiated at the contact surface and near the trailing edge of contact. Other small secondary cracks were also observed. However, these did not grow enough to cause any specimen failure. Several specimens were sectioned to investigate the primary and secondary crack orientations and locations. The experimentally observed primary crack orientations angles were either -45^0 or 45^0 with a variation of $\pm$ 15^0 from a perpendicular to the loading direction [26]. Figure 3 shows a typical initiation orientation and propagation path of the primary crack. The primary crack in this specimen was initially oriented at the interface at an angle, θ = -30° initially and then at θ = -43°, where negative angles are measured in clockwise direction from the perpendicular to   the direction of applied axial

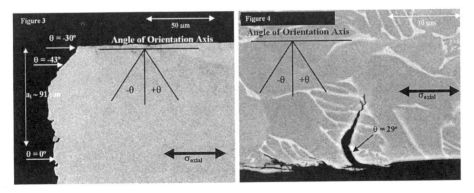

Figure (3)   Figure (4)

Figures 3 and 4 - *Primary (Fig 3) and Secondary (Fig. 4) Crack Growth Path in Specimens Tested with 101.6 mm and 50.8 mm Cylindrical Pads respectively.*

stress. The crack propagated in a tortuous manner until it reached a depth of about 91 μm. At this location the crack orientated perpendicular to the axial stress, θ = 0°. Figure 4 also shows a typical secondary crack, which orientated at the contact interface at an angle, θ = 29°. It should be noted that the cracks shown in Figures 3 and 4 have initiated/propagated in both intergranular and transgranular manners. Similar features were noted with other specimens. Thus, the grain boundaries were not the preferred location/direction for crack initiation/propagation in the tested specimens. Further, this behavior about crack initiation and orientation was independent of the fretting pad geometry.

Results And Discussion

The fretting fatigue data, i.e. the applied stress versus fatigue life are compared to the plain fatigue life data in Figure 5. Both sets of these data were generated at different stress ratios. It is well documented in the fatigue literature that there is a mean stress or stress ratio effect on the fatigue life or strength. There are several ways this effect can be accounted for. In this study, a method suggested by Walker was used [32], which is as follows:

$$\sigma_{effective} = \sigma_{max}(1-R)^m \qquad (1)$$

where $\sigma_{effective}$ is the effective stress taking into account the stress ratio effect, σ_{max} is the maximum applied stress on the specimen, R is the stress ratio and m is a fitting parameter, which was determined to be 0.45 in a previous study [26]. The resulting fit to the plain fatigue life data is shown as a solid line in Figure 5. The plain fatigue data,

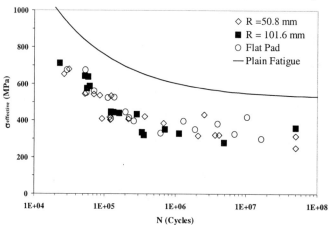

Figure 5 - *Effective Stress versus Life Relationships for Fretting Fatigue from Three Pad Geometries and Plain Fatigue*

obtained from R values ranging from −1.0 to 0.5, are not shown for the sake of clarity. This relationship can be expressed as an exponential function containing two terms as:

$$\sigma_{\text{effective}} = C_1 N_i{}^{C_2} + C_3 N_i{}^{C_4} \qquad (2)$$

where C_1, C_2, C_3 and C_4 are fitting parameters. The constants in equation (2) were obtained by using the best fit to the plain fatigue data, which provided $C_1 = 4.51 \times 10^5$ MPa, $C_2 = -0.720$, $C_3 = 246$ MPa and $C_4 = 5.81 \times 10^{-3}$. This comparison clearly shows that there is a reduction in the fatigue life under the fretting condition as expected. However, this type of comparison is of a simplistic nature, as it does not provide any mechanistic basis for estimating life debit due to fretting. Further, this direct comparison based on the stress applied to the specimen does not account for the two very important factors; a) the stress magnification in the contact region under the fretting condition and b) the multiaxial loading effects also under the fretting condition. Therefore, multiaxial fatigue parameters are considered next.

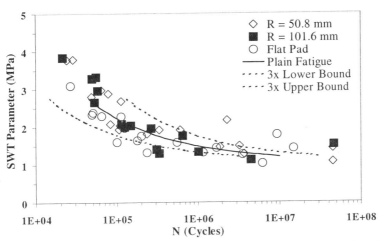

Figure 6 - *SWT Parameter versus Life Relationships for Fretting Fatigue from Three Pad Geometries and Plain Fatigue*

Smith-Watson-Topper (SWT) Critical Plane Approach

Szolwinski and Farris [21] modified the Smith-Watson-Topper parameter for the fretting fatigue crack initiation. This modified parameter assumed that crack initiation occurs on the plane where the product of normal strain amplitude, ε_a and maximum normal stress, $\sigma_{\max}$ is the maximum. Using the computed stresses and strains from the finite element analysis of the fretting fatigue experiments, this parameter was calculated at all planes ranging from $-90° \leq \theta \leq 90°$ in increments of $0.1°$, which provided this parameter's maximum value, location and critical plane orientation. The measured fretting fatigue life data were then plotted as a function of SWT parameter along with the counterpart from the plain fatigue condition (Figure 6). This clearly shows that the majority of the fretting fatigue crack initiation life data lie within $\pm 3N_i$ scatter bands of the plain fatigue life data. In other words, the fretting fatigue crack initiation life can be predicted from the SWT versus fatigue life relationship of the plain fatigue. In addition, the predicted crack location was determined to be near the trailing edge of the contact based on this approach, which was in good agreement with experimental observations. Further, the predicted angle of crack orientation based on this approach at the contact surface was within $+5°$ to $+8°$ of the perpendicular to the applied loading direction, and these were not in agreement with experimental crack orientations. This clearly shows that this parameter has the limitation in this respect.

Shear Stress Range Critical Plane Approach

In this approach, the maximum and minimum shear stresses were computed on all planes ranging from $-90° \leq \theta \leq 90°$ in $0.1°$ using the computed stresses and strains obtained from the finite element analysis. From this, the critical plane and magnitude of the maximum shear stress range, $\Delta\tau_{crit} = \tau_{max} - \tau_{min}$ was obtained. Then the shear stress ratio effect on the critical plane was accounted for by incorporating a technique proposed by Walker [32], which is expressed as:

$$\Delta\tau_{crit,effective} = \tau_{max} (1 - R_\tau)^m \qquad (3)$$

where τ_{max} is the maximum shear stress on the critical plane, R_τ is the shear stress ratio on the critical plane, and m is a fitting parameter, which was determined to be 0.45 from the plain fatigue data. The measured fretting fatigue life data are plotted as a function of the parameter, $\Delta\tau_{crit,effective}$ (Figure 7). As can be seen, the majority of the fretting fatigue data lie within $\pm 3N_i$ scatter bands of the plain fatigue data. Hence, this parameter can be also used to predict the fretting fatigue life from the plain fatigue data in conjunction with an analysis.

It should be noted that the maximum shear stress range occurs on two planes, one in the positive quadrant (i.e. from $0°$ to $90°$ from the perpendicular to the loading direction, Figure 3) and the other in the negative quadrant (i.e. from $0°$ to $-90°$ from the perpendicular to the loading direction). For each state of stress, there are thus two critical shear stress planes, which are at $90°$ to each other and either of the two orientations is equally possible, however local variation in microstructural properties may cause one orientation to be preferred. The orientation of primary cracks at the contact surface predicted from this approach was either from $+45°$ to $+50°$ or from $-45°$ to $-50°$ which correlated well with the experimental observations. Furthermore, the predicted crack initiation location agreed with the experimental results and it was determined to be near the trailing edge of the contact.

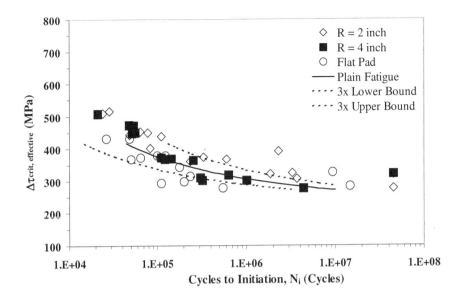

Figure 7 - *Shear Stress Range Critical Plane Parameter versus Life Relationships for Fretting Fatigue from Three Pad Geometries and Plain Fatigue*

Findley Parameter

The crack initiation in the multiaxial fatigue loading should be influenced by both normal and shear stresses (strains). The previous approach is primarily a shear stress based approach. In order to investigate the effect of the normal stress on the critical plane, a multiaxial fatigue approach on the critical plane, where normal stresses (strains) is included explicitly, was also considered in this study. One of these is the Findley parameter, (FP) [33]. In this approach, crack initiation is assumed to be governed by both the maximum shear stress amplitude, $\tau_a = (\tau_{max} - \tau_{min})/2$ and maximum stress normal to orientation of the maximum shear (σ_{max}) multiplied by an influence factor, k, as shown in the following. The k was determined to be equal to 0.35 from the plain fatigue data [25].

$$FP = \tau_a + k\sigma_{max} \qquad (4)$$

The Findley parameter, FP was calculated at all planes ranging from -90° $\leq \theta \leq$ 90° in 0.1° from the computed stresses and strains obtained from the finite element analysis. These calculations provided the critical plane where this parameter was the maximum. The measured fretting fatigue crack initiation life data were plotted as a function of this parameter (Figure 8), along with the corresponding relationship for the plain fatigue data. In this case, the FP versus fatigue life relationships for the plain fatigue and fretting fatigue did not agree with each other showing that this parameter can not be used to predict the fretting fatigue lives from the plain fatigue data. However, fretting fatigue data from different geometries were in agreement with each other within a scatter band. Further, the predicted angle of crack orientation based on this approach at the contact surface was within $+5^0$ to $+25^0$ of the perpendicular to the applied loading direction . The predicted crack initiation location was near the trailing edge. This shows that this parameter has these two limitations, i.e. it did not predict the fretting fatigue life from the plain fatigue data well, and the predicted crack orientations were different from than those observed experimentally.

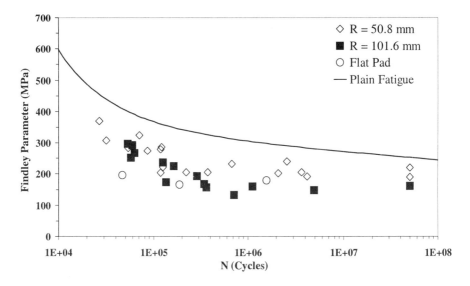

Figure 8 - *Findley Parameter versus Life Relationships for Fretting Fatigue from Three Pad Geometries and Plain Fatigue*

Modified Shear Stress Range (MSSR) Critical Plane Approach

To overcome the above shortcoming of the Findley parameter, shear stress range critical plane parameter was modified in the form similar to Findley parameter to account the normal stress that generally aids in opening the crack surfaces. This modified version of shear stress range critical plane parameter, MSSR is expressed as follows:

$$MSSR = A\Delta\tau^{B}_{crit} + C\sigma^{D}_{max} \qquad (5)$$

In this approach, the first term, $\Delta\tau_{crit}$ is the same as in equation 3 and the second term is the maximum normal stress on the critical plane. The constants, A, B, C and D were obtained by curve fitting. The measured fretting fatigue life data are plotted as a function of the MSSR parameter (Figure 9). As can be seen, the majority of the fretting fatigue data lie within $\pm 3N_i$ scatter bands of the plain fatigue data. Hence, this parameter can be also used to predict the fretting fatigue life from the plain fatigue data in conjunction with an analysis. Furthermore, this modified parameter predicted angle of crack orientation and crack location which were same as in the case of the shear stress range critical plane parameter, and these were in agreement with their experimental counterparts as stated earlier. Finally, this modified parameter explicitly included the effects of the shear stress as well as normal stress as it is should be the case in multiaxial fatigue loading.

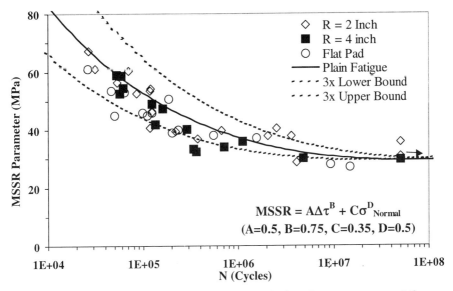

Figure 9 - *Modified Shear Stress Range Critical Plane Parameter versus Life Relationships for Fretting Fatigue from Three Pad Geometries and Plain Fatigue*

Summary

This study examined several critical plane based multiaxial fatigue models/criteria/parameters to investigate fretting fatigue crack initiation behavior in titanium alloy, Ti-6Al-4V. These were evaluated on their ability to predict the number of cycles to fretting fatigue crack initiation, crack initiation location and crack orientation angle along the contact surface. The normal stress or tensile-cracking based Smith-Watson-Topper critical plane parameter predicted the cycles to fretting fatigue crack initiation and crack location, which were in agreement with their experimental counterparts. However, its prediction of the crack orientation was not in agreement with the experimental counterparts. On the other hand, a shear stress or shear-cracking based critical plane parameter predicted the cycles to fretting fatigue crack initiation, crack initiation location, and crack orientation which were in agreement with their experimental counterparts. Further, the Findley parameter, a critical plane approach involving both normal and shear stresses (or both tensile and shear cracking) predicted the cycles to fretting fatigue crack initiation and crack orientation which were not in agreement with their experimental counterparts. Hence, a critical plane parameter involving both normal and shear stresses (or both tensile and shear cracking) is proposed that predicted the cycles to fretting fatigue crack initiation, crack initiation location, and crack orientation which were in agreement with their experimental counterparts. From these observations, it appears that that fretting fatigue crack mechanism in the tested titanium alloy is governed by the shear stress on the critical plane. However, the role of the normal stress on the critical plane upon the fretting fatigue crack initiation behavior appears to be unclear from this study, and hence more experiments involving various pad geometries and loading conditions are needed to investigate this phenomenon.

Acknowledgment

The authors would like to gratefully acknowledge the computer support for this study provided by The Ohio Supercomputer Center (OSC).

References

[1]   A. B. Cowles, "High Cycle Fatigue in Aircraft Gas Turbines - An Industry Perspective", *International Journal of Fracture*, Vol. 80, 1996, pp. 147-163.

[2]   T. Nicholas, "Critical Issues in High Cycle Fatigue", *International Journal of Fatigue*, Vol. 21, 1999, pp. S221-S231.

[3]   D. W. Hoeppner, V. Chandrasekaran, C. B. Elliot, Editors. "Fretting Fatigue: Current Technologies and Practices", *ASTM STP 1367*, American Society for Testing and Materials, West Conshohocken, PA, 2000.

[4]   D. Rayaproula and R. Cook, "A Critical Review of Fretting Fatigue Investigations at the Royal Aerospace Establishment", Standardization of Fretting Fatigue Test Methods and Equipment, *ASTM STP 1159*, M. Helmi and R. Waterhouse, Eds. American Society for Testing and Materials, Philadelphia, 1992, pp. 129-152.

[5]   T. Lindley and K. Nix, "Fretting Fatigue in the Power Generation Industry: Experiments, Analysis, and Integrity Assessment", Standardization of Fretting Fatigue Test Methods and Equipment, *ASTM STP 1159*, M. Helmi and R. Waterhouse, Eds. American Society for Testing and Materials, Philadelphia, 1992, pp. 153-169.

[6]   R. Waterhouse, "Fretting Fatigue", *International Materials Review*, Vol. 37, No. 2, 1992, pp. 77-97.

[7]   S. Adibnazari and D. Hoeppner, "Study of Fretting Fatigue Crack Nucleation in 7075-T6 Aluminum Alloy", *Wear*, Vol. 159, 1992, pp. 257-264.

[8]   R. Antoniou and T. Radtke, "Mechanisms of Fretting Fatigue of Titanium Alloys", *Materials Science and Engineering*, A237, 1997, pp. 229-240.

[9]   T. Lindley and K. Nix, "The Role of Fretting in the Initiation and Early Growth of Fatigue Cracks in Turbo Generator Materials", *Multiaxial Fatigue*, ASTM, 1985, pp. 340-360.

[10] K. Nix and T. Lindley, "The Application of Fracture Mechanics to Fretting Fatigue", *Fatigue and Fracture of Engineering Materials and Structures*, Vol. 8, No.2, 1985, pp. 143-160.

[11] A. E. Giannakopoulus, T. C. Lindley and S. Suresh, "Aspects of Equivalence Between Contact Mechanics and Fracture Mechanics: Theoretical Connections and a Life-Prediction Methodology for Fretting Fatigue", *Acta Materialla*, Vol. 46, 1998, pp. 2955-2966.

[12] S. Faanes and U. Fernando, "Life Prediction in Fretting Fatigue using Fracture Mechanics", *Fretting Fatigue*, Mechanical Engineering Publications, London, 1994, pp. 149-159.

[13] C. Ruiz, P. H. B. Boddington and K. C. Chen, "An investigation of fatigue and fretting in a dovetail joint", *Experimental Mechanics*, Vol. 24, 1984, pp. 208-217.

[14] D. Nowell and D. A. Hills, "Crack Initiation Criteria in Fretting Fatigue", *Wear*, Vol. 136, 1990, pp. 329-343.

[15] C. D. Lykins, S. Mall and V. K. Jain, "An evaluation of parameters for predicting fretting fatigue crack initiation", *International Journal of Fatigue*, 20, 2000, pp. 703-716.

[16] M. Kuno, R. B. Waterhouse, D. Nowell, and D. A. Hills, "Initiation and growth of fretting fatigue cracks in the partial slip regime", *Fatigue and Fracture of Engineering Materials and Structures* 12/5, 1989, pp. 387-398.

[17] C. H. Goh, R. W. Neu, and D. L. McDowell, "Shakedown, Ratchetting, and Reversed Cycle Plasticity in Fretting Fatigue of Ti-6Al-4V Based on Polycrystal Plasticity Simulation", *presented at the 6th National Turbine Engine High Cycle Fatigue Conference*, Jacksonville, FL, 5-8 March 2001.

[18] A. Krgo, A. R. Kallmeyer, and P. Kurath, "Evaluation of HCF Multiaxial Fatigue Life Prediction Metrologies for Ti-6Al-4V", *Proceedings of the 5th National Turbine Engine High Cycle Fatigue Conference*, Arizona, 2000.

[19] J. E. Shigley and C. R. Mischke, "Mechanical Engineering Design", McGraw-Hill, New York, 1989.

[20] J. Morrow, "Fatigue Design Handbook", *SAE AE-6*, J. A. Graham, Ed., Society of Automotive Engineers, 1968, pp. 17-58.

[21] M. Szolwinski and T. Farris, "Mechanics of fretting fatigue crack formation", *Wear*, Vol. 198, 1996, pp. 93-107.

[22] R. N. Smith, P. Watson, and T. H. Topper, "A Stress-Strain Function for the Fatigue of Metals", *Journal of Materials*, JMLSA, Vol. 5, 1970, pp. 767-778.

[23] R. Neu, J. Pape, D. Swalla-Michaud, "and Methodologies for Linking Nucleation and Propagation Approaches for Predicting Life Under Fretting Fatigue", *Fretting Fatigue: Current Technology and Practices*, ASTM 1367, D. Hoeppner, V. Chandrasekaran and C. Elliot, Eds. American Society for Testing and Materials, 1999.

[24] A. Fatemi and D. Socie, "A critical plane approach to multiaxial fatigue damage including out of phase loading", *Fatigue and Fracture of Engineering Materials and Structures*, Vol. 11, 1988, pp. 149-165.

[25]   C. D. Lykins, S. Mall and V. K. Jain, "Combined experimental-numerical investigation of fretting fatigue crack initiation", *International Journal of Fatigue*, Vol. 23, 2001, pp. 703-711.

[26]   C. D. Lykins, S. Mall and V. K. Jain, "A shear stress based parameter for fretting fatigue crack initiation", *Fatigue and Fracture of Engineering Materials and Structures*, Vol. 24, 2001, pp. 461-473.

[27]   K. D. Van and M. H. Maitournam, "On a New Methodology for Quantitative Modeling of Fretting Fatigue", *Fretting Fatigue: Current Technologies and Practices, ASTM STP 1367*, D. W. Hoeppner, V. Chandrasekaran, C. B. Elliot, eds, American Society Testing and Materials, Philadelphia, 2000, pp. 538-552.

[28]   R. Cortez, S. Mall and J.R. Calcaterra, "Interaction of High Cycle and Low Cycle Fatigue on Fretting Behavior of Ti-6-4", *Fretting Fatigue: Current Technologies and Practices, ASTM STP 1367*, D. W. Hoeppner, V. Chandrasekaran, C. B. Elliot, eds, American Society Testing and Materials, Philadelphia, 2000, pp. 183-198.

[29]   K. Iyer and S. Mall, "Analysis of contact pressure and stress amplitude effects on fretting fatigue life", *ASME Journal of Engineering Materials and Technology*, Vol. 123, No.1, 2001, pp. 85-93.

[30]   C. T. Tsai and S. Mall, "Elasto-plastic finite element analysis of fretting stresses in pre-stressed strip in contact with cylindrical pad", *Finite Elements in Analysis and Design*. 36, 2000, pp. 171-187.

[31]   ABAQUS Standard User's Manual, Hibbit, Karlsson and Sorensen Inc., 1995

[32]   K. Walker, "The effect of stress ratio during crack propagation and fatigue for 2024-T3 and 7075-T6 Aluminum", *Effects of Environment and Complex Load History on Fatigue Life, STP 462*, American Society for Testing and Materials, West Conshohocken, PA, 1970, pp. 1-14.

[33]   W. N. Findley, "Fatigue of Metals Under Combinations of Stresses", *Transactions, ASME*, Vol. 79, 1957, pp. 1337-1348.

S. Shirai,[1] K. Kumuthini,[1] Y. Mutoh,[1] and K. Nagata[2]

Fretting Fatigue Characteristics of Titanium Alloy Ti-6Al-4V in Ultra High Cycle Regime

REFERENCE: Shirai, S., Kumuthini, K., Mutoh, Y. and Nagata, K., "Fretting Fatigue Characteristics of Titanium Alloy Ti-6Al-4V in Ultra High Cycle Regime," *Fretting Fatigue: Advances of Basic Understanding and Applications, ASTM STP 1425*, Y. Mutoh, S. E. Kinyon, and D. W. Hoeppner, Eds., ASTM International, West Conshohocken, PA, 2003.

ABSTRACT: In the present study, fretting fatigue behavior of a Ti-6Al-4V alloy under high stress ratios in the ultra high cycle regime (up to 10^9 cycles) has been investigated. The slope of S-N curve decreased with increasing stress ratio. The slope was steeper in short fatigue life region and became flatter in longer life region. The test results under various stress ratios followed the modified Goodman diagram for both the fatigue strengths defined at 10^7 cycles and 10^8 cycles. The so-called non-propagating cracks, which were often observed in steels and aluminum alloys, could not be found in the run-out specimens tested up to 10^7 cycles. In the further loading cycles up to 10^9 cycles at the same stress amplitude, no crack initiation and propagation could be observed, which indicated that the conventional fretting fatigue strength defined 10^7 cycles was still effective at least up to 10^9 cycles. Different behavior between the titanium alloy used and the other materials mainly results from the different crack initiation behavior.
KEYWORDS: fretting fatigue, mean stress, stress ratio, ultra high cycle, titanium alloy

Introduction

Titanium alloy has been increasingly applied to steam turbine components as well as gas turbine components due to its excellent specific strength, corrosion resistance, etc. In these applications, the components such as blades suffer a high centrifugal force with vibrating loads, which produces a high cycle fatigue condition with high mean stress. It is also known that endurance lives in service are 10^8-10^9 cycles for car engines, 10^9 cycles for high-speed trains and 10^{10} cycles for aircraft engines, which are further longer than the endurance life of 10^7 cycles in conventional fatigue tests. Therefore, fatigue behavior under high stress ratio as well as in the ultra high cycle regime should be understood for estimating fatigue lives of these components in service.

[1] Nagaoka University of Technology, Nagaoka-shi, 940-2188 Japan.
[2] Toshiba Corp., Fuchu-shi, Japan.

Since 1970, many researches have been devoted in fretting fatigue of titanium alloys [1–11]. Fretting fatigue behavior of titanium alloys has been discussed in detail from both microstructural and mechanical points of view. However, detailed fretting fatigue processes, such as how long the crack initiation stage is, have not yet been clear. Effect of high mean stress on fretting fatigue behavior has also not yet been obvious. Since no report on fretting fatigue test over 10^8 cycles has been available, fretting fatigue behavior in the ultra high cycle regime has not yet been understood. There are several topics to be solved in fretting fatigue in the ultra high cycle regime: existence of fatigue limit, change of fatigue fracture mechanisms, non-propagating crack, etc.

In the present study, fretting fatigue tests at stress ratios of –1, 0.5 and 0.7 up to 10^8 cycles and at a stress ratio of –1 up to 10^9 cycles were conducted. Fretted surfaces and longitudinal cross-sections of tested specimens were observed in detail to discuss the effect of high stress ratio and the fretting fatigue crack initiation and propagation behaviors in the ultra high cycle regime.

Experimental Procedures

The material used for the specimen was a Ti-4Al-6V alloy, which was also used for the contact pad. Chemical composition and mechanical properties of the material are shown in Tables 1 and 2, respectively. Figure 1 shows a microstructure of the material.

Table 1-*Chemical composition of Ti-6Al-4V*

Al	V	Fe	O	C	N
6.62	4.31	0.28	0.165	<0.005	0.003

Table 2-*Mechanical properties of Ti-6Al-4V*

Yield Strength [MPa]	Tensile Strength [MPa]	Elongation [%]
926	1013	17.6

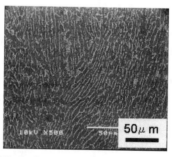

Fig.1- *Microstructure of Ti-6Al-4V*

Shapes and dimensions of the fretting fatigue specimen and the contact pad are shown in Fig. 2. A pair of contact pads was clamped against the opposite flats of the fretting fatigue specimen using a calibrated proving ring, as schematically illustrated in Fig. 3. The contact pads were mounted in the clamping jig. The contact pressure of 100MPa was applied to the contact pads. The clamping load was measured with strain gages bonded to the proving ring. The relative slip amplitude between the contact pad and the specimen was monitored by calibrated extension meters of the type shown in Fig. 3. The frictional force (tangential force) between the contact pad and the specimen was also measured by strain gages bonded at the neck part of the clamping jig. The tests were performed in air at frequencies of 20 Hz for $\sim 10^7$ cycles and 30-47 Hz for $\sim 10^8$ cycles using a conventional servo-hydraulic testing machine. The test up to 10^9 cycles was conducted at a frequency of 340 Hz using a newly developed electro-dynamic fatigue machine. The effect of frequency was checked by comparing three kinds of fretting fatigue test results at the stress amplitude of 270 MPa under R=-1: the fretting fatigue life at a frequency of 340 Hz was 9.2 × 10^4 cycles and that of frequency alternating test for

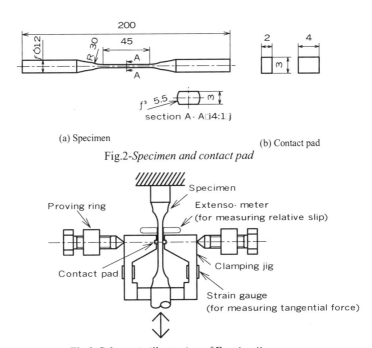

(a) Specimen

(b) Contact pad

Fig.2-*Specimen and contact pad*

Fig.3-*Schematic illustration of Fretting jig*

each 10^4 cycles between 30 Hz and 340 Hz was 7.9×10^4 cycles, which were in good agreement with that at 30 Hz (1.2×10^5 cycles).The stress ratios were -1, 0.5 and 0.7 to investigate the fretting fatigue strengths at high stress ratios. The cracks on the fretted surface and the fretting fatigue crack path on the longitudinal cross-section of the specimens tested were observed using a scanning electron microscope (SEM) in detail.

Results

S-N curve

The S-N curves for the Ti-6Al-4V alloy are shown in Fig. 4. From the figure, the following characteristics can be pointed out:
(1) The slope of S-N curve was different between regions below and above 10^7 cycles: The steeper slopes in the region below 10^7 cycles were observed, while the slopes in the region above 10^7 cycles were almost flat.
(2) In the region below 10^7 cycles, the slope decreased with increasing the stress ratio.
(3) The fretting fatigue strengths defined at 10^7 cycles were 130MPa for R=-1, 100MPa for R=0.5 and 80MPa for R=0.7. Those defined at 10^8 cycles were 130MPa for R=-1, 90MPa for R=0.5 and 70MPa for R=0.7. (The fatigue strength is defined as follows: if the two data points at the same stress amplitude are available, the fatigue strength is the stress where one is failure and the other run-out. If only one data point is available, the fatigue strength is the stress for run-out.)
(4) The specimen tested at 130MPa, which was the fatigue strength defined at 10^7 cycles, survived up to 10^9 cycles.

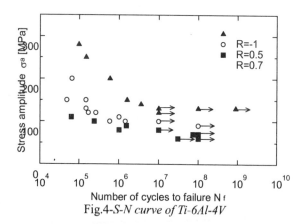

Fig.4-*S-N curve of Ti-6Al-4V*

The modified Goodman diagram obtained from the S-N curves is shown in Fig. 5. It can be seen from the figure that the data points are well laid along a straight line of the Goodman diagram for both the fatigue strengths defined at 10^7 cycles and at 10^8 cycles.

Tangential Force and Relative Slip Amplitude

Both tangential force and relative slip amplitude were unstable up to initial few hundreds cycles. However, after that, they are stable and almost constant during fretting fatigue tests until failure or run-out. Relationship between tangential force and stress amplitude is shown in Fig. 6, where stable values of tangential force are plotted. From the figure, it seems that tangential force depends on stress amplitude regardless of stress ratio.

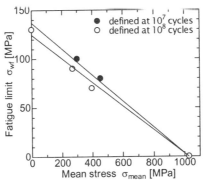

Fig.5-*Goodman's diagram of Ti-6Al-4V*

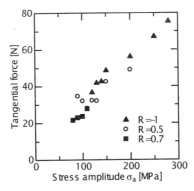

Fig.6 *Tangential force vs. stress amplitude*

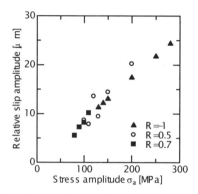

Fig.7— *Relative slip amplitude vs. stress amplitude*

Relationship between relative slip amplitude and stress amplitude is shown in Fig. 7, where stable values of relative slip amplitude are also plotted. As can be seen from the figure, relative slip amplitude depends on stress amplitude regardless of stress ratio. These behaviors result from that mean stress is static and only stress amplitude induces relative movement between the specimen and the contact pad.

Observations of Fretted Surface and Longitudinal Cross section of the Specimens Tested

Figure 8 shows observations of fretted surface damage and surface roughness measurements of the specimens tested up to 10^8 cycles under stress ratios of -1, 0.5 and 0.7. It is found that fretted damage and surface roughness decrease with increasing the stress ratio. This may result from the decrease in stress amplitude, which consequently results in the decrease of the relative slip amplitude, with increasing the stress ratio. Figure 9 shows observations of fretted surface damage and surface roughness measurements of the specimens tested at 130MPa with a stress ratio of -1 up to 10^7, 10^8 and 10^9 cycles. It is obvious that fretted surface damage and surface roughness increase with increasing the number of cycles.

Surface cracks were carefully observed on the fretted surfaces. An example of the initiated surface crack is shown in Fig. 10. However, such a crack could be found on the fretted surfaces of only five specimens among 29 specimens tested, sizes of which were from 10μm to 100μm on the fretted surface. The depths of these cracks were tried to measure on the longitudinal cross-section of the specimens. However, an initiated crack could be observed in only one specimen, as shown in Fig. 11. In other cases, no cracks could be found on the longitudinal cross-section because of too small size of the cracks.

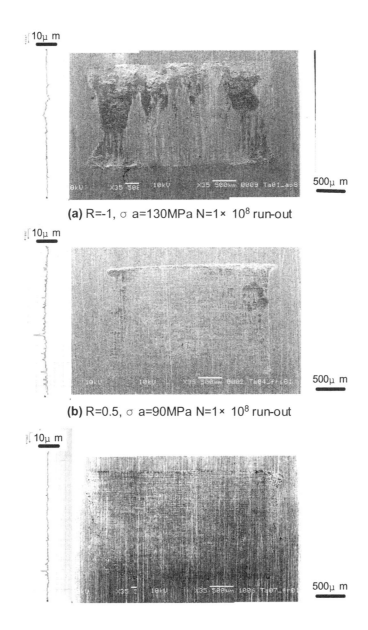

(a) R=-1, σ a=130MPa N=1× 10^8 run-out

(b) R=0.5, σ a=90MPa N=1× 10^8 run-out

Fig.8 - *Fretted surfaces on the Ti-6Al-4V specimens tested up to N=1× 10^8 cycles.*

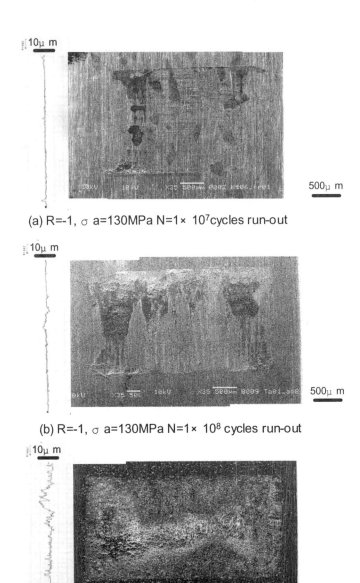

(a) R=-1, σ a=130MPa N=1× 10^7cycles run-out

(b) R=-1, σ a=130MPa N=1× 10^8 cycles run-out

(c) R=-1, σ a=130MPa N=1× 10^9cycles run-out

Fig.9 - *Fretted surfaces on the Ti-6Al-4V specimens tested at R=-1 and a=130MPa.*

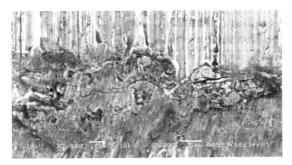

Fig. 10 *Cracks on the fretted surface for R=-1, σa=130MPa, N=1×10⁷cycles.*

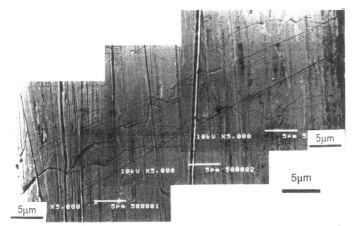

Fig.11 *Initiated crack on the longitudinal cross-section for R=-1 N=10⁷cycles.*

Discussions

Fretting Fatigue Behavior in the Ultra High Cycle Regime

The transition of crack initiation site from surface to subsurface in high strength alloys has been often observed at 10^6–10^8 cycles under high cycle fatigue without fretting [12–19]. This corresponds to the change of crack initiation mechanism from the extrusion-intrusion to the internal defect, which induces high local stress concentration. Selection of the two mechanisms is competitive depending on stress level, size and shape of defect, cyclic deformation behavior, etc. The transition also introduces a knee followed by a rather sharp drop in S-N curve [18]. Subsurface crack origins of titanium alloys are often cleavage facets because of their high purity [14–19], while those of high strength steels are often non-metallic inclusions [12,13].

There is no report on fretting fatigue test over 10^8 cycles, while a limited number of reports on fretting fatigue tests up to 10^8 cycles is available. The available test results up to 10^8 cycles indicate no transition of crack initiation site and no change of the shape of S-N curve [20–24]. The present fretting fatigue test results up to 10^9 cycles for a Ti-6Al-4V alloy also show no particular change of S-N curve and fretting fatigue crack initiation mechanism. This may result from high stress concentration at the fretting contact edge even at low stress amplitudes. The contact edge region with high stress concentration due to fretting action is always an initiation site of fretting fatigue crack even for materials with large defects, such as sintered porous steels [25], cast iron [26], etc.

Another problem in fretting fatigue in the ultra high cycle regime will be the non-propagating crack. It has been often reported that non-propagating cracks have been observed in run-out specimens for steels and aluminum alloys [4]. This kind of non-propagating crack problem is very similar to the case of notched specimen. Although it is not clear whether the non-propagating crack observed at 10^7 cycles can be still arrested in the following cycles, it should be reasonable to consider that the crack can propagate in the following cycles, as can be seen in Ref. [27]. Since this non-propagating crack problem in fretting fatigue includes the small crack and mix mode crack problems, further detailed investigation is required to make this complicated problem clear. In Ti-6Al-4V alloy, initiated cracks were rarely found in the fractured specimens as well as the run-out specimens. In another words, so-called non-propagating cracks could not be observed in the run-out specimens. Therefore, it is speculated that once a crack initiates, it grows rapidly to final failure. It is also found that the crack initiation stage is very late of fatigue life, which is significantly different from steels and aluminum alloys with the early stage of crack initiation [4] and more like to intermetallics with the very late crack initiation [28].

Shape of S-N curve

The slope of S-N curve in fretting fatigue of titanium alloy, which is steeper in shorter life region, becomes more flat in longer life region, as can be seen from Fig. 4. The similar change of slope in fretting fatigue has been found in aluminum alloys [22,23], a rotor steel [20] and a carbon steel [29]. The deformation of contact edge between the specimen and the contact pad can be categorized in three types: a) sticking, b) slipping and c) gapping. Based on the detailed analysis [30], no singular behavior at the contact edge can be found in case of gapping. Stronger stress singularity at the contact edge was found in case of slipping compared to the case of sticking. The deformation type depends on many parameters: rigidity of contact pad, contact pressure, stress amplitude, relative slip amplitude, etc. When the stress amplitude decreases, the deformation type is changing from slipping to sticking, that is the stress singularity at contact edge becomes weaker. This change of deformation type and then the change of stress singularity at contact edge will be one of the reasons for changing the slope of S-N curve in fretting fatigue.

Conclusions

Fretting fatigue behavior of a Ti-6Al-4V alloy under high stress ratios and in the very high cycle regime was investigated. The main conclusions obtained are summarized as follows.

1) Fretting fatigue strengths with various stress ratios defined both 10^7 cycles and 10^8 cycles were followed the modified Goodman diagram.
2) Surface cracks on the fretted surface of both the fractured and the run-out specimens could rarely be found in Ti-6Al-4V alloy, suggesting that once a crack nucleates, it grows rapidly to final failure.
3) No changes of crack initiation mechanism and initiation site were found for fretting fatigue in the present range of fatigue cycle up to 10^9 cycles, contrary to the case of fatigue without fretting. The high stress concentration at the contact edge in fretting fatigue will be a possible reason for these phenomena.
4) In Ti-6Al-4V alloy, non-propagating crack could not be found in the survived specimen at 10^7 cycles. In the further loading cycles up to 10^9 cycles at the same stress amplitude, no crack initiation and propagation could be observed, which indicated that the conventional fretting fatigue strength defined at 10^7 cycles was still effective at least up to 10^9 cycles.

References

[1] Milestone, W. D., "A New Apparatus for Investigating Friction and Metal-to-Metal Contact in Fretting Joints," Effects of Environment and Complex Load History on Fatigue Life, ASTM STP 462, 1970, pp. 318–328.
[2] Salkind, M. J., Lucas, J. J., "Fretting Fatigue in Titanium Helicopter Components," NACE International Conference on Corrosion Fatigue: Chemistry, Mechanics and Microstructure, 1971.
[3] Waterhouse R. B. and Dutta, M. K., "The Fretting Fatigue of Titanium and Some Titanium Alloys in a Corrosive Environment," *Wear*, Vol. 25, 1973, pp. 171–175.
[4] Goss, G. L. and Hoeppner, D. W., "Normal Load Effects in Fretting Fatigue of Titanium and Aluminum Alloys," *Wear*, Vol. 27, 1974, pp.153–159.
[5] R. B. Waterhouse and M. H. Wharton, "The Behavior of Three High-Strength Titanium Alloys in Fretting Fatigue in a Corrosive Environment," *Lubrication Engineering*, Vol. 32, 1976, pp. 294–298.

[6] Lutynski, C., Simansky, G., and McEvily, A. J., "Fretting Fatigue of Ti-6Al-4V Alloy," *Materials Evaluation Under Fretting Conditions, ASTM STP 780*, 1982, pp. 150–164.

[7] R. Cortez, R. Mall and J. R. Calcaterra "Interaction of High-Cycle and Low-Cycle Fatigue on Fretting Behavior of Ti-6-4," *Fretting Fatigue: Current Technology and Practices, ASTM STP 1367*, 2000, pp. 183–198.

[8] Satoh, T., "Influence of Microstructure on Fretting Fatigue Behavior of Near-alpha Titanium," *Fretting Fatigue: Current Technology and Practices, ASTM STP 1367*, 2000, pp. 295–307.

[9] Huston A. L. and Nicholas, T., "Fretting Fatigue Behavior of Ti-6Al-4V Against Ti-6Al-4V Under Flat-on-Flat Contact with Blending Radii," *Fretting Fatigue: Current Technology and Practices, ASTM STP 1367*, 2000, pp. 308–321.

[10] S. E. Kinyon and D. W. Hoeppner, "Spectrum Load Effects on the Fretting Fatigue Behavior of Ti-6Al-4V," *Fretting Fatigue: Current Technology and Practices, ASTM STP 1367*, 2000, pp. 100–115.

[11] Anton, D. L., Lutian, M. J, Favrow, L. H., Logan D., and Annigeri, B., "The Effects of Contact Stress and Slip Distance on Fretting Fatigue Damage in Ti-6Al-4V/17-4PH Contacts," *Fretting Fatigue: Current Technology and Practices, ASTM STP 1367*, 2000, pp. 119–140.

[12] Masuda, C., Nishijima, S., Sumiyoshi, H., Tanaka Y., and Ishii, A., "Mode II Crack Observed at the Origin of Fish Eye for SCr 420 Carburized Steel," *J. Soc. Mater. Sci. Jpn.* (Zairyo) Vol. 34, 1985, pp. 664–669.

[13] Abe T. and Kanazawa, K., "Fatigue strength and fatigue crack initiation and propagation of high strength steels," *J. Soc. Mater. Sci., Jpn.* (Zairyo), Vol. 40, 1991, pp. 1447–1452.

[14] Neal D. F. and Blenkinsop, P. A., *Acta Metall.*, Vol. 24, 1976, p. 59.

[15] Ruppen, J., Bhowal, P., Eylon, D., and McEvily, A. J., *Fatigue Mehcanisms, ASTM STP 675*, 1979, p. 47.

[16] Eylon, D., *J. Mater. Sci.,* Vol. 14, 1979, p. 1914.

[17] Ruppen J. A., Eylon, D., and McEvily, A. J., *Metall. Trans.,* 11A, 1980, p. 1072.

[18] Atrens, A. Hoffelner, W. Duerig, T. W. and Allison, J. E., *Scr. Metall.* Vol. 17, 1983, p. 601.

[19] Hagiwara M., Kaieda Y., Kawabe, Y., and Miura, S., *Tetsu-to-Hagane*, Vol. 76, 1990, p. 2182 (in Japanese).

[20] King, R. N., Lindley, T. C. "Fretting Fatigue in a 3.5NiCrMoV Rotor Steel," *Proc. of Fifth Int. Conf. on Fracture*, 1980, pp. 631–640.

[21] Spink, G. M., "Fretting Fatigue of A2.5%NiCrMoV Low Pressure Turbine Shaft - The Effect of Different Contact Pad Materials and of Variable Slip Amplitude," *Wear*, Vol. 136, 1990, pp. 281–297.

[22] Rayaprolu, D. B., Cook, R., "A Critical Review of Fretting Fatigue Investigations at

the Royal Aerospace Establishment," *ASTM STP 1159*, 1992, pp. 129–152.

[23] Lindley, T. C., Nix, K. J., "Fretting Fatigue in the Power Generation Industry: Experiments, Analysis, and Integrity Assessment, *ASTM STP 1159*, 1992, pp. 153–169.

[24] Fischer, G., Grubisic, V., Buxbaum, O. "The Influence of Fretting Corrosion on Fatigue Strength of Nodular Cast Iron and Steel under Contact Amplitude and Load Spectrum Tests," *ASTM STP 1159*,1992, pp. 178–189.

[25] Nishida, T., Takeuchi, M., and Mutoh, Y., "Fretting Wear and Fatigue Properties of High Strength Sintered Steels," SURFACE MODIFICATION TECHNOLOGIES VII, *Proc. of Seventh Int. Conf.,* Sanjo, Niigata, Japan, T. S. Sudarshan, Ed., *The Inst. of Mater.,* 1994, pp. 11–24.

[26] Mutoh Y. and Tanaka K., "Fretting Fatigue in Several Steels and A Cast Iron," *Wear,* Vol. 125, 1988, pp. 175–191.

[27] Shirai, S., Mutoh, Y., and Nagata, K., "Fretting Fatigue Behavior of Ti-6Al 4V and Structural Steel in the Very High Cycle Regime," *Proc. of Int. Conf. on Fatigue in the Very High Cycle Regime*, Vienna, July, 2001, pp. 295–302.

[28] Hansson, T., Kamaraj, M., Mutoh, Y., and Petterson, B., "High Temperature Fretting Fatigue Behavior in an XDTM γ-base TiAl," Fretting Fatigue: Current Technology and Practices, ASTM STP 1367, 2000, pp. 65–79.

[29] Endo K. and Goto H., "Initiation and Propagation of Fretting Fatigue Cracks," *Wear,* Vol. 38, 1976, pp. 311–324.

[30] Mutoh, Y., Kondoh, K., and Xu, J. Q., " Observations and Analysis of Fretting Fatigue Crack Initiation and Propagation," presented at Third International Symposium on Fretting Fatigue, Nagaoka, 15–18 May, 2001, to be published in ASTM STP, 2001.

SURFACE TREATMENT

Tomohisa Nishida,[1] Junnosuke Mizutani,[2] Yoshiharu Mutoh,[3] and Masatsugu Maejima [4]

Effect of Lubricating Anodic Film on Fretting Fatigue Strength of Aluminum Alloy

---

REFERENCE: Nishida T., Mizutani J., Mutoh Y., and Maejima M., "Effect of Lubricating Anodic Film on Fretting Fatigue Strength of Aluminum Alloy," *Fretting Fatigue: Advances in Basic Understanding and Applications, STP 1425*, Y. Mutoh, S. E. Kinyon, and D. W. Hoeppner, Eds., ASTM International, West Conshohocken, PA, 2003.

ABSTRACT: Wear, scratch, plain fatigue and fretting fatigue tests were carried out using aluminum alloy JIS 6063-T5 specimens with anodic film and lubricating anodic film. Though the surface hardness of lubricating anodic film was lower than that of anodic film, lubricating anodic film had excellent wear resistance and crack resistance compared to anodic film. Plain fatigue strength of the specimen with anodic film was lower than that of the specimen with lubricating anodic film, which almost coincided with that of the specimen without film. The low fatigue strength of the specimen with anodic film may result from the brittleness and tensile residual stress of the anodic film, which attribute to the cracking at lower applied stress. Cracking of the lubricating anodic film during fatigue tests hardly occurred due to its ductile nature and low residual stress and consequently the fatigue strength was not degraded. Fretting fatigue strength of the specimen with lubricating anodic film was higher than that of the specimen with anodic film, which was higher than that of the specimen without film. The lubricating anodic film had the lowest coefficient of friction and a higher cracking resistance, which contributed to the retardation crack initiation and the low crack growth rates. The anodic film also had low coefficient of friction compared to the specimen without film and prevented meta-to-metal contact, which may result in higher fatigue strength compared to the specimen without film. The lubricating anodic film is of significant benefit to

---

[1]Associate Professor, Department of Mechanical Engineering, Numazu College of Technology, Numazu 410-8501, Japan.
[2]Associate Professor, Department of Electronic Control Engineering, Toyama College of Maritime Technology, Shinminato 933-0293, Japan.
[3] Professor, Department of Mechanical Engineering, Nagaoka University of Technology, Nagaoka 940-2188, Japan.
[4]Manager, Materials Research Laboratory, Fujikura, Limited , Tokyo 135-0042, Japan.
fretting fatigue strength without degradation of plain fatigue strength.

fretting fatigue strength without degradation of plain fatigue strength.

KEYWORDS: plain fatigue, fretting fatigue, lubricating anodic film, anodic film, in-situ observation, tangential force coefficient

Introduction

Aluminum alloy, which has superior mechanical properties, low cost, light weight and reliability, has been widely used for automobile parts, aircraft parts, air and oil compressors and other components. However, aluminum alloy has problems of surface damage due to its softness and corrosion. Therefore, improvement of surface properties is required in practical applications. Anodic oxide treatment, which is so-called anodic film treatment is widely used for this purpose. Anodized film impregnated with fluorine resin or molybdenum sulfide has also been used for automobile parts and oil compressors because of high wear resistance. Joining of components is common in such machines and structures. Once one of the components or both the two components suffer a load, fretting fatigue can be a potential failure mode. Therefore, not only plain fatigue properties but also fretting fatigue properties are important for designing machines with high reliability. Although some research works on fatigue of aluminum alloys with anodic film have been reported [1-4], those on fretting fatigue have rarely been done. In the present study, plain fatigue and fretting fatigue tests were carried out to investigate fretting fatigue properties of aluminum alloy JIS 6063-T5 specimens with anodic film and lubricating anodic film. Furthermore, fretting fatigue cracks of the specimen were observed in-situ using a fatigue testing machine with SEM (Scanning Electron Microscope).

Experimental Procedure

Material

The material used for specimens was an aluminum alloy (JIS 6063-T5), of which the chemical compositions and mechanical properties are shown in Tables 1 and 2, respectively. The material used for the contact pad was a chromium-molybdenum steel (JIS SCM420). Three types of the specimens were used: the specimen without film, the specimen with anodic film (alumite), and the specimen with lubricating anodic film. After degreasing an aluminum specimen surface with a neutral degreasing agent, it was anodized for 35 min in a bath containing 20 wt% sulfuric acid solution with constant current density of 2.5 A/dm^2, to obtain an oxide film with 30 µm thickness. This film was re-anodized in aqueous solution of ammonium thiomolybdate with constant current density of 0.1 A/dm^2 for 6 min and then molybdenum sulfide was impregnated into oxide film pores.

Wear and Scratch Test

Wear and scratch tests were carried out to investigate the fundamental properties of the anodic film prior to the fatigue tests. The wear test was carried out using a ball-on-disk type wear tester (material of ball: a high carbon chromium bearing steel (JIS SUJ2), diameter of ball: 5 mm, sliding distance per revolution: 3.14 cm) under sliding distance of 62 m, load of 1.5 kN and sliding speed of 5 cm/sec in air. The scratch test was also carried out with a sapphire needle. Film surface was horizontally scratched by a 0.2 mm diameter sapphire needle at a normal load of 4.9 N and test speed of 50 mm/min. The resultant cracks were observed using an optical microscope.

Residual Stress of Anodic Film

Since the anodic oxide film is amorphous, the measurement of residual stress by X-ray diffraction method is very difficult. The residual stress was estimated by the following method.

After degreasing 99.8 % pure aluminum foil (50 μm×25 mm×100 mm) with a neutral degreasing agent, it was perfectly annealed (400 ℃, 1 hr). After one surface of the foil was masked by an organic adhesive tape, 2,5,8,10,15,20 and 24 μm thickness of anodic oxide films were formed on the other non-masked surface in a bath containing 15 wt% sulfuric acid solution at 15 ℃ bath temperature with constant current density of 2 A/dm^2. In 0.1 wt% aqueous solution of ammonium thiomolybdate (($NH_4)_2MoS_4$), the stainless steel (JIS SUS304) plate was arranged as the cathode and the anodic oxide film was arranged as anode. Electrolysis was processed at 20 ℃ bath temperature with a constant current density of 50 mA/ dm^2 to 120 V. After anodizing, aluminum board was well rinsed and masking tape on it was peeled carefully with organic solvent, dried in hot air at 60 ℃ and held at room temperature. The residual stress of anodic film was calculated using equation of Barklie & Davies [5] with the measured radius of curvature.

Table 1 — *Chemical composition (wt%).*

	Si	Fe	Cu	Mn	Mg	Cr	Zn	Ti
6063-T5	0.44	0.20	0.03	0.02	0.53	0.01	0.01	0.02

	C	Si	Mn	P	S	Cu	Ni	Cr	Mo
SCM420	0.21	0.23	0.70	0.016	0.014	0.13	0.07	1.01	0.15

Table 2 — *Mechanical properties.*

	Tensile strength σ_B(MPa)	Elongation ϕ (%)	Young's modulus E(GPa)	Hardness HV(MPa)
6063-T5	230	21	65	770
SCM420	940	16	206	2850

Plain Fatigue and Fretting Fatigue Test

The shapes and dimensions of the plain fatigue specimen, fretting fatigue specimen and bridge-type contact pad are shown in Fig. 1 and Fig. 2. The gage parts of all the fatigue specimens and pad contact feet were polished using successively finer grade of emery paper and the final polishing was conducted longitudinally with 1500 grade emery paper. Fretting was induced by clamping a pair of bridge-type pads onto both sides of the specimen (Fig. 1(b)), using a proving ring [6]. The clamping pressure was adjusted before and during the tests to give a constant value of 50 MPa. The tangential force between the specimen and the pad during the tests was measured by strain gages attached underneath the central part of the pad. The plain fatigue and fretting fatigue tests were carried out using a servo-hydraulic fatigue test machine with a capacity of 98 kN under a stress ratio R of −1 at frequencies ranging from 10 to 20 Hz. Fracture surfaces of the specimens tested were observed using a scanning electron microscope and surface roughness of the fretted region was measured using a surface roughness tester.

In-situ and Continuous Observation of Fretting Fatigue Crack

Fretting fatigue cracks were continuously observed using small fretting fatigue specimens and a fatigue testing machine with SEM (Scanning Electron Microscope). The shapes and dimensions of the fretting fatigue specimen and the bridge-type contact pad are shown in Fig. 3. Fretting fatigue tests were carried out under a stress ratio R of −1 and a clamping pressure of 50 MPa, which was applied by using a proving ring, as shown in Fig. 4.

Results and Discussion

Fundamental Properties of Anodic Film and Lubricated Anodic Film

Figure 5 shows the relationship between frictional coefficient and sliding distance. The anodic film shows stable value of friction coefficient, 0.4, after about 10 m running (300 revolutions). On the other hand, the lubricating anodic film shows stable value of friction coefficient, 0.2, after about 50 m running (1500 revolutions). The frictional coefficient of the lubricating anodic film was approximately half of that of the anodic film.

Figure 6 shows the observations of scratched surface of 24 μm thickness film. It was found that number of cracks by scratched test in the lubricating anodic film was less than that in the anodic film, which indicated that the lubricating anodic film had higher resistance for cracking compared to the anodic film.

Figure 7 shows the relationship between hardness and distance from the surface of film. The hardness test was carried out under a load of 0.5 N and hold time of 30 seconds. Surface hardness of the anodic film and lubricating anodic film was 0.55 GPa and 0.37 GPa, respectively. The hardness of base material was about 0.1 GPa. This value, HV=0.1 GPa is the same in comparison with aluminum alloy before anodizing.

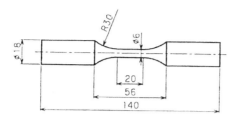

Figure 1 — *Plain fatigue specimen.*

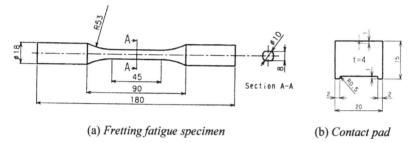

(a) *Fretting fatigue specimen*   (b) *Contact pad*

Figure 2 — *Fretting fatigue specimen and contact pad.*

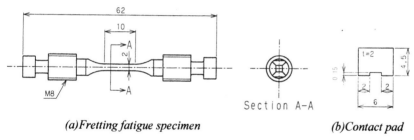

(a)Fretting fatigue specimen   *(b)Contact pad*

Figure 3 — *Fretting fatigue specimen and contact pad for SEM Servo.*

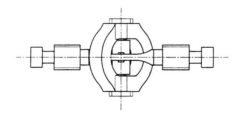

Figure 4 — *Schematic illustration of fretting fatigue test.*

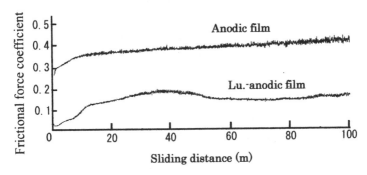

Figure 5 — *Relationship between frictional force coefficient and sliding distance.*

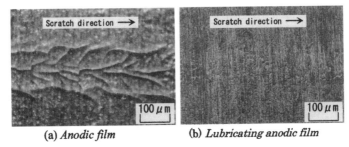

(a) *Anodic film*   (b) *Lubricating anodic film*

Figure 6 — *Observations of surface crack induced by the scratch test.*

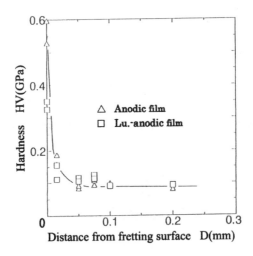

Figure 7 — *Relationship between hardness and distance from fretting surface.*

Though the surface hardness of lubricating anodic film was lower than that of anodic film, the former exceeds the latter in frictional property and crack resistance.

Residual Stress of Anodic Film

It is well known that the volume of anodic oxide film increases about 1.49 times of that of aluminum bulk metal, if perfect alumina crystal structure is assumed as explanation of Kabasbersky's volume ratio (oxide film/metal) [7]. Therefore, compressive residual stress might be induced in anodic film. When the volume ratio is less than 1.0, tensile residual stress may be caused [7], which is known to influence mechanical properties and corrosion resistance in most materials.

In the anodic oxidizing process, dense layer with 5 % porosity, which is almost equivalent to a volume ratio of 1.49 [8,9], is formed at the initial stage, while the volume ratio decreases with increasing thickness of film. On the average, the through-thickness porosity will be 20 %. Near surface layer of the film will have much higher porosity.

According to the simple calculation, the volume ratio at the initial stage (2~5 μm, thickness) is 1.40 (1.49×(1-0.05)=1.40). When the porosity is assumed to be about 20 %, the volume ratio of the film is estimated to be 1.18. Therefore, the compressive residual stress in the film will be considerably reduced [10].

The residual stress in the film is influenced not only by porosity but also by other factors, such as solution, intermetal compounds, etc. It is difficult to measure the residual stress by X-ray method, because the anodic film is amorphous.

Figure 8 shows curvatures of the foil specimen with various thickness of anodic film, which is produced on the one side of fully annealed aluminum foil. The curvature increases with increasing film thickness. The same result was observed in the case of lubricating anodic film. Figure 9 shows the difference in the curvature between anodic film and lubricating anodic film. By impregnation of molybdenum sulfide in lubricating anodic film, the compressive residual stress increased and the curvature was reduced. Based on these observations, the residual stress of anodic film and lubricating anodic film is estimated according to Borklie & Davies [5] equation:

$$\sigma = h^2 E / 6rd(1-d/h),$$

where, h: foil thickness in mm, E: Young's modulus of the aluminum foil in MPa, r: radius of curvature in mm, d : anodic oxide film thickness in mm. Figure 10 shows the relationship between residual stress and thickness of anodic oxide film. As can be seen from the figure, the tensile residual stress of lubricating anodic film was reduced by about 20 MPa compared to that of anodic film.

Plain Fatigue Strength

The S-N curves for plain fatigue of anodic film and lubricating anodic film aluminum alloys are shown in Fig.11. The plain fatigue strengths at 10^7 cycles were around 80 MPa for the specimen with anodic film and around 90 MPa for the specimen with lubricating anodic film and also around 90 MPa for the specimen without film.

Shiozawa [1] has reported that the plain fatigue strength at 10^7 cycles using bending

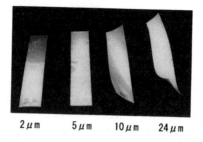

2 μm   5 μm   10 μm   24 μm

Figure 8 — *Foil curvatures for varying thickness of anodized coatings.*

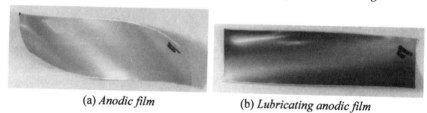

(a) *Anodic film*   (b) *Lubricating anodic film*

Figure 9 — *Difference in deformation of anodic film between non-impregnation and impregnation, which indicates the difference in residual stress.*

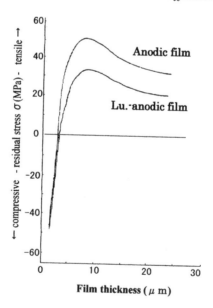

Figure 10 — *Residual stress in anodic film and lubricating anodic film.*

rotational fatigue tester for the aluminum alloy specimen (JIS A2014-T6 and JIS A6151-T6) with anodic film was lower 20~30 % than that without film. Hirata et al. [2] have reported that though the plain fatigue life in high stress region for the aluminum alloy specimen (JIS A1080) with anodic film was shorter than that without film, the life in low stress region was longer than that without film. These results show that cracking of anodic film during fatigue tests results in low fatigue strength.

Fig.12 shows a SEM observation of plain fatigue crack initiation after the specimen failed. Though the fatigue cracks of the specimen without film were initiated by associating with slipping on the surface as shown in Fig.12(a), it seemed that the fatigue cracks of the specimen with anodic film were initiated from fracture region of the film as shown in Fig.12(b). On the other hand, it seemed that the fatigue cracks of the specimen with lubricating anodic film were initiated inside base material as shown in Fig.12(c).

Fretting Fatigue Strength

The S-N curves for fretting fatigue are shown in Fig.11. The fretting fatigue strengths of 10^7 cycles were around 30 MPa for the specimen without film, around 40 MPa for the specimen with anodic film and around 50 MPa for the specimen with lubricating anodic film.

Tangential Force Coefficient

The relationship between tangential force coefficient measured after getting stable condition and relative slip amplitude is shown in Fig.13. The tangential force coefficient is given as $\mu = Fa/P$, where Fa is the amplitude of tangential force and P is the contact pressure. The tangential force coefficient increased with increasing relative slip amplitude for all materials and attained a constant value. Relative slip amplitude was the calculated value as rigid to contact pad. The constant value of the frictional force coefficient of the specimen without film, anodic film and lubricating anodic film were 0.85, 0.7 and 0.6 respectively. The small difference in frictional force coefficient between two anodic films in fretting tests compared to the case of ball-on-disk test may result from the severe test condition including contact piece geometry and contact load.

Surface Roughness

The maximum roughness of the fretted surface was also measured in order to examine the fretting surface damage. The relationship between maximum surface roughness and relative slip amplitude is shown in Fig.14. The maximum surface roughness increased with increasing relative slip amplitude for all materials and was saturated over 10~20 μm of relative slip amplitude. The maximum surface roughness values were around 45 μm for the specimen without film, around 20 μm for the specimen with anodic film and around 10 μm for the specimen with lubricating anodic film. The maximum surface roughness values for the specimen with lubricating anodic film were significantly lower than those of other materials. From the results of tangential force coefficient and surface roughness, though the difference was not so significantly

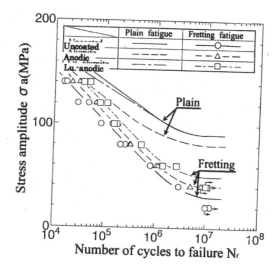

Figure 11 — S-N curves.

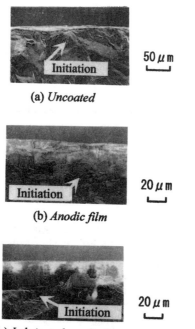

(a) Uncoated

(b) Anodic film

(c) Lubricated anodic film

Figure 12 — Crack initiation under the plain fatigue.

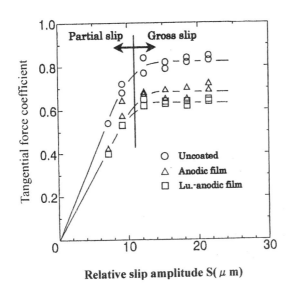

Figure 13 — *Relationship between tangential force coefficient
and relative slip amplitude.*

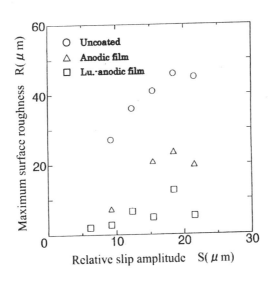

Figure 14 — *Relationship between maximum surface roughness
and relative slip amplitude.*

large as the case of the frictional coefficient on the wear test, lubricating anodic film excelled anodic film in the fretting wear characteristics.

SEM Observations of Fretting Region

In all materials, fretting fatigue cracks were initiated near the edge of the contact region. A longitudinal cross section of the fretting fatigue specimen is shown in Fig.15. Fig.15(a) shows that the fretting crack on the specimen without film grows to 45° to fretting surface. This is almost the same with the case of other materials without film [*11 -13*]. The observation near the edge of the contact region is shown in Fig.15(b). The damage of film was significantly severe and that there were lots of small fretting cracks. The damage of lubricating anodic film was very small as shown in Fig.15(c). The

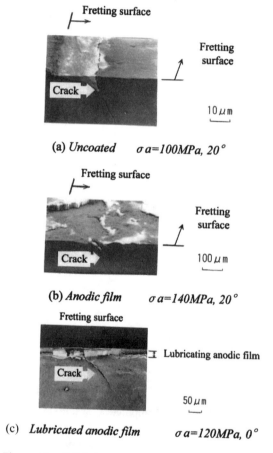

(a) *Uncoated* $\sigma a=100MPa, 20°$

(b) *Anodic film* $\sigma a=140MPa, 20°$

(c) *Lubricated anodic film* $\sigma a=120MPa, 0°$

Figure 15 — *SEM observations of fretting cracks on the longitudinal cross-section of the specimens.*

lubricating anodic film had the lowest coefficient of friction and a higher cracking resistance, which contributed to the late crack initiation and the low crack growth rates.

In-situ Observations of Fretting Fatigue Cracks

Fretting fatigue crack growth behavior in anodic film is shown in Fig.16. Fretting fatigue cracks in anodic film were initiated in the early stage of fatigue life. In some cases, the cracks were initiated just when contact pressure was applied. However, the cracks in the anodic film generally initiated over 5×10^2 cycles and they grew into base material over 3×10^3 cycles.

Figure 17 shows the fretting fatigue crack growth behavior in anodic film and lubricating anodic film at the same stress level. In both materials, wear powder produced during the test was observed in the late stage of life but was less in quantity compared to the case of material without film. The surface hardness was improved due to the hard film, which contributed to the wear resistance. The cracks in the anodic film initiated over 5×10^2 cycles, and the cracks grew into the base material over 3×10^3 cycles. On the other hand, the cracks in the lubricating anodic film initiated over 1.2×10^3 cycles, and the cracks grew into the base material over 6.4×10^3 cycles. Lubricating anodic film contributed to the retardation crack initiation and also to the low crack growth rate due to the lower tangential force coefficient than that of the bulk material without film.

Effect of Anodic film on fatigue strength

The effects of anodic film on fatigue strength are summarized from results obtained as follows.

Plain fatigue strength of the specimen with anodic film was lower than that of the specimen with lubricating anodic film, which almost coincided with that of the specimen without film. The low fatigue strength of the specimen with anodic film may result from the brittleness and tensile residual stress of the anodic film, which are attributed to the cracking at lower applied stress. On the other hand, cracking of the lubricating anodic film during fatigue tests hardly occurred due to its ductile nature and low residual stress and consequently the fatigue strength was not degraded.

Fretting fatigue strength of the specimen with lubricating anodic film was higher than that of the specimen with anodic film, which was higher than that of the specimen without film. The lubricating anodic film had the lowest coefficient of friction and a higher cracking resistance, which contributed to the retardation crack initiation and the low crack growth rates. The anodic film also had low coefficient of friction as compared with the specimen without film and so prevents meta-to-metal contact, which may result in higher fatigue strength compared to the specimen without film.

The lubricating anodic film is of significant benefit for fretting fatigue strength without degradation of plain fatigue strength.

Conclusions

Fundamental properties of aluminum alloy JIS 6063-T5 specimens with anodic film

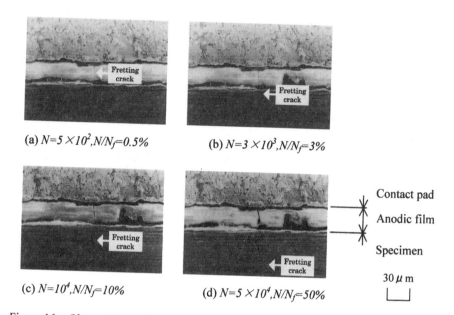

(a) $N=5\times10^2, N/N_f=0.5\%$ (b) $N=3\times10^3, N/N_f=3\%$

(c) $N=10^4, N/N_f=10\%$ (d) $N=5\times10^4, N/N_f=50\%$

Figure 16—*Observations of fretting crack on the side surface of the specimen in real time with SEM(anodic film : $\sigma_a=120MPa, P=50MPa, N_f=10^5$).*

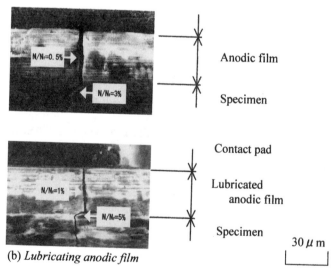

(b) *Lubricating anodic film*

Figure 17—*Observations of fretting crack on the side surface of the specimen with SEM ($\sigma_a=120MPa, P=50MPa$).*

and lubricating anodic film were investigated. Plain fatigue and fretting fatigue tests were also carried out to investigate fretting fatigue properties of the aluminum alloyspecimens with anodic film and lubricating anodic film.

Furthermore, fretting fatigue cracks of the specimen were observed in-situ using small fretting fatigue specimens and a fatigue testing machine with SEM (Scanning Electron Microscope).

The main results obtained are summarized as follows:

(1) Though the surface hardness of lubricating anodic film was lower than that of anodic film, the lubricating anodic film had excellent frictional property and crack resistance compared to the anodic film.

(2) Plain fatigue strength of the specimen with anodic film was lower than that of the specimen with lubricating anodic film, which almost coincided with that of the specimen without film. The low fatigue strength of the specimen with anodic film may result from the brittleness and tensile residual stress of the anodic film, which are attributed to the cracking at lower applied stress. Cracking of the lubricating anodic film during fatigue tests hardly occurred due to its ductile nature and low residual stress and consequently the fatigue strength was not degraded.

(3) Fretting fatigue strength of the specimen with lubricating anodic film was higher than that of the specimen with anodic film, which was higher than that of the specimen without film. The lubricating anodic film had the lowest coefficient of friction and a higher cracking resistance, which contributed to the retardation crack initiation and the low crack growth rates. The anodic film also had low coefficient of friction as compared with the specimen without film and prevents meta-to-metal contact, which may result in higher fatigue strength compared to the specimen without film.

The lubricating anodic film is of significant benefit for fretting fatigue strength without degradation of plain fatigue strength.

References

[1]   Shiozawa, K., "Effect of Fatigue Strength on Anodized Aluminum Alloy" *Journal of the Japanese Anodizing Association,* Vol.3-324, 1999, pp.3–10.

[2]   Hirata, M., Maejima, M., Saruwatari, K., Shigeno, H., and Takaya, M., "Rotational Bending Fatigue of Anodized Coating of Aluminum," *Aluminum SurfaceTechnology,* Vol. 47-4, 1996, pp.376–377.

[3]   Cree, A. M., Weidmann, G. W., and Hermann, R., *Journal of Material Science Letters,* Vol. 4, 1995, pp.1505–1507.

[4]   Design Data for Fatigue Strength of Metal Materials, (III) Effect of Environment, *Japan Society of Mechanical Engineers,* 1974, p.142.

[5]   Barklie, R. H., and Davies, J., *Institute of Mechanical Engineering,* 1957, p. 731.

[6]   Nishida, T., Mizutani, J., Mutoh, Y., Yoshii, K., Ebihara, O., and Miyashita, S., "Plain Fatigue and Fretting Fatigue Strengths of AC4CH Aluminum Alloy" *Journal of Japan Institute of Light Metals,* Vol. 49-10, 1999, pp. 493–498.

[7]   Baba, N., *Journal of the Japanese Anodizing Association,* Vol. 281-3, 1994, p.1.

[8]   Pilling, N. B., and Bedworth, R. E., *Journal Inst. Metals,* Vol. 29, 1923, p.529.

[9]   Nelson, J. C., and Oriani, R. A., *Corrosion Science,* Vol. 34, 1993, p. 307.

[10] Maejima, M., Saruwatari, K., Isawa, K., and Takaya, M., "Curvatures of Anodized Coatings of Aluminum" *Surface technology,* Vol.46-9, 1995, pp. 856–859.

[11] Tanaka, K., Mutoh, Y., and Sakoda, S., "Effect of ContactMaterials on Fretting Fatigue in a Spring Steel" *Japan Society of Mechanical Engineers,* Vol.51-464, 1986, pp.1200–1207.

[12] Mutoh, Y., Nishida, T., and Sakamoto, I., "Effect of Relative Slip Amplitude and Contact Pressure on Fretting Fatigue Strength" *Japan of the Society of Material Science,* Vol.37-417, 1988, pp.649–655.

[13] Waterhouse, R. B., "Fretting Fatigue in Aqueous Electrolytes," *Fretting Fatigue, Applied Science Publication,* 1981, pp.159–175.

Masaki Okane,[1] Kazuaki Shiozawa,[2] Masaharu Hiki,[3] and Kazutaka Suzuki[4]

Fretting Fatigue Properties of WC-Co Thermal Sprayed NiCrMo Steel

REFERENCE: Okane, M., Shiozawa, K., Hiki, M., Suzuki, K., "**Fretting Fatigue Properties of WC-Co Thermal Sprayed NiCrMo Steel**," *Fretting Fatigue: Advances in Basic Understanding and Applications, ASTM STP 1425,* Y. Mutoh, S. E. Kinyon, and D. W. Hoeppner Eds., ASTM International, West Conshohocken, PA, 2003.

ABSTRACT: Fretting fatigue tests of NiCrMo steel (JIS SNCM439) sprayed with tungsten carbide with additive of 12% cobalt (WC-Co) by HVOF (High Velocity Oxygen Fuel) were carried out to study the effect of WC-Co coating on fretting fatigue behavior. Since the fatigue strength of the present alloy steel was reduced to about 27% by the effect of fretting, fretting fatigue strength was improved by approximately 30% by WC-Co spraying onto the specimen. The tangential force coefficient, which is tangential force measured during the fretting fatigue tests divided by contact load, behaviors of WC-Co sprayed specimen were different form those of the non-sprayed specimen. Especially, the tangential force of WC-Co specimen in an early stage of the test indicated significant low value compared with that of non-sprayed specimen and this was caused from good wear resistance of WC-Co layer. Therefore, the initiation of a fretting crack in case of WC-Co sprayed specimen delayed, and crack propagation rate was lower than that of non-sprayed specimen. These were main reasons for the improvement of fretting fatigue strength by WC-Co spray.

KEYWORDS: fretting fatigue, NiCrMo steel, spray, WC-Co, HVOF, fretting wear

Introduction

Fretting fatigue, the characteristic fatigue process caused by surface damage induced as the result of microscopic movement between contacting two surfaces, is well known as one of the most important problems in designing or maintenance of many industrial machineries. Many investigators have paid a lot of effort to understand the

[1] Research Associate, Department of Mechanical and Intellectual Systems Engineering, Faculty of Engineering, Toyama University, 3190 Gofuku, Toyama, 930-8555, Japan
[2] Professor, Toyama University, Japan
[3] Graduate Student, Toyama University, Japan
[4] Executive Director, Ohsuzu Giken Co. Ltd., 773-1 Wada-cho, Hamamatsu, 435-0016, Japan

fretting or improvement of fretting fatigue strength [1-4].

Various provisions for fretting fatigue fracture have been considered, and most of them were introduced by Waterhouse [5] and Hirakawa [6]. Surface treatment technologies were applied for resisting the fretting fatigue, such as nitriding [7], carbonizing [8], tufftriding [9], shot peening [10,11] and so on. They were applied for preventing the initiation of fretting fatigue crack and for retardation of the crack propagation, by hardening the surface and by introducing residual compression stress near the surface. The authors have previously reported [12,13] the fretting fatigue properties of TiN coated carbon steel by PVD (Physical Vapor Deposition), and it has been clarified that the TiN coating was effective for improving fretting fatigue strength, by the retardation of fretting fatigue crack initiation.

The PVD or CVD (Chemical Vapor Deposition) technique produces "thin layer" only few microns in thickness. On the other hand, spraying technologies can laminate thicker layer compared with PVD or CVD, further the manufacturing process and equipment of spraying may be simple compared with PVD or CVD. Application of hard layer which has excellent wear resistance by spraying is expected to not only improve the fretting fatigue strength but also save cost.

High Velocity Oxygen Fuel (HVOF) spraying, which is a relatively new spraying method, enables to generate denser sprayed coating compared with the process with conventional plasma spraying, because of higher particle velocity during the spraying process (over 1000 m/sec.) [14]. Furthermore, since the flame temperature is comparatively low (less than 3000K), HVOF must be extremely effective for spraying of carbide powders, such as tungsten carbide (WC), which is expected as one of the hard-wearing materials but weak against heat [14]. In other words, HVOF can construct the higher quality carbide sprayed coating which is unable to be produced by the plasma spraying. Nowadays, the mainstream of carbide spraying has been HVOF [14].

Major applications of WC coating by HVOF are for the sliding parts, and a lot of investigations about wear properties of WC coating by HVOF were conducted [15-18]. They generally reported good wear resistance of the WC coating. Very few studies on fatigue behaviors of the WC sprayed steel by HVOF are conducted [19,20], and improvement of fatigue strength by the coating has been reported.

As described above, WC sprayed coating has both good wear and fatigue resistance, it is expected to improve fretting fatigue strength caused by cracking at contacting surface, but no report has been published regarding that. In this study, fretting fatigue tests were carried out using WC-Co sprayed NiCrMo steel to investigate the basic fretting fatigue properties and its performance for fretting resistance. The fretting fatigue fracture process of the WC-Co sprayed steel was also discussed.

Experimental Procedure

Fatigue Specimen and Contact pad

In this study, Nickel-Chromium-Molybdenum steel JIS SNCM439 was used for the substrate material and for the contact material. The chemical compositions of the material used are shown in Table 1. The specimen was machined from a bar with a diameter of 22mm after the heat treatments as follows, at 1123K (850°C) for 1.7h O.Q. (Oil

Table1 - *Chemical composition of the substrate material (wt.%)*

C	Si	Mn	P	S	Cu	Ni	Cr	Mo
0.37	0.28	0.75	0.021	0.017	0.05	0.74	0.81	0.15

Table2 - *Mechanical properties of the substrate material* [21]

Tensile strength σ_B [MPa]	1120
Proof stress σ_y [MPa]	1060
Young's modulus E [GPa]	210
Elongation δ [%]	22
Reduction of area Φ [%]	58
Hardness HV	301

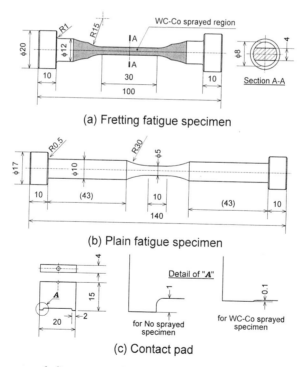

(a) Fretting fatigue specimen

(b) Plain fatigue specimen

(c) Contact pad

Fig.1 - *Shapes and dimensions of the (a) fretting fatigue specimen, (b) plain fatigue specimen and (c) contact pad. (Unit : mm)*

Quenching) then at 893K (620°C) for 2h W.C. (Water Cooling). The mechanical properties of the material after heat treatment are shown in Table 2.

The shapes and dimensions of the specimens used are shown in Fig.1: (a) shows the fretting fatigue specimen, (b) the plain fatigue specimen and (c) the contact pad. The gauge parts of both fatigue specimens (fretting fatigue specimen: 30mm long, plain fatigue specimen: 10mm long) were polished with emery paper up to grade #1500 longitudinally and final polishing was conducted for the plain fatigue specimen by using alumina slurry with grain size of 0.3 μm. The contact pad was cut from a bar with a diameter of 60mm by wire-EDM (Wire Electric Discharge Machining) into designed shape and two contact area of each contact pad (2mm×4mm face) were finalized by the flat-surface grinding machine. Two kinds of contact pad, as shown in Fig. 1(c) were used for the fretting fatigue tests. One was, with 1mm foot height, for the fretting fatigue test of non-sprayed specimen and the other was, with 0.1mm foot height, for the test of WC-Co sprayed specimen. Detail explanation will be given later in this paper.

Spray Conditions by HVOF

Tungsten carbide with additive of 12wt.% cobalt (WC-12%Co, Grain size: 6–38μm) was sprayed onto surface of the fretting fatigue specimen by HVOF (High Velocity Oxygen Fuel) process. The DJ (Diamond Jet [22]) gun was used for HVOF process. The chemical compositions and spraying conditions were summarized in Table 3 and Table 4, respectively. Before HVOF process, blast treatment was applied onto the specimen surface to enhance the bonding between substrate metal and WC-Co layer. The blast

Table3 - *Chemical composition of the spray powder used (wt.%)*

Co	TC	Fe	W
12.7	5.4	0.2	bal.

Table4 - *Spraying conditions by HVOF*

Gas pressure [MPa]	Oxygen	1.03
	C_3H_8	0.69
	Air	0.52
Gas flow [m³/s]	Oxygen	4.77×10^{-3}
	C_3H_8	1.31×10^{-3}
	Air	5.84×10^{-3}
Powder delivery rate [N/s]	0.3626~0.3724	
Spraying distance [m]	0.17	
Specimen Temperature [K]	Pre-heating	333
	After spraying	413~433

Table5 - *Blast conditions*

Blast material	Alumina grid #46
Pressure [MPa]	0.44
Blasting distance [m]	0.1
Diameter of the blast nozzle [m]	0.008

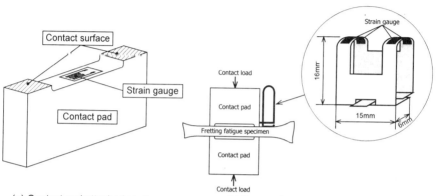

(a) Contact pad attached strain gauge   (b) Set up of the extensometer

Fig.2 - *Measuring procedures of the tangential force and the relative slip.*

conditions were shown in Table 5. Surface of the WC-Co layer was ground perpendicularly by using diamond grind wheel, and after that final polish was conducted longitudinally by using diamond emery paper up to grade of #1500. The thickness of WC-Co layer was 100μm and the micro vickers hardness of the layer HV was about 1152 (about 11.3 GPa).

Fretting Fatigue Tests and Plain Fatigue Tests

A fatigue test machine with 49 kN capacities was used for fretting fatigue tests and plain fatigue tests. All the tests were carried out under a load-controlled sinusoidal wave form condition with a stress ratio R ($\sigma_{min}/\sigma_{max}$)=-1, and frequency f=20Hz in air at room temperature. For the fretting fatigue tests, two contact pads were pressed onto both sides of the gauge section, and a contact pressure was controlled at constant value of 100 MPa by using proving ring.

Tangential force between specimen and contact pad during the fretting fatigue test was measured using a strain gauge attached on the bottom of the contact pad as shown in Fig. 2(a). And relative slip amplitude between them was also measured by using small extensometer as shown in Fig. 2(b).

After the tests, fretted surfaces were observed in detail by using a scanning electron microscope (SEM). The cross section of the specimen around the fretted region was also

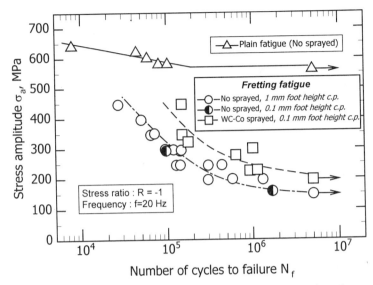

Fig. 3 - *S-N curves for fretting fatigue of WC-Co sprayed steel.*

observed using SEM to discuss the initiation and growth of the fretting cracks.

Results and Discussions

S-N Curves (Relationship between fretting fatigue lives and applied stresses)

The relationship between applied stress amplitude and number of cycles to failure (S-N curve) is shown in Fig. 3. In the figure, data points with arrow show that no failure occurred up to the testing cycles of 5×10^6 cycles. It can be seen from the figure that the S-N curve for plain fatigue of the substrate material ($\triangle$) was almost horizontal and the fatigue strength at 5×10^6 cycles was around σ_a=560 MPa. The fretting fatigue strength at 5×10^6 cycles of both no-sprayed ($\bigcirc$) and WC-Co sprayed ($\square$) specimen were significantly reduced compared with the plain fatigue strength. The fretting fatigue strength of the no-sprayed specimen at 5×10^6 cycles was σ_a=150 MPa which was about 27% comparing with the plain fatigue strength and that of the WC-Co sprayed specimen was σ_a=200 MPa. About 30% increase of fretting fatigue strength was observed in the WC-Co sprayed specimen compared with the no-sprayed specimen.

As mentioned before, two kinds of contact pads as shown in Fig. 1(c) were used in this study. The contact pad with 1 mm foot height was used for the fretting fatigue test of no-sprayed specimen and 0.1 mm foot height contact pad was used for the experiment of WC-Co sprayed specimen. This is because that when the higher foot contact pads were used for the experiments of WC-Co sprayed specimen, the contact pads were broken at their foot roots before the specimen failure occurred. Therefore, contact pads were used for the tests of WC-Co sprayed specimen with foot height of 0.1 mm for reducing stress concentrations at their roots. It is well known that the fretting fatigue life and strength are

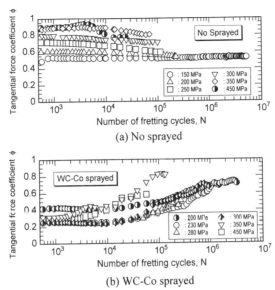

(a) No sprayed

(b) WC-Co sprayed

Fig. 4 - *Variations of tangential force coefficients ϕ during the fretting fatigue tests.*

significantly affected by the shape and dimension of the contact pad [23,24]. For confirmation, several fretting fatigue tests of the no-sprayed specimens using 0.1 mm foot height contact pads were conducted. Results of the tests were indicated in Fig. 3 by half-solid circles. No significant differences were found in fretting fatigue lives and strength between the test results with 1 mm and 0.1 mm foot height contact pads. And also no significant difference was found in the results of tangential force coefficient and relative slip amplitude between them as described later (Fig. 6). Therefore, the difference in the shapes and dimensions of two kinds of contact pads used in this study seems to have no influences on fretting fatigue behaviors of the present material. All experimental results of WC-Co sprayed specimen shown in Fig. 3 were the results using 0.1 mm foot height contact pads.

Tangential Force Coefficient and Relative Slip Amplitude

Variations of the tangential force coefficient ϕ during the fretting fatigue tests are shown in Fig. 4. The tangential force coefficient ϕ was defined as a ratio of the tangential force F to the contact load P, $\phi=F/P$. In case of no-sprayed specimen, tangential force coefficients were almost constant during the fretting fatigue tests, which increased with increasing stress amplitude. This tendency is similar to that of many metallic materials observed by many investigators previously [1-4]. On the other hand, in case of WC-Co sprayed specimen tangential force coefficients at early stage of the experiments, were about 0.2 to 0.4 and have no dependence on the applied stress level. They were clearly low compared to those of the no-sprayed specimens. Thereafter, the tangential force coefficients increased gradually with the number of fretting cycles. In this study, a pair of contact pads were used for each fretting fatigue test and the tangential force was

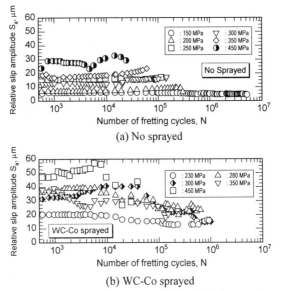

(a) No sprayed

(b) WC-Co sprayed

Fig. 5 - *Variations of relative slip amplitude S_a during the tests.*

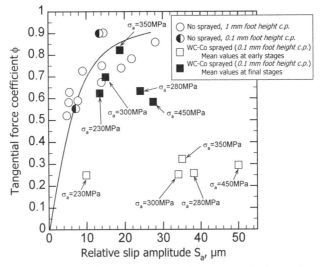

Fig. 6 - *Relationship between tangential force coefficient ϕ and relative slip amplitude S_a.*

measured at each contact pad respectively. In case of no-sprayed specimen, since the two values of tangential forces were almost equal, the mean values of them were plotted in Fig. 4(a). However, in case of WC-Co sprayed specimen, there were some cases that significant difference was observed between two measured tangential forces especially in latter stage of the experiment. In that case, the tangential force at the side where the main crack initiated was plotted in Fig. 4(b).

Variations of the relative slip amplitude S_a during the fretting fatigue tests are shown in Fig. 5. In case of no-sprayed specimen, the relative slip amplitude S_a was almost constant during the test and increased with increasing stress level. On the other hand, in case of WC-Co sprayed specimen S_a was larger than that of no-sprayed specimen at early stage of the test, which increased with increasing stress level, and became smaller with the number of fretting cycles. Variations of the relative slip amplitude almost synchronized with those of the tangential force as mentioned before.

Relationship between the tangential force coefficient and the relative slip amplitude is shown in Fig. 6. All plots of no-sprayed specimens ($\bigcirc$) are the mean values of the tangential force coefficient and relative slip amplitude measured during the tests. The results in case of WC-Co specimen separated into two plots in the figure, which are the mean values at early stage ($\square$), i.e. the mean values of them up to cycles of about 5×10^3, and the final values ($\blacksquare$). In case of no-sprayed specimen, tangential force coefficient increased with increasing relative slip amplitude and became constant value above the certain value of relative slip amplitude as observed in many metallic materials [1-4]. In case of WC-Co sprayed specimen, the lower tangential force coefficient and the larger relative slip amplitude at early stage became almost similar values with those of no-sprayed specimen at final stage.

Observations of Fretted Surfaces and Fretting Fatigue Cracks

Fretted specimen surfaces after the experiments were observed by SEM. The

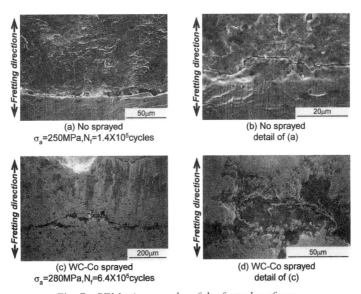

(a) No sprayed
σ_a=250MPa,N_f=1.4X10⁵cycles

(b) No sprayed
detail of (a)

(c) WC-Co sprayed
σ_a=280MPa,N_f=6.4X10⁵cycles

(d) WC-Co sprayed
detail of (c)

Fig. 7 - *SEM micrographs of the fretted surfaces.*

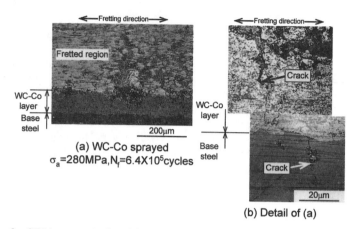

Fig. 8 - *SEM micrographs of fretting cracks in the WC/Co sprayed specimen.*

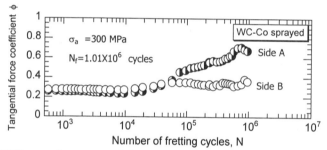

Fig. 9 - *Difference in tangential force coefficient caused by difference of fretting wear condition.*

results are shown in Fig. 7. Figure 7(a) shows a fretted surface around the contact edge of no-sprayed specimen and a detailed observation of Fig. 7(a) is shown in Fig. 7(b). It is clearly shown that there are some cracks near an edge of contact induced by fretting action. All of no-sprayed specimens fractured from the cracks as shown in Fig. 7(a) and 7(b). Figure 7(c) shows a fretted surface around the contact edge of WC-Co sprayed specimen. Damage on the WC-Co sprayed specimen was not so severe as the fretting damage of no-sprayed specimen. Some cracks are observed near the contact edge where the wear of WC-Co layer is severe as shown in Fig. 7(d), and all WC-Co sprayed specimens fractured from these cracks. Wear and spalling of WC-Co layer occurred at whole of contact region. However, since the full-thickness wear and attrition of the WC-Co layer did not occur, the layer was remaining until the final fracture occurred.

Cross section of the WC-Co sprayed specimen tested, which was cut along longitudinal direction, was observed by using SEM. The SEM micrographs are shown in Fig. 8. Although cracks in the WC-Co layer are not so clear to find out as those in substrate metal, it can be seen that a crack initiated at specimen surface by the fretting

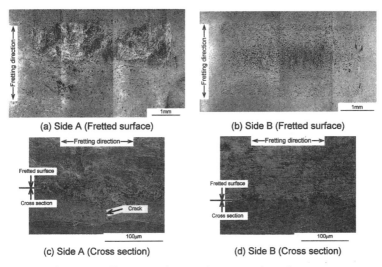

(a) Side A (Fretted surface)   (b) Side B (Fretted surface)

(c) Side A (Cross section)   (d) Side B (Cross section)

Fig. 10 - *Difference in fretting damage and crack initiation.*

action propagates through the WC-Co layer with slight angle. The crack has further propagation into the substrate metal after slight growth along the boundary between the layer and the substrate.

Fretting Fatigue Fracture Process in WC-Co Sprayed Specimen

From the experimental results, it was found that fretting fatigue lives of WC-Co sprayed specimen increased 2-10 times compared to those of no-sprayed specimen. And the fretting fatigue strength of WC-Co sprayed specimen significantly increased comparing with that of no-sprayed specimen. Therefore, WC-Co spray by HVOF seems to become an effective method for improving fretting fatigue strength. In this section, fretting fatigue fracture process will be discussed.

As mentioned before, there were some cases in which significant difference in tangential force variation of WC-Co sprayed specimen was observed. The most significant case is shown in Fig. 9. For convenience, each fretted region is called "Side A" and "Side B" respectively in this study. The tangential force coefficient at "Side A" increased gradually with increasing number of fretting cycles after the middle stage of the test. However, the tangential force coefficient at the opposite side did not increase, as much as "Side A", it seems to be constant. The principal crack occurred at "Side A" in this case. Figure 10 shows fretted surfaces and cross sections of "Side A" and "Side B" observed by using SEM. It can be clearly found from Fig. 10(a) and 10(b) that the damage on the surface of "Side A" is severer than that of "Side B". It can be also seen from Fig. 10(c) and 10(d) that there is fretting crack at "Side A", but no crack was found at "Side B". These results mean that in case of the present WC-Co sprayed specimen, fretting crack is not induced under a condition that the tangential force coefficient is lower than approximately 0.4, which is the final value of "Side B". The larger value of

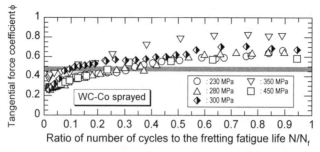

Fig. 11 - *Variations of the tangential force coefficient as a function of the fatigue life ratio N/N_f.*

tangential force coefficient than 0.4 seems to be essential to initiate the fretting cracks. In the case shown in Fig. 9 and 10, fretting cracks seemed to initiate on "Side A" at a certain stage when the tangential force coefficient became larger than approximately 0.4. The initiation condition of fretting crack is thought to be depending on not only the tangential force (tangential force coefficient) but also applied stress level. However, the tangential force coefficients at early stage were almost equal irrespective of stress level, and the tangential force, rather than the applied stress, affected local stress condition around the crack initiation site significantly [25,26]. Therefore, the tangential force coefficient at the point of fretting crack initiation does not seem to depend on the applied stress.

Relationship between the tangential force coefficient and the fatigue life ratio of WC-Co sprayed specimen, i.e. the ratio of number of fretting cycles to fretting fatigue life N/N_f, is shown in Fig. 11. If the fretting crack is assumed to initiate when the tangential force coefficient becomes 0.45-0.5 (the range is indicated in Fig. 11 by a gray colored band), based on the previous discussion, the fretting crack initiation life is estimated about 10-30% of whole fatigue life as shown in Fig. 11. In many metallic materials, the fretting fatigue crack initiates at very early stage, which is less than 25% of the life, and almost the whole of fretting fatigue life is spent in propagating the fretting crack [27]. Thus, no significant difference in crack initiation life is found between the case of present WC-Co sprayed specimen and the cases of many metallic materials. Since total fretting fatigue lives of WC-Co sprayed specimen are 2-10 times longer than those of no-sprayed specimen, the crack initiation cycles of WC-Co sprayed specimen are thought to be larger number than those of no-sprayed specimen. This is considered one of the reasons that the fretting fatigue life was improved by WC-Co spraying.

Furthermore, the tangential force coefficient at crack initiation in WC-Co sprayed specimen is smaller compared with that in no-sprayed specimen under the condition of same stress level. Since fretting fatigue crack propagation rate strongly depends on the magnitude of tangential force coefficient i.e. the tangential force, the propagation rate of WC-Co specimen seemed to be lower than that of no-sprayed specimen. Thus, the propagation lives of WC-Co sprayed specimen were thought to be longer than those of no-sprayed specimen as a consequence. This is considered as the other reason why the fretting fatigue life was improved by WC-Co spraying. Further experiment and discussion will be needed for clarifying the mechanism.

The reason that the fretting fatigue life was improved by WC-Co spraying is summarized as follows. The tangential force coefficient of WC-Co sprayed specimen was significantly lower than that of no-sprayed specimen, because of excellent fretting wear resistance of WC-Co, especially in the early stage of the test. Therefore, initiation of the fretting fatigue crack in WC-Co sprayed specimen was delayed compared with that of no-sprayed specimen, and the crack propagation rate in WC-Co specimen was lower compared with that of no-sprayed specimen. The cooperation of these two mechanisms may give the improvement in fretting fatigue life and strength in WC-Co sprayed specimen. In other word, the fretting fatigue properties of WC-Co sprayed specimen strongly depend on fretting wear properties of WC-Co, which is affected significantly by spraying process and quality of the powder. Another series of fretting fatigue test of WC-Co sprayed specimen should be conducted with controlling the spraying process systematically.

Conclusions

Fretting fatigue tests were carried out using NiCrMo steel (JIS-SNCM439) sprayed with tungsten carbide with additive of 12wt.% cobalt (WC-12%Co) by High Velocity Oxygen Fuel (HVOF) process. The main results obtained are summarized as follows:
1. The fatigue strength of the NiCrMo steel was reduced into 1/4 by fretting.
2. The fretting fatigue strength of the NiCrMo steel was improved about 30% by WC-Co spraying.
3. The tangential force coefficients of no-sprayed specimens were almost constant during the test, and those values depended on the applied stress amplitude.
4. The tangential force coefficients of WC-Co sprayed specimens were low at early stage of the tests, which were about 0.3 irrespective of stress level, and after that, they increased gradually with the number of fretting cycles.
5. The WC-Co sprayed specimen was fractured by the fretting crack which was initiated at the surface of WC-Co layer and propagated into the substrate steel.
6. The improvement in fretting fatigue strength by WC-Co spraying was given by interaction of retardation of fretting fatigue crack initiation and its lower propagation rate.

Acknowledgments

This study was supported by a Grant-in-Aid for Encouragement of Young Scientists from the Ministry of Education, Science and Culture, Japan. Some of the materials used were supplied for fretting fatigue round robin tests of the Japan Society of Mechanical Engineers (JSME). The authors greatly thank Mr. N. Kadonyu and Mr. T. Itoh for their collaboration in experiments.

References

[1] Edited by R.B.Waterhouse, "Fretting Fatigue", *Applied Science Publishers Ltd. London*, (1981).
[2] Edited by M.H.Attia and R.B.Waterhouse, "Standardization of Fretting Fatigue Test Method and Equipment", *ASTM STP 1159*, (1992).

[3] Edited by R.B.Waterhouse and T.C.Lindley, "Fretting Fatigue", *ESIS 18, Mechanical Engineering Publications, London*, (1994).

[4] Edited by D.W.Hoeppner, V.Chandrasekaran and C.B.Elliott, "Fretting Fatigue: Current Technology and Practices", *ASTM STP 1367*, (2000).

[5] R.B.Waterhouse, "Fretting Fatigue", *Applied Science Publishers Ltd. London*, (1981), pp.221-240.

[6] K.Hirakawa, "Case Histories and Prevention of Fretting Fatigue Failure", *Proceeding of the "Symposium, Fretting – Wear and Fatigue"*, (1993), pp.111-124 (in Japanese).

[7] D.Yunsh, Z.Baoyu and L.Weili, "The Fretting Behaviour of a Nitrided Steel 38CrMoAl", *Wear 125-1*, (1988), pp.193-204.

[8] S.M.Kudva and D.J.Duquette, "Effect of Surface Residual Stresses on the Fretting Fatigue of a 4130 Steel", *ASTM STP 776*, (1982), pp.195-203.

[9] K.Nishioka and H.Komatsu, "Researches on Increasing the Fatigue Strength of Press-Fitted Axles (3rd Repot, Effect of Tufftriding)", *Trans. Jpn. Soc. Mech. Eng.*, 36-291, (1970), pp.1805-1811 (in Japanese).

[10] T.Satoh, K.Machida, Y.Mutoh, K.Tanaka and E.Tsunoda, "Improvement of High-Temperature fretting Fatigue Strength by Shot Peening", *Trans. Jpn. Soc. Mech. Eng. (A)*, 56-528, (1990), pp.1784-1791 (in Japanese).

[11] Y.Mutoh, T.Satoh and E.Tsunoda, "Improving Fretting Fatigue Strength at Elevated Temperatures by Shot Peening in Steam Turbine Steel", *ASTM STP 1159*, (1992), pp.199-209.

[12] M.Okane, K.Shozawa and T.Ishikura, "Fretting Fatigue Behavior of Low Carbon Steel with TiN Coating by PVD Method", *Trans. Jpn. Soc. Mech. Eng. (A)*, 65-632, (1999), pp.827-832 (in Japanese).

[13] M.Okane, K.Shozawa and T.Ishikura, "Fretting Fatigue Behavior of TiN-Coated Steel", *ASTM STP 1367*, (2000), pp.465-476.

[14] T.Itsukaich, in *Thermal Spraying Technique*, 16-3, (1997), pp.52-57 (in Japanese).

[15] H.Liao, B.Normand and C.Coddet, "Influence of Coating Microstructure on The Abrasive Wear Resistance of WC/Co Cernet Coatings", *Surface and Coating Technology*, 124, (2000), pp.235-242.

[16] J.Voyer and B.R.Marple, "Sliding wear behavior of high velocity oxy-fuel and high power plasma spray-processed tungsten carbide-based cermet coatings", *Wear 225-229*, (1999), pp.135-145.

[17] D.A.Stewart, P.H.Shipway and D.G.McCartney, "Abresive wear behaviour of conventional and nanocomposite HVOF-sprayed WC-Co coatings", *Wear 225-229*, (1999), pp.789-798.

[18] L.Jacobs, M.M.Hyland and M.De Bonte, "Study of the Influence of Microstructural Properties on the Sliding-Wear Behavior of HVOF and HVAF Sprayed WC-Cermet Coatings", *Journal of Thermal Spray Technology*, 8(1), (1999), pp.125-132.

[19] T.Ogawa, K.Tokaji, T.Ejima, Y.Kobayashi and Y.Harada, "Evaluation of Fatigue Strength of WC Cermet-and 13Cr Steel-Sprayed Materials", *J. Soc. Mater. Sci. Jpn.*, 46-10, (1997), pp.1124-1129 (in Japanese).

[20] J.U.Hwang, T.Ogawa, K.Tokaji, T.Ejima, Y.Kobayashi and Y.Harada, "Fatigue Strength and Fracture Mechanisms of WC Cermet-Sprayed Steel", *J. Soc. Mater. Sci. Jpn.*, 45-8, (1996), pp.927-932 (in Japanese).

[21] T.Nishida, Y.Mutoh and S.Niwa, in *Proceeding of the 48th Annual Meeting of Japan Society of Materials Science,* (1999), pp.117-117 (in Japanese).

[22] Sulzer Meteco Ltd.

[23] Y.Mutoh, T.Nishida and I.Sakamoto, "Effect of Relative Slip Amplitude and Contact pressure on Fretting Fatigue Strength", *J. Soc. Mater. Sci. Jpn.*, 37-417, (1988), pp.649-655 (in Japanese).

[24] K.Nakazawa, M.Sumita and N.Maruyama, "Effects of Fretting Pad Shape and Clamping Condition on Fretting Fatigue in SNCM439 Steel", *Proceeding of 1998 JSME Annual Meeting,* 98-1, (1998), pp.161-162 (in Japanese).

[25] Y.Mutoh, "Mechanisms of Fretting Fatigue", *JSME Inter. Journal, A*, 38-4, (1995), pp.405-415.

[26] R.B.Waterhouse, "Fretting Fatigue", *International Materials Reviews*, 37-2, (1992), pp.77-97.

[27] K.Tanaka, Y.Mutoh and S.Sakoda, "Effect of Contact Materials on Fretting Fatigue in a Spring Steel", *Trans. Jpn. Soc. Mech. Eng. (A)*, 51-464, (1985), pp.1200-1207 (in Japanese).

CASE STUDIES AND APPLICATIONS

Trevor C. Lindley

Fretting Wear and Fatigue in Overhead Electrical Conductor Cables

Reference: Lindley, T. C. "**Fretting Wear and Fatigue in Overhead Electrical Conductor Cables,**" *Fretting Fatigue: Advances in Basic Understanding and Applications, ASTM STP 1425*, Y. Mutoh, S. E. Kinyon, and D. W. Hoeppner, Eds., ASTM International, West Conshohocken, PA, 2003.

Abstract: Fretting wear and fatigue damage induced by wind vibration in overhead power conductors has been investigated by metallographic and fractographic examination. It has been concluded that wear, corrosion and fatigue can each play a role in the degradation of cable, sometimes leading to complete failure.

Keywords: Overhead electrical conductors, wear, fatigue, corrosion

Introduction

Wind induced vibrations can cause wear, sometimes leading to cracking and possibly complete failure in overhead electricity supply lines [1,2]. For power lines situated near the coast, the problem can be further exacerbated by corrosion, wind borne sea salt spray being carried inland for many miles.

The present paper is concerned with fretting wear and fatigue damage in the clamp-conductor region of ACSR (Aluminium Conductor Steel Reinforced) multi-layer cables. The wind induced vibration modes which can subject a conductor to fatigue loading are either sub-conductor oscillation or aeolian vibration. Aeolian is small amplitude vibration which causes fretting between individual wires making up the cable. Sub-conductor oscillation can cause much more severe bending forces and is induced by strong winds

Professor, Imperial College of Science, Technology and Medicine, Department of Materials, Prince Consort Road, London SW7 2BP, UK

and gales which may even cause cable "galloping". Despite both types of vibration modes being wind induced, sub-conductor oscillation promotes displacements which are essentially restricted to a horizontal plane whereas aeolian vibration gives rise to a vertical motion of the conductor. The two vibration modes also differ in frequency response and often in magnitude of induced stress. Sub-conductor oscillation has a vibration frequency of about 1Hz and is typically associated with stress amplitudes values of +/-40 MPa. Aeolian vibration typically occur at frequencies up to 20Hz at stresses of +/- 10 MPa. The present work is particularly concerned with the aeolian vibration which induces fretting damage by promoting small oscillatory displacements between individual strands making up the conductor cable. The conductors involved in the present study had been in service for more than 20 years.

The paper describes a detailed metallurgical investigation including fractography of cables damaged by fretting where wear, corrosion and fatigue can each play a role in the degradation of the cable.

Experimental Method

Macroscopic and microscopic examination was carried out in order to satisfy the following objectives:
1) to note the location of significant wear and to assess the extent of wear between individual strands and relate this wear to the observation of cracking.
2) to record the occurrence and orientation of all cracks and to identify the sites of crack initiation.
3) to determine the mechanism of cracking
4) if the cracking was due to fatigue, determine the operative vibration mode (sub-conductor or aeolian).
5) to determine the extent to which corrosion was important in the failures.

Experimental Results

Cable Construction

Each conductor comprised 54 strands in three layers of commercial purity aluminium re-inforced by a central core of seven steel strands (figure 1). These steel strands support the weight of the conductor with any stress in the aluminium strands being relieved by primary creep. The conductors are suspended in groups of four which are held in a square configuration by Andre spacers (figure 2). Each conductor was gripped by a carbon loaded neoprene bush fitted into the aluminium alloy spacer.

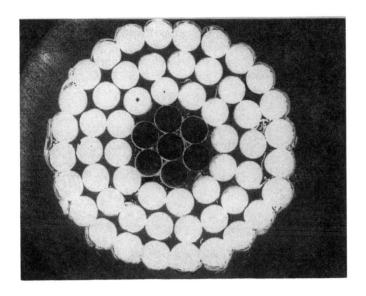

Figure 1 Cross Section through ACSR Conductor

General Examination

In-service damage was located in regions close to the Andre spacers (figures 3 and 4) where the cable exits a support bush. The four conductors can be conveniently identified by their positions in the spacer as 'top inner' (TI), 'top outer' (TO), 'bottom inner' (BI) and 'bottom outer' (BO). After dismantling, the position previously occupied by the spacer clamp is easily defined by two dark witness marks (figure 3a) , sometimes showing evidence of pronounced corrosion in some regions (figure 3b). Limited corrosion was noted on TO conductors but patches of corrosion was found on TI, BI and BO conductors. The corrosion present on some strands was sufficient to give a significant reduction in cross section (figure 4) but such corrosion was never associated with either strand cracking or complete failure. In all cases, cracking occurred close to the dark witness mark (near the edge of the clamp) and ran diagonally across each strand when viewed from the outside of the conductor (see figure 5a and 5b). In each conductor, all cracked or fractured strands were at the same end of the clamp. Cracking never occurred at both ends of a clamp.

Locations of Cracking and Corrosion

The outer layers of the conductor samples were examined optically and the locations of all fractured, cracked or severely corroded strands recorded as shown in figure 6. The cracked or fractured strands tended to be grouped together in any one conductor rather than being randomly distributed around the circumference. There was apparently no relationship between the occurrence of cracking in a strand and the presence of corrosion (compare figures 7 and 8). It was noted that corrosion tended to occur in strands positioned toward the bottom of the conductor whereas cracking tended to occur at all points on the circumference of the conductor with the exception of the very bottom strand, none of which showed cracks.

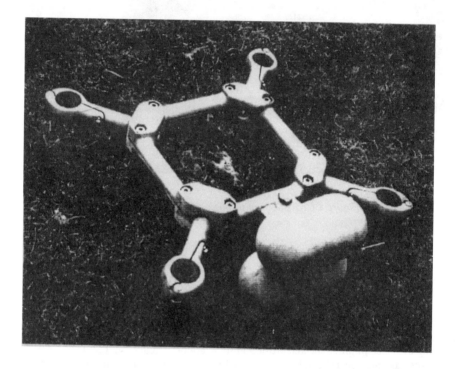

Figure 2 The Andre Spacer (ignore the weights in the foreground which were absent in the present study)

Wear in Conductor Strands in Adjacent Layers

A conductor was partly dismantled in order to investigate the wear between strands in the outer and second layers. Some of the outer layer strands were removed, exposing lens shaped wear scars on both the second layer strands and on the matching faces of the outer layer strands as shown in figures 9. The lens markings are bright in appearance with relatively little surface oxidation for positions near to a clamp. At distances greater than about 100mm from a clamp, a few lens markings were found to be covered with patches of a dark surface oxide (figure 10). Significant quantities of white oxide found between

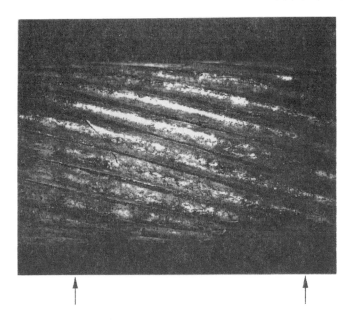

Figure 3 General Views of the Conductors in the Region of the Spacer Clamps
(a) the TO position showing two dark witness marks (arrowed)

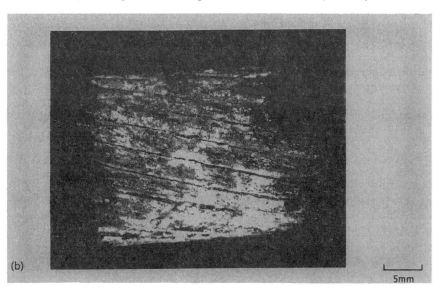

(b) the TI position showing heavily corroded region beneath the clamp

9mm

Figure 4 detail of severely corroded region beneath clamp (conductor BO)

the strands was identified by X-ray analysis as a mixture of the Gibbsite and Bayerite forms of aluminium hydroxide. Metallographic cross-sections taken through a strand at a lens marking showed a relatively smooth surface with little sub-surface damage. The lens type wear was quantified by measuring the length of individual lens scar markings around the circumference of a conductor section. Lens scar length measurements were made over a range of distances from the centre of the clamp (figure 11). The wear between strands was a maximum close to the edges of a clamp and gradually reduced with distance away from the clamp.

Wear Between Adjacent Strands in the Outer Layer

In addition to the lens shaped scars on the underside of the outer strands, flats were apparent on the side faces of these strands (figure 12a) i.e. at the point of contact between adjacent strands in the outer layer. This form of wear was completely absent at the centre of the clamp, with flats developing from a point very close to the edge of the clamp (figure 12b). Figure 13 shows the variation in the width of the flat with distance from the centre of the clamp, plotted on both sides (labelled left and right) of the clamp.

(a)

5mm

Figure 5 General View of Conductor in the region of the Spacer Clamp
(a) conductor BO

(b)

5mm

(b) conductor BI

Mechanisms of Wear

As already noted, the wear between adjacent strands resulted in the formation both of lens type markings (on the underside of outer layer strands) and flats (on the side faces of the strands). Both forms of wear result from relative sliding between adjacent layers. The factors controlling the amount of wear include the contact pressure between strands, the amplitude of relative sliding and the number of sliding cycles. Both types of wear show a maximum near to the edge of the clamp (figures 11 and 13), a consequence of both a high contact pressure due to the proximity of the clamp together with significant relative sliding. At the centre of the clamp, the lens type wear was somewhat reduced since sliding between the outer and second layers was restrained by the clamp. Wear resulting in flats was completely absent at this point suggesting that relative sliding was not possible as a consequence of the outer layer strands being locked together by the clamp.

Both forms of wear show a reduction in magnitude with distance from the clamp although the lens type wear was present over the whole of the conductor sections examined whereas no flats were present at positions remote from the clamp. This would indicate that the formation of flats was due to a wear mechanism more specific to the clamp, unlike the formation of lens markings which are more of a long range process increased in magnitude by the presence of the clamp.

In the region adjacent to the clamp, the appearance of the wear damage produced by the two forms of wear is a manifestation of two extremes of repeated sliding wear behaviour. The lens type wear associated with little sub-surface damage and large quantities of white oxide deposits is typical of wear at a relatively low contact pressure and high values of relative slip, promoting material removal by a grinding type action. Conversely, the flats with their adherent oxide and evidence of sub-surface damage is typical of fretting wear at a relatively high contact pressure (greater than 15MPa) and a low range of relative slip (<30µm). In this later case, surface material is removed by the repeated shear and fracture of asperities.

Cracking and Failure of Conductor Strands

The presence of an angled crack in a single strand from the outer conductor layer is shown in a view of the side face in figure 14. Wear between adjacent strands in the outer layer has caused the introduction of a flat and subsequently the angled crack. In other cases, two cracks were found which had initiated from both flats on either side of the strand. Crack initiation at a flat was confirmed by scanning electron microscopy (SEM) as shown in figure 15. Note that this crack did not grow to fracture and the final fatigue crack tip position before breaking open is indicated. The fracture surface near to the crack initiation site was heavily oxidised, consistent with the crack developing very slowly in its early stages.

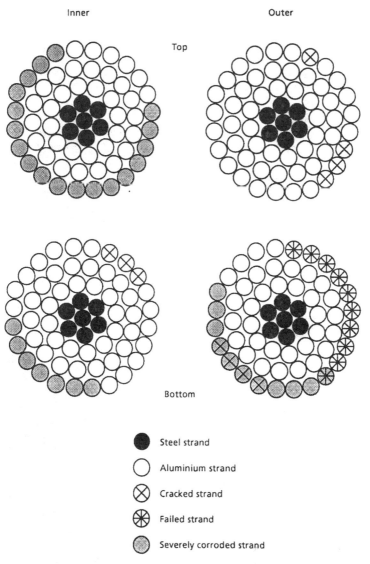

Figure 6 Schematic showing locations of cracked, failed and severely corroded strands.

In the present investigation, all cracked strands (figure 6) showed crack initiation at a flat at a position close to the outer edge of the clamp. Each crack then grows into the strand at a shallow angle and in the direction beneath the clamp. Such a crack trajectory is typical of a crack developing by fretting. Crack initiation by a fretting mechanism was further supported by taking metallographic axial sections through the strands

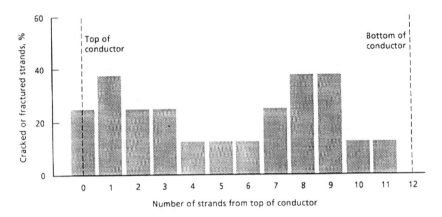

Figure 7 Relationship between cracking in strands and position in conductor

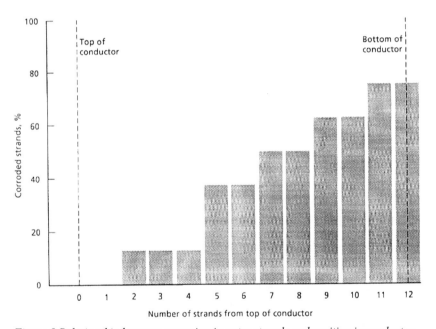

Figure 8 Relationship between corrosion in outer strands and position in conductor

perpendicular to flats in regions adjacent to the major cracks. The section in figure 16 shows multiple initiation of angled cracks, typical of fretting fatigue.

Two outer layer strands which failed completely are shown in figure 17, one of which failed from a single fatigue crack whilst the other failed from two cracks initiating on either side of the strand. Fatigue cracks often grew to occupy about 80% of the strand

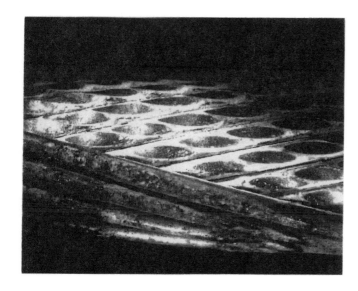

5mm

Figure 9 Partially dismantled conductor showing wear between outer and second layer strands responsible for lens shaped markings

1mm

Figure 10 Lens shaped marking approximately 200mm from edge of clamp showing patchy surface oxide

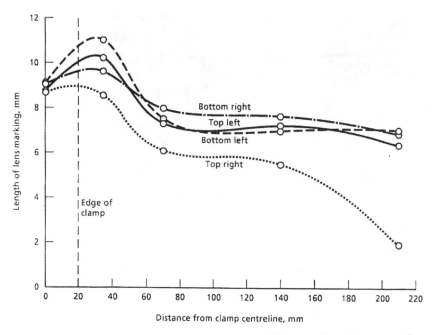

Figure 11 Length of lens shaped markings in terms of distance from clamp centreline

cross section prior to failure. The final failure occurs by a ductile failure giving a fracture surface which is approximately perpendicular to the strand surface.

Note that in the present survey, cracks were not found at the lens shaped scars (arising from contact between strands in adjacent layers). Here, it is possible that under the prevailing contact conditions, the wear rate might be sufficiently high to remove any incipient cracks.

Conclusions

1. The cracking observed in outer layer strands of the conductors initiated by fretting fatigue at a position close to the edge of the spacer clamp.
2. Fretting fatigue occurred due to the contact between the side faces of adjacent strands in the outer conductor layer. This fretting also resulted in the formation of wear flats at a location specific to the neighbourhood of the clamp.
3. A second wear mechanism was apparent which took place between strands in the outer and second layers. This wear was less specific to the clamp region, being

(a)

1mm

Figure 12 Flats on side faces of outer layer strands formed by wear against adjacent outer layer strand (a) fully developed flat about 20mm from edge of clamp

(b)

1mm

(b) development of flat close to edge of clamp

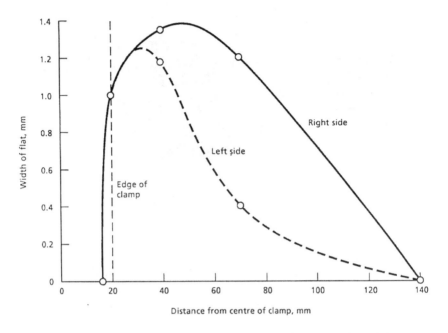

Figure 13 Variation in width of flat with distance from centre line of clamp

1mm

Figure 14 Angled crack in flat on side face of an outer layer strand

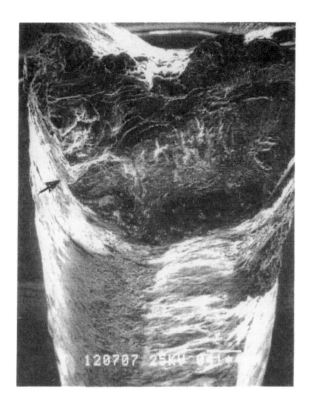

1mm

Figure 15 Low power SEM micrograph of cracked strand broken open to reveal fracture surface (final position of fatigue crack tip arrowed)

apparent to a distance of at least 0.5m from the clamps. This wear resulted in the formation of lens shaped markings on the underside of outer layer strands. The wear was more severe than that between outer layer strands due to a greater amount of relative sliding between strands, further exacerbated by the formation of a large amount of fretting debris. There were no incidences of cracking with this type of wear.

4. Considerable amounts of corrosion were present on some strands. This corrosion was concentrated in the region beneath the clamps to the underside of the conductor. This corrosion sometimes caused large reductions in the cross section of the strands but no cracking was associated with corrosion so that the fretting fatigue was not related to the corrosion.

5. Microcracks formed by fretting subsequently developed by mechanical fatigue to eventually occupy about 80% of the cross section, final strand failure then occurring by ductile fracture. The inclined orientation of the cracks indicated that the loading

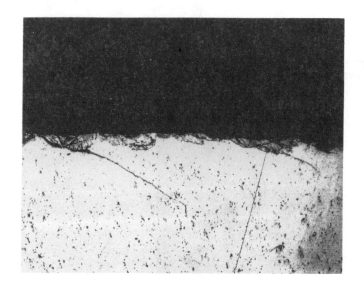

0.1mm

Figure 16 Cross section through strand at a flat showing multiple initiation of angled cracks typical of fretting fatigue

developed in the strands was significantly influenced by local contact stress in the proximity of the clamp.
6. There was no evidence to indicate that which mode of conductor loading (subconductor oscillation or aeolian vibration) was responsible for the cracking.

Acknowledgements

The author is extremely grateful to Drs K J Nix and A Camyab who each made significant contributions to the work reported.

References

[1] Cardou,A., Leblond, A., Goudreau, S. and Cloutier, L., "Electrical Conductor Bending Fatigue at Suspension Clamps: a Fretting Fatigue Problem", *Fretting Fatigue*, ESIS 18, R.B.Waterhouse and T.C.Lindley, Eds, European Structural Integrity Society, Mechanical Engineering Publications, London, 1994, pp 257-266.
[2] Cardou, A., Cloutier, L., St-Louis, M. and Leblond, A., "ACSR Electrical Conductor Fretting Fatigue at Spacer Clamps", *Standardisation of Fretting Fatigue Test Methods*

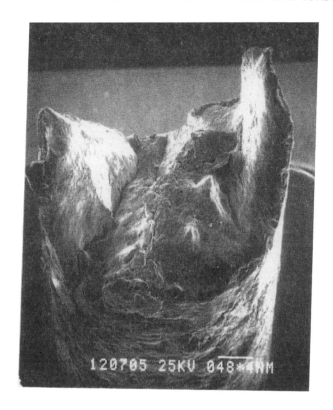

1mm

Figure 17a SEM micrograph of fractured outer layer strand failing from two fatigue cracks on either side of strand

*and Equip*ment, ASTM 1159, M.H.Attia and R.B.Waterhouse, Eds., American Society Testing and Materials, West Conshohocken, PA, 1992, pp 231-242.

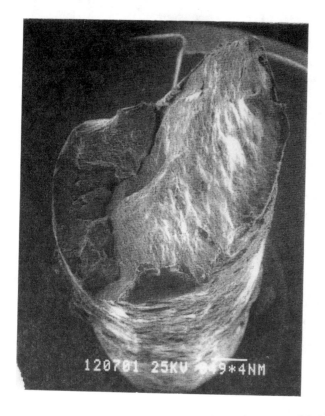

1mm

Figure 17b SEM micrograph of fractured outer layer strand failing from a single fatigue crack

List of Figures

Figure 1. Cross section showing the construction of the conductor.

Figure 2. The André spacer (ignore the weights shown in the foreground which were absent in the present study).

Figure 3. General views of the conductors in the region of the spacer clamps
(a) the TO position showing two dark witness marks (arrowed)
(b) the TI position showing heavily corroded region beneath the clamp.

Figure 4. Detail of severely corroded region beneath clamp (conductor BO).

Figure 5. General views of conductors in the region of the spacer clamp showing major cracking (a) conductor BO (b) conductor BI.

Figure 6. Schematic showing locations of cracked, failed and severely corroded strands. (orientation of conductors as during service).

Figure 7. Relationship between cracking in outer strands and position in conductor.

Figure 8. Relationship between corrosion in outer strands and position in conductor.

Figure 9. Partially dismantled conductor showing wear between outer and second layer strands responsible for lens shaped markings.

Figure 10. Lens marking approximately 200mm from edge of clamp showing patchy surface oxide.

Figure 11. Length of lens shaped markings in terms of distance from clamp centreline.

Figure 12 Flats on side faces of outer layer strands formed by wear against adjacent outer layer strand.
(a) fully developed flat about 20mm from edge of clamp.
(c) development of flat close to edge of clamp.

Figure 13. Variation in width of flat with distance from centre line of clamp.

Figure 14. Angled crack in flat on side face of an outer layer single strand.

Figure 15 Low power SEM micrograph of cracked strand broken open to reveal fracture surface (final position of fatigue crack tip arrowed).

Figure 16. Cross section through strand at a flat showing multiple initiation of angled cracks typical of fretting fatigue.

Figure 17. SEM micrographs of two fractured outer layer strands
(a) failing from two fatigue cracks from either side of the strand
(b) failing from a single fatigue crack

Toshihiko Yoshimura,[1] Takashi Machida,[2] and Toshio Hattori[3]

Evaluating Fatigue Life of Compressor Dovetails by Using Stress Singularity Parameters at the Contact Edge

REFERENCE: Yoshimura, T., Machida, T., and Hattori, T., "Evaluating Fatigue Life of Compressor Dovetails by Using Stress Singularity Parameters at the Contact Edge," *Fretting Fatigue: Advances in Basic Understanding and Applications, STP 1425,* Y. Mutoh, S. E. Kinyon, and D. W. Hoeppner, Eds., ASTM International, West Conshohocken, PA, 2003.

Abstract: Fretting fatigue life of compressor dovetail for gas turbine was investigated using stress singularity theory. It is found that the crack initiation of fretting fatigue between blade dovetail and wheel dovetail occurs at the contact edge of blade dovetail. The parameters of stress singularity obtained from maximum principal stress distribution are capable to evaluate the fretting fatigue strength with the cyclic load having a constant angle against the contact surface under a variable contact pressure. The ratio of between the intensity range of stress singularity and the critical intensity range of stress singularity, $\Delta H1 / \Delta Hc$, is an effective parameter to predict the fretting fatigue life of compressor dovetail.

Keywords: fretting fatigue, dovetail, stress singularity, critical intensity range of stress singularity

Introduction

An increase in the length of gas-turbine compressor blades with an increase in the compression ratio and the volume of gas passing through may cause fretting fatigue problems because of an increase in the blade centrifugal force and contact pressure between the blade and wheel dovetails. Constant contact stress at the edge between the blade and wheel dovetails leads fretting fatigue. In order to estimate the life-time of gas turbine compressor dovetail, a fatigue experiment was conducted and the stress singularity theory was applied to estimate the fretting fatigue strength of a compressor dovetail.

[1]Senior researcher, Mechanical Engineering Research Laboratory, Hitachi, Ltd., 3-1-1, Saiwai, Hitachi, Ibaraki 317-0073, Japan.
[2]Chief researcher, Mechanical Engineering Research Laboratory, Hitachi, Ltd., 3-1-1, Saiwai, Hitachi, Ibaraki 317-0073, Japan.
[3]Chief researcher, Mechanical Engineering Research Laboratory, Hitachi, Ltd., 502, Kandatsu, Tsuchiura, Ibaraki 300-0073, Japan.

Experiment

Figure 1 shows the fatigue experiment conditions. 15-Cr steel for the blade and NiCrMoV steel for the wheel were used. The blade dovetail was inserted into the wheel dovetail. During the fatigue experiment, the maximum axial force, which included the blade centrifugal force, was applied to both the dovetails. The cyclic force corresponding to the gas-bending force of the compressor blade was then superimposed on the axial force. The loading conditions for the fatigue experiment are shown in Fig. 2. Stress ratio R ranged from 0–0.91. The number of crack initiation cycles were detected based on the output variation from a pair of strain gauges, which were installed on both sides of the blade dovetail. As shown in Fig. 3, to clarify the effect of the contact angle between the blade and wheel dovetails (dovetail shoulder angle) on the fatigue life of the compressor blade, the kinds of dovetail shoulder angles of 30, 35, and 40 degrees were chosen.

Blade dovetail
(15Cr steel)

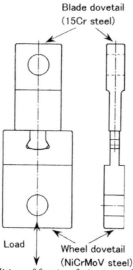

Load
Wheel dovetail
(NiCrMoV steel)

Figure 1—*Condition of fretting fatigue test for compressor dovetail.*

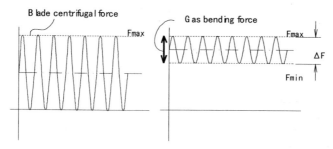

Figure 2 —*Loading condition of fatigue (Fmin/Fmax=0 ∼0.91)*

Analysis

The stress around a contact edge is represented by the following equation [1].

$$\sigma \ (r) = H / r^{\lambda} \tag{1}$$

Here, σ is the stress (MPa), r is the distance from the contact edge, H is the intensity of stress singularity, and λ is the order of stress singularity.

In this study, the crack initiation due to fretting fatigue was evaluated by using parameters H and λ. The order of stress singularity, λ, was obtained from the stress distribution calculated by a field-element method, and λ was obtained by calculations using the contact angle of the both dovetails. The dovetail region around the contact edge was divided into 100-μ m-long meshes as shown in Fig. 4. The stress distribution was analyzed by using the successive-finite-element method and the intensity of stress singularity, H, and the order of stress singularity, λ were calculated. The circumferential stress distribution and the maximum principal stress distribution were obtained by defining the contact edge as the origin of the polar coordinate.

Results

Figure 5 shows the dovetail contact region before fretting fatigue crack initiation. As can be seen in the figure, many abrasion particles appeared around the contact area. Crack initiation was observed at the contact edge between the blade and wheel dovetails. These results suggest that the friction between the blade and wheel dovetails, which was added to the contact pressure due to a high centrifugal force, leads to fretting fatigue. The cracks due to fretting fatigue tend to grow along the direction of the maximum shear stress.

Figure 6 shows the relationship between the number of crack initiation cycles, N_f, and the ratio of the repetition load to the maximum load, $\Delta F/F_{max}$. The time of crack initiation due to fretting fatigue depends on the dovetail shoulder angle, which means that the life of a dovetail with a shoulder angle of $30°$ will be longer than that of a dovetail with a shoulder angle of 40.

The parameters of stress singularity obtained from the circumferential stress distribution and the maximum principal stress distribution by the finite-element method are summarized in Table 1. The circumferential stress distribution was obtained along the direction of crack propagation by defining the contact edge as the origin of the polar coordinate. The order of stress singularity, λ_{θ}, λ_1, tended to increase with an increase in the dovetail shoulder angle. While the intensity of stress singularity, H_{θ}, from the circumferential stress distribution increased with an increase in the dovetail shoulder angle, the intensity of stress singularity, H_1, from the maximum principal stress distribution did not show any significant increase. The reason of such a discrepancy might be due to the stress distribution along the direction of the maximum shear stress. In the case of the circumferential stress, the force normal to contact surface, which directly affects the circumferential stress, decreases with the dovetail shoulder angle. On the other hand, in the case of maximum principal stress, the maximum principal stress may not depend on the dovetail shoulder angle, if Mohr's circle for stress around the contact area is taken into consideration.

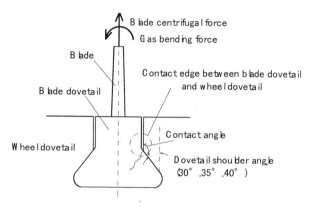

Figure 3 —*Structure of compressor dovetail.*

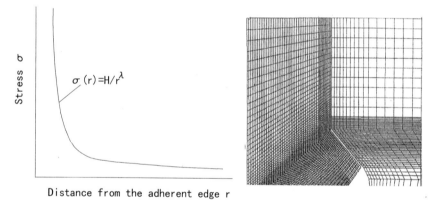

Figure 4—*FEM mesh for dovetail.*

Table 1—*Stress singularity parameters for each dovetail angle.*
(Load; $\sigma m=140MPa$)

Dovetail shoulder angle (deg)	Circumferential stress		Maximum principal stress	
	$\lambda \theta$	$H\theta$	$\lambda 1$	$H1$
30	0.238	167	0.278	322
35	0.285	208	0.365	322
40	0.309	259	0.404	335

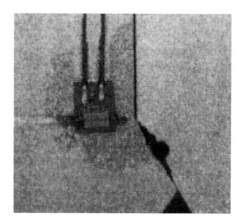

Figure 5 — *Dovetail specimen before fretting fatigue crack initiation.*

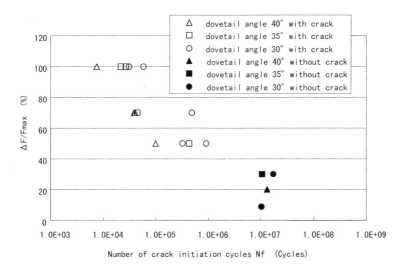

Figure 6 — *Crack initiation life of dovetail fatigue test.*

Discussion

The predictability of crack initiation due to fretting fatigue under arbitrary contact conditions was investigated by comparing the intensity of stress singularity, H, and the order of stress singularity, λ, obtained by the finite element method, with the critical intensity of stress singularity, H_c.

Critical intensity of stress singularity H_c is the intensity of stress singularity that causes crack initiation when the stress of a specific distance from the contact edge reaches its limit. The fatigue limit (smoothed test piece) , σ_{wo}, of the 15Cr steel used for the blade was 780 MPa (R=0), and the threshold stress intensity factor range, ΔK_{th}, was 3.1 MPa$\sqrt{}$m=98 MPa$\sqrt{}$mm (R=0). These values are plotted in this order in Fig.7. Fatigue limit σ_{wo} does not depend on the distance from the contact edge and is constant. In contrast, threshold stress intensity factor range ΔK_{th} depends on the (depth)$^{-1/2}$. The intersection point of σ_{wo} and ΔK_{th} is considered to define a specific distance, which corresponds to the crack initiation distance [2]. This leads to the following equation.

$$H_c = 780 \times 0.0025^\lambda \qquad (2)$$

The relationship between the critical intensity of stress singularity, H_c, and the order of stress singularity, λ, is graphically shown in Fig. 8. The symbol at λ =0 corresponds to the fatigue limit of a [smoothed test piece], while the symbol at λ =0.5 corresponds to threshold stress intensity factor range ΔK_{th}. When λ =0.5, the following relationship is formed.

$$H_c\ (\lambda=0.5)\ = \Delta K_{th} / \sqrt{2\pi} \qquad (3)$$

To compare ΔH corresponding to the cyclic force shown in Fig. 2 with ΔH_c, we used R=0 for $\Delta \sigma_{wo}$ and ΔK_{th}. In the case of 15Cr steel, $\Delta \sigma_{wo}$ was higher and ΔK_{th} was lower than these values for HT60 steel ($\Delta \sigma_{wo}$=547 MPa (R=0), ΔK_{th}=7.5 MPa$\sqrt{}$m (R=0)) [3] and for Ni-Cr-Mo steel ($\Delta \sigma_{wo}$=360 MPa (R=-1), ΔK_{th}=12.4 MPa$\sqrt{}$m (R=-1)), which lead to a decrease in the specific distance from the contact edge needed to induce crack initiation [3].

The relationship between the intensity of stress singularity and the order of stress singularity obtained from the circumferential stress distribution (H_θ and λ_θ, respectively) is plotted in Fig. 8. The solid circle, solid square, and solid triangle correspond to dovetail shoulder angles of 30, 35, and 40, respectively. The intensity ranges of stress singularity, ΔH_θ, calculated by using the ratio of the cyclic force, ΔF, to the maximum axial force, F_{max}, are shown in Fig. 9. The open symbols show the specimens with crack initiation, while the solid symbols show the specimens without crack initiation. It is recognized that the crack initiation occurs on even the specimens with intensity of stress singularity ΔH_θ being lower than ΔH_c. The relationship between the ratio of the intensity range of stress singularity to the critical intensity range of stress singularity, $\Delta H_\theta / \Delta H_c$, and the number of crack initiation cycles, N_f, is shown in Fig. 10. The open symbols show the specimens with crack initiation. The solid symbols show the specimens without crack initiation.

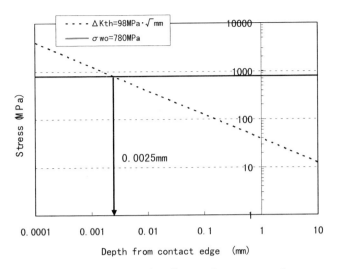

Figure 7—Crack initiation diatance from contact edge.

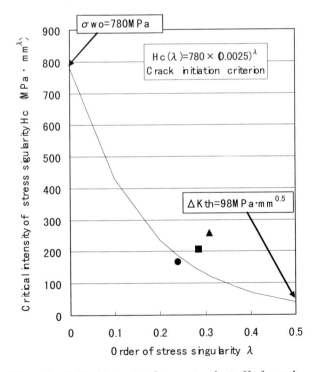

Figure 8—*Critical intensity of stress singularity Hc for each*
order   stress singularity. (30°,   35°,   :40°:
from circumferential   stress distribution)

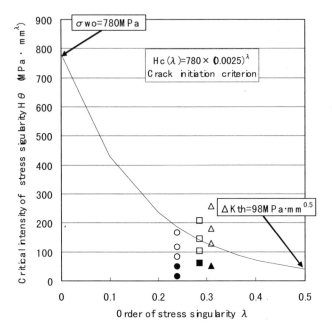

Figure 9—*Intensity range of stress singularity from circumferential stress distribution for critical intensity of stress singularity.*
(30°,   35°,   40°: with crack,
30°,   35°,   40°: without crack)

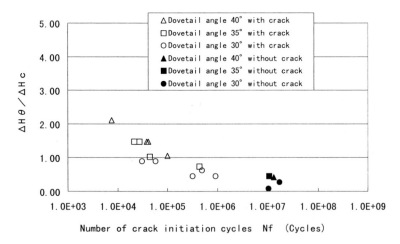

Figure 10—*Relationship between ratio of Δ Hθ/Δ Hc and number of crack initiation cycles Nf.*
(Δ Hθ : from circumferential stress distribution)

As can be seen in the figure, the deviation is smaller than that in the relationship between the number of crack initiation cycles, N_f, and the ratio of the cyclic load to the maximum load, $\Delta F/F_{max}$, as shown in Fig. 6. When $\Delta H_\theta/\Delta H_c$ was more than 0.5, crack initiation occurred.

The relationship between the intensity of stress singularity and the order of stress singularity obtained from the maximum principal stress distribution (H_1 and λ_1, respectively) is plotted in Fig. 11. The solid circle, solid square, and solid triangle correspond to dovetail shoulder angles of 30, 35, and 40, respectively. Here, the ratio of H_1/H_c was higher than that of H_θ/H_c. The intensity ranges of stress singularity, ΔH_1, calculated by using the ratio of cyclic force, ΔF, and the maximum axial force, F, are shown in Fig. 12. As for the circumferential stress distribution, the open symbols indicate the specimens with crack initiation, while the solid symbols indicate the specimens without crack initiation. We found that the specimens with intensity of stress singularity ΔH_1 lower than ΔH_c did not show crack initiation. Figure 13 shows the relationship between the ratio of intensity range of stress singularity to the critical intensity range of stress singularity, $\Delta H_1/\Delta H_c$, and the number of crack initiation cycles, N_f. The open symbols indicate the specimens with crack initiation, while the solid symbols indicate the specimens without crack initiation. We found that the deviation was smaller than the deviation in the relationship between Nf and the ratio of $\Delta F/F_{max}$, as shown in Fig. 6 for $\Delta H_\theta/\Delta H_c$. Crack initiation occurred when $\Delta H1/\Delta Hc$ was more than 1.0. These results suggest that the ratio of $\Delta H_1/\Delta H_c$ is effective in predicting fretting fatigue life. One of our specimens had crack initiation at a ratio of $\Delta H_1/\Delta H_c$ slightly greater than 1.0.

Earlier, T. Hattori and T. Watanabe [3] investigated fretting fatigue by using stress singularity parameters and a cyclic load parallel to the contact surface under a constant pressure on the contact surface. In this study, the cyclic load had a constant angle against the contact surface under a variable pressure on the contact surface. The results of this study indicate that under these conditions, fretting fatigue can be evaluated using the stress singularity theory. Thus, while Hattori and Watanabe used parameters of stress singularity obtained from a circumferential stress distribution to evaluate the fretting fatigue of a cyclic load parallel to the contact surface under a constant pressure, we used the maximum principal stress distribution to evaluate the fretting fatigue of a cyclic load with a constant angle against the contact surface under a variable contact pressure. When R=0.91, which corresponds to a constant gas-bending force of a compressor blade with a ratio of $\Delta H_1/\Delta H_c$ bellow 1.0, there will be no fretting fatigue. Thus, to ensure reliability of gas turbines, the ratio of $\Delta H_1/\Delta H_c$ must be bellow 1.0 in the start-up, shutdown, and abnormal operation of gas turbines.

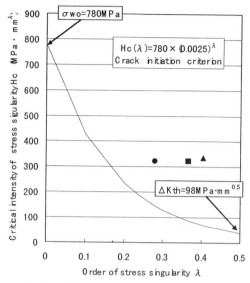

Figure 11–*Critical intensity of stress singularity Hc for each order stress singularity. (30°,   35°,   :40°: from maximum principal   stress distribution)*

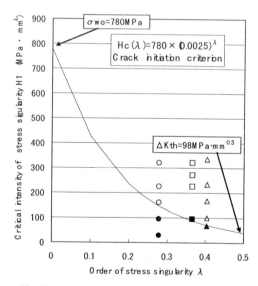

Figure 12–*Intensity range of stress singularity from maximum principal stress distribution for critical intensity of stress singularity.*
(○30° , □35° , △40° : with crack,
●30° , ■35° , ▲40° : without crack)

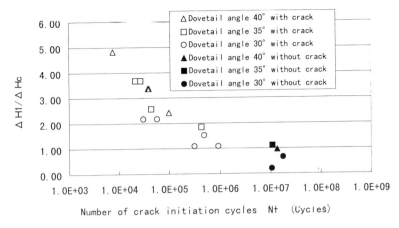

Figure 13 — *Relationship between ratio of Δ H1/Δ Hc and number of crack initiation cycles Nf.*
(Δ H1 : from maximum principal stress distribution)

Conclusion

The relationship between the dovetail shoulder angle and the fretting fatigue life of a compressor dovetail for a gas turbine by using the singularity theory was investigated.

(1) Crack initiation due to fretting fatigue occurred at the contact edge of the blade dovetail between the blade and wheel dovetails, and progressed to the side of the blade dovetail.

(2) The parameters of stress singularity obtained from the maximum principal stress distribution can be used to evaluate the fretting fatigue of a cyclic load with a constant angle against the contact surface under a variable contact pressure.

(3) The ratio of the intensity range of stress singularity to the critical intensity range of stress singularity, $\Delta H_1 / \Delta H_c$, is an effective parameter to predict the fretting fatigue life.

References

[1] Bogy, D. B., Journal of Applied Mechanics, Vol. 38, (1971) p.377.

[2] Hattori, T., Journal of the Marine Engineering Society in Japan, Vol. 33, No.3 (1998-3) pp.203-211.

[3] Hattori, T. and Watanabe, T., The Japan Society of Mechanical Engineers, Proc. of M&M2000 (2000-10) pp.413-414.

Masanobu Kubota, [1] Hidenori Odanaka,[2] Chu Sakae,[3] Yoshihiro Ohkomori,[4] and Yoshiyuki Kondo[5]

The Analysis of Fretting Fatigue Failure in Backup Roll and its Prevention

Reference: Kubota, M., Odanaka, H., Sakae, C., Ohkomori, Y., and Kondo, Y., "**The Analysis of Fretting Fatigue Failure in Backup Roll and its Prevention,**" *Fretting Fatigue: Advances in Basic Understanding and Applications, STP 1425*, Y. Mutoh, S. E. Kinyon, and D. W. Hoeppner, Eds., ASTM International, West Conshohocken, PA, 2003.

Abstract: In the present research, a fretting fatigue failure in a backup roll (BUR) was examined. Fretting fatigue properties of the roll material were experimentally obtained and the prevention of the fretting fatigue failure was proposed. BUR is one of the most important components in a steel making rolling mill for the control of plate thickness and flatness. The BUR was initially supported by oil film bearings. Recently, oil film bearings have been rapidly replaced by roller bearings for the purpose of high-precision rolling. After the replacements, fretting fatigue failures of BUR have begun to occur. Rotating bending fatigue tests were performed with a 40 mm diameter shrink-fit axle to evaluate the fretting fatigue strength of the BUR material and to examine the improvement of the fretting fatigue strength of the holding part of the BUR. The effects of roller machining and overhanging of the contact edge on the fretting fatigue strength were investigated. Based on the results of this study, roller machining was applied to the holding part of the BUR. Furthermore, the diameter of the holding part of the BUR was increased. No failure occurred due to fretting in the BUR for more than four years since these countermeasures were applied.

Keywords: fretting fatigue, failure analysis, steel making rolling mill, backup roll, shrink-fitted axle, surface rolling, overhanging of the contact edge

Introduction

The backup roll (BUR) is one of the most important components in the steel making rolling mill with regard to the quality of rolled products. A steel making rolling mill is shown in Fig. 1. The rolling mill consists of BURs, intermediate rolls and work rolls.

---

[1] Technical Assistant, Department of Intelligent Machinery and Systems, Kyushu University, 6-10-1 Hakozaki, Higashi-ku, Fukuoka, 812-8581 Japan

[2] Graduate School of Kyushu University

[3] Associate Professor, Department of Intelligent Machinery and Systems, Kyushu University

[4] R & D Department, Japan Casting & Forging Corp., 46-59 Nakahara Sakinohama, Tobata-ku, Kitakyushu, 804-8555, Japan

[5] Professor, Department of Intelligent Machinery and Systems, Kyushu University

The BUR has an important role in controlling with high precision the thickness and flatness of the steel plate under the rolling load of more than 10 MN [1]. In the middle of 1990s, a succession of BUR failure accidents occurred. The failure occurred at the axle part of the BUR as shown in Fig. 1, where the inner race of the roller bearing is fitted. The failure was due to fretting fatigue. Since the working conditions of BUR have become severer in response to demands for improving productivity and precision of the rolled products, the problem of fretting fatigue in the BUR has risen in recent years. The present study shows a failure analysis of the BUR, results of the fretting fatigue tests and changes made to the actual BUR based upon the fretting fatigue tests.

Failure Analysis of Steel Making Backup Roll

The need for improving work efficiency in rolling has tended toward extended campaign cycles and higher rolling loads [2]. An oil film bearing has been used in the steel making rolling mill. Recently, oil film bearing has been replaced by roller bearing for high-precision rolling in recent years [1]. Thus, the problem of fretting fatigue has occurred in the BUR. The BUR axle is assembled by a shrink fit with the inner race of the roller bearing. Figure 2 shows the fitted part of the BUR.

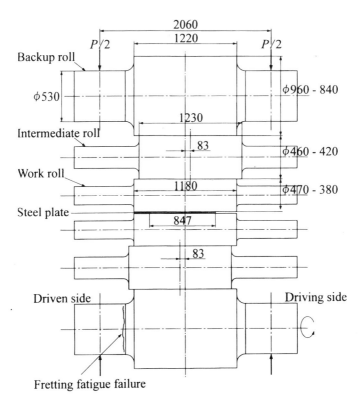

Figure 1 — *Structure of the Steel Making Rolling Mill*

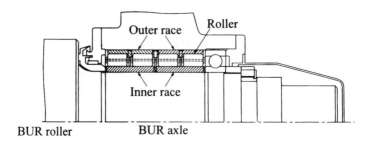

Figure 2 — *Shrink-Fitted Part of BUR with Roller Bearing*

Oil film bearings from 13 BURs were replaced by roller bearings. One of these BURs with roller bearings failed within three years due to cracking at the BUR axle. Five similar failures followed the first one within 6 months. Two of them were completely broken. It was estimated that the number of cycles to failure was approximately 4×10^7.

The surface of the BUR, which suffers from wear and spalling, is ground at regular interval to remove small cracks and irregularities. About 10 to 20 years of BUR service life is usually determined by the decrease in the diameter by the grinding. The BUR axle must endure more than 10^8 cycles of rolling load during service.

Figures 3 and 4 show a cracked BUR. Since unusual severe vibration occurred in the operation, the BUR was removed from the rolling mill. When the inner races of the

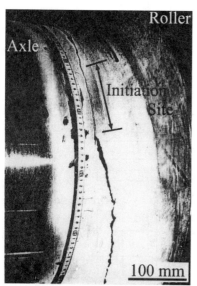

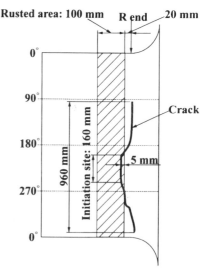

(a) *Penetration Testing*   (b) *Development of Cracked Part*

Figure 3 — *Cracked Backup Roll*

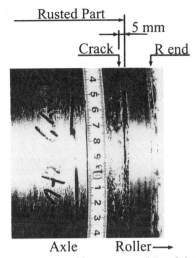

Figure 4 — *Macroscopic Observation of Crack Initiation Site*

roller bearing were separated from the BUR axle, several large cracks were found along with rust. The rust did not cover the entire surface where the inner race was fitted but covered only the contact edge. This appearance shows the same for typical fretting fatigue failure found in fitted axles such as railway axles. The axial length of the rusted area was about 100 mm. The fatal cracks were initiated within the rusted area which was located 5 mm inside from the contact edge. The length of the largest crack was about one-third the circumference. Figure 5 shows the fracture surface of the BUR. There are distinct traces of fatigue crack growth. No material faults and manufacturing defects were detected at the crack initiation site.

Figure 6 shows the rusted surface, which was observed by SEM after removal of the rust. Many small cracks were observed with many small pits. Figure 7 is a photo of these cracks in the axial section. The cracks propagated at an angle to the surface. The characteristics of the cracks and the pits were consistent with these observed for typical long-life fretting fatigue [3].

The maximum bending stress at the cracked part of the BUR axle was estimated to be less than 130 MPa. Taking the failure life, the observations of the surface of the axle and cracks and the stress condition into consideration, it was concluded that the cause of the failure of the BUR axle was fretting fatigue.

Fretting Fatigue Test

A study on the fatigue strength of a fitted-axle assembly, which has its origin in the studies of the wheelset for rolling stock carried out by Whöler [4, 5], has been studied for more than 140 years. Nowadays, many measures to improve fatigue strength of the fitted axle are known. However, the fatigue strength of the BUR axle has not been clarified [1]. Also the effective measures to prevent fretting fatigue failure depend on the structures and materials of the machines [3]. Fretting fatigue tests were performed using

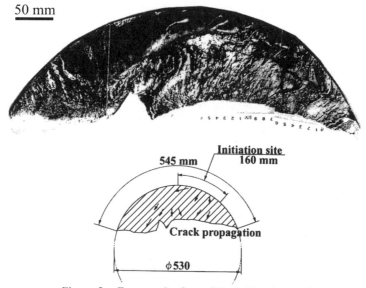

Figure 5 — *Fracture Surface of Failed Backup Roll*

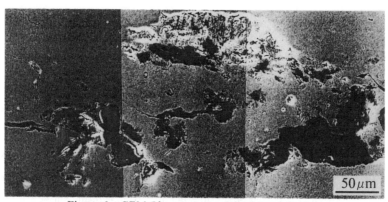

Figure 6 — *SEM Observations of the Rusted Surface*

Figure 7 — *Sectional View of Microcracks*

shrink-fitted axle assemblies with a 40 mm diameter in order to examine the cause of the failure and to establish the measures to prevent BUR axle fretting fatigue failure.

Experimental procedure

Rotating bending fatigue test machines, whose load capacity is 4.4 kN and rotating speed is 1600 rpm, were used for the fretting fatigue tests. The test machines have a mechanism to align automatically the axle center. Usually the test machines are used to evaluate the fatigue strength of high-speed railway axles [6, 7].

Figure 8 shows a drawing of the shrink-fitted axle assembly used in the experiments. The assembly consists of an axle and boss. Both ends of the axle are supported by two ball bearings on the test machine. The boss is midway between the bearings and is given the bending load by the servo-hydraulic loading equipment.

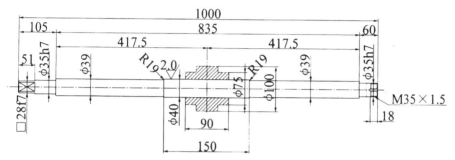

Figure 8 — *Drawing of Shrink-Fitted Axle Assembly Used in the Fretting Fatigue Tests*

Three kinds of axles were used in the experiments. One was manufactured by the same method as the actual BUR axle. The others were finished by surface rolling which introduces a surface compressive residual stress to the fitted part. One of the rolled axles had two grooves cut at the contact edges. The grooves were introduced to make the boss edges overhang. The rolled and overhung axle is simply named "the overhung axle". The condition of surface rolling and the shape of the groove are shown in Fig. 9. Compressive residual stress prevents growth of fatigue cracks [3, 8]. Overhanging of the boss edge decreases the relative slip amount [9, 10]. Therefore, it is expected that the fretting fatigue strength of the BUR axle may be improved by the surface rolling and the overhanging. Since the actual BUR has more than a 500 mm diameter at the axle part and weighs over 11 tons, the methods for improving the fretting fatigue strength are very limited.

The contact pressure between the BUR axle and inner race of the roller bearing is approximately 12 MPa. Interference of the fit at such a low contact pressure is too small for the 40 mm-diameter axle with regard to machining tolerance; therefore, the contact pressure of the axle assembly used in the experiment was 24 MPa. The fretting fatigue limit in the experiment was defined as the fretting fatigue strength at 2×10^7 cycles considering that the fracture of the BUR axles occurred at more than 10^7 cycles.

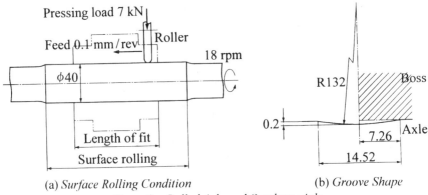

(a) *Surface Rolling Condition*   (b) *Groove Shape*

Figure 9 — *Rolled Axle and Overhung Axle*

The axle was made with SKD 6 forged steel. The boss was made with SFNCM alloy steel. Both materials are defined by Japanese Industrial Standards. The chemical compositions are shown in Table 1. The axle material was taken from the actual BUR. The boss material was the same as the actual inner race. Table 2 shows the mechanical properties of the materials used. Carburizing and quenching were done to the boss in order to make the hardness equal to that of the actual inner race. The boss is twice as hard as the axle in Vickers hardness.

The distribution of the axial residual stress of the surface-rolled axle is shown in Fig. 10. The maximum compressive stress was approximately 1000 MPa at the surface. The extent of the compressive residual stress was from the surface to an approximately 1.5 mm deep.

Results of Fretting Fatigue Tests

The results of the fretting fatigue tests are shown in Fig. 11. The fretting fatigue

Table 1 — *Chemical Compositions of the Materials (wt%)*

	C	Si	Mn	P	S	Ni	Cr	Mo	V
Axle (SKD6)	0.45	0.57	0.54	0.006	0.010	0.19	4.92	0.98	0.391
Boss (SFNCM)	0.27	0.80	0.3	0.006	0.001	3.65	1.84	0.46	0.103

Table 2 — *Mechanical Properties of the Materials*

	0.2%Proof Strength (MPa)	Tensile Strength (MPa)	Elongation at Fracture (%)	Reduction of Area (%)	Hardness (HV)
Axle	867	1066	19.0	53.1	339
Boss	733*	858*	21.6*	64.0*	670**

*Before Carburizing and Quenching, **After Carburizing and Quenching

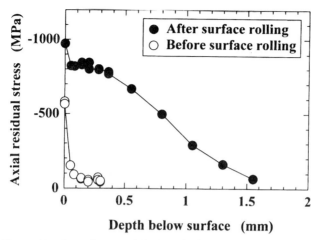

Figure 10 — *Distribution of the Residual Stress of the Rolled Axle*

limit based on the stress to break specimen, σ_{wf2}, of the normal axle was in the range from 155 MPa to 172 MPa. Since the fatigue limit of the BUR material obtained by the rotating bending fatigue test with the 12 mm-diameter specimens was 550 MPa [2], the fatigue limit decreased approximately one-third due to fretting. Since the fretting fatigue strength of fitted axles is significantly influenced by the diameter of the axle [11. 12], the fretting fatigue strength of the actual BUR axle is assumed to be lower than that obtained by the experiment.

The σ_{wf2} values of the rolled axle and the overhung axle were more than 182 MPa. The improvement in the σ_{wf2} of both axles were considered to be attained by the restraint of the fatigue crack growth due to the residual stress induced by surface rolling.

The fretting fatigue limit based on a stress to initiate cracks, σ_{wf1}, of the overhung axle was more than 97 MPa. The σ_{wf1} values of the normal and rolled axle were less than 97 MPa. The overhanging of the contact edge has a positive effect on the improvement of fatigue strength of the crack initiation.

The relative slip amount of the normal axle was measured by the displacement sensor attached near the contact edge. Since the contact edge of the overhung axle was covered by the overhung boss edge, the relative slip amount of the overhung axle could not be directly measured. Therefore, the relative slip amount of the overhung axle was evaluated by a finite element analysis [1]. Figure 12 shows the relative slip range obtained by the finite element analysis and the experiments. The parameter controlling the friction between the axle and the boss in the analysis was determined so that the analytical slip amount coincided with the experimental one. Since the axial stress near the contact edge was decreased by the groove, the relative slip amount of the overhung axle is small in comparison with that of the normal axle. A decrease in the relative slip amount leads to a decrease in the tangential force coefficient [13]. Therefore, it can be considered that the σ_{wf1} value of the overhung axle was improved.

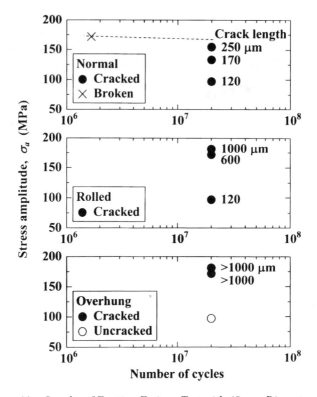

Figure 11 — *Results of Fretting Fatigue Test with 40 mm Diameter Axle*

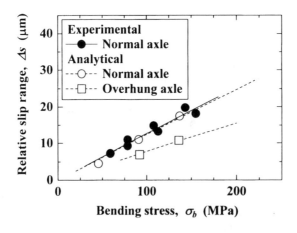

Figure 12 — *Relative Slip Range of 40 mm Diameter Axle*

Fretting fatigue cracks

Figure 13 shows the fretting fatigue cracks. The initiation site of the fretting fatigue cracks was not at the contact edge but at the fretted area inside the fitted part, which was from 0.5 mm to 2 mm from the contact edge. The observations of the fretting wear and fretting fatigue cracks show the same characteristics as the failed BUR. Since severe wear wore away minute surface cracks [13], the fretting fatigue cracks were observed at the fretted surface where the wear depth gradually decreased toward the center of the fitted part.

The crack lengths measured are shown in Fig. 11. Although the σ_{wfl} value of the overhung axle was higher than that of the rolled axle, the overhung axle had large cracks above 172 MPa in stress amplitude. When the overhung axle was manufactured, the grooves were cut after the surface rolling in order to keep the groove edge sharp. Therefore, the compressive residual stress near the groove was presumed to decrease. This could be the reason why the large cracks existed in the overhung axle.

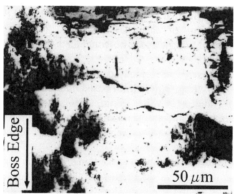

Figure 13 — *Fretting Fatigue Cracks (Normal Axle, σ_a = 97 MPa, N = 2 × 10⁷)*

Application to the Actual BUR

On the basis of this study, surface rolling was applied to the actual BUR. Furthermore, the diameter of the BUR axle was increased in order to decrease the stress at the fretted part. Four years have passed and no BUR has failed after these changes were made. To increase the diameter of the BUR axle is a difficult task, since reconstructing the steel making rolling mill should be done. Although the rolled axle endured 2×10^7 cycles with the higher stress amplitude compared with the normal axle in the experiment, many non-propagating cracks, which are as much as 1 mm or more, existed in the rolled axle. It should be noted that the BUR should endure more than 1×10^8 cycles of repetition load during its service life and that there is a kind of size effect on the fretting fatigue strength [11]. This is the reason why the increase in the BUR axle diameter for decreasing stress was achieved in cooperation with the roll user. Since the test result of the overhung axle indicated that the improvement of the fretting fatigue strength was less than expected, the overhanging of the contact edge was not adopted.

Conclusion

Recently, to comply with the demand of increased production efficiency, rolling load of the steel making rolling mill is steadily increasing. The backup roll (BUR) of the rolling mill was initially supported by oil film bearings. Nowadays, the oil film bearings have been rapidly replaced by roller bearing for the purpose of high-precision rolling. After the replacements, six backup rolls successively failed. The present failure analysis study of the BUR and fretting fatigue tests developed countermeasures for the actual BUR based upon the experiments. The results obtained are as follows.

1. The appearance of the surface of the failed BUR axle where the inner race of the roller bearing was fitted showed typical fretting fatigue of the fitted axles as in railway wheelsets. The failure life was estimated to be approximately 4×10^7 cycles. The cracks led the BUR axle to failure initiated within the rusted area, which is located 5 mm inside from the contact edge in the fitted part. Many small cracks, which propagated obliquely to the surface, were observed together with many small pits. Stress amplitude at cracked part of the BUR axle was estimated to be less than 130 MPa, which is considerably lower than the plain fatigue limit of the BUR material. Taking the above results of investigations into consideration, it could be concluded that the cause of the BUR axle failure was fretting fatigue.

2. Fretting fatigue tests were performed using shrink-fitted axle assemblies with a diameter of 40 mm. The fretting fatigue limit based on a stress to break a specimen, σ_{wf2}, was between 155 MPa and 172 MPa. Taking the size effect on the fretting fatigue strength of fitted axle into consideration, it can be considered that the fatigue strength of the BUR axle was not enough to survive the stress amplitude of the BUR axle caused by the rolling load.

3. The σ_{wf2} value of the rolled axle and the overhung axle were more than 182 MPa. The σ_{wf2} value of the both axles were considered to be improved because of the restraint of the fatigue crack propagation due to the residual stress induced by surface rolling.

4. The overhanging of the contact edge improved the fatigue strength based on the stress to initiate cracks σ_{wf1}. The relative slip amount of the overhung axle, which was estimated by the finite element analysis, was lower than that of the normal axle; therefore, the improvement of σ_{wf1} of the overhung axle is presumed to be due to decreasing of the tangential force coefficient with the decreasing amount of the relative slip. However, the crack length of the overhung axle was larger than that of the rolled axle. Since the grooves were cut after the surface rolling in order to keep the groove edge sharp, the compressive residual stress near the groove could be released.

5. The surface rolling was applied to the actual BUR in order to improve fretting fatigue strength of the BUR axle. The diameter of the BUR axle was also increased in order to decrease the stress amplitude at the fretted part. Since the test results of the overhung axle indicated that the improvement of the fretting fatigue strength was less than expected, the overhanging of the contact edge was not adopted. Four years have passed and no BUR has failed since the above countermeasures were applied.

References

[1]   Ohkomori, Y., Odanaka, H., Kubota, M., and Sakae, C., "The Analysis and Prevention of Failure in Steel Making Backup Roll," *Preprint of Japan Society of Mechanical Engineers Kyushu Branch Summer Meeting, The Japan Society of Mechanical Engineers*, No. 008-2, 2000, pp. 5-6, Japanese.

[2]   Ohkomori, Y., and Nagamatsu, T., "Fretting Fatigue Strength Properties of Journal Portion of Steel Making Backup Roll," *Preprint of Regular Meeting, The Iron and Steel Institute of Japan*, CAMP-ISIJ Vol. 12, 1999, pp. 1172, Japanese.

[3]   Hirakawa, K., "Case Histories and Prevention of Fretting Fatigue Failure," *Sumitomo Metals*, Vol. 46, No. 4, 1994, pp. 4-16, Japanese.

[4]   Whöler, A., "Versuche über Biegung und Verdrehung von Eisenbahnwagen-Achsen Während der Fahrt," *Zeit. Bauwesen*, Vol. 8, 1858, pp. 641-652.

[5]   Whöler, A., "Versuche über die Relative Festigkeit von Eisen," *Stahl und Kupfer*, Vol. 16, 1866, pp. 67-84.

[6]   Makino, T., Yamamoto, M., and Hirakawa, K., "Fracture Mechanics Approach to the Fretting Fatigue Strength of Axle Assemblies," *ASTM STP 1367*, D. W. Hoeppner, V. Chandrasekaran and C. B. Elliott III, Eds., American Society for Testing and Materials, West Conshohocken, PA, 2000, pp. 509-522.

[7]   Makino, T., Yamamoto, M., and Hirakawa, K., "The Fretting Fatigue Crack Initiation Behavior at Press-Fitted Axle Assembly with Variable Stress Loading (Effects of Stress Cycle Ratio, Number of Load Levels and Axle Size)," *Proceedings of 12th International Wheelset Congress*, 1998, pp.147-152.

[8]   Horger, O. J., and Maulbetsch, J. L., "Increasing the Fatigue Strength of Press-Fitted Axle Assemblies by Surface Rolling," *Journal of Applied Mechanics*, Vol. 3, 1936, pp. A-91-A-98.

[9]   Hirakawa, K., Toyama, M., and Kubota, M., "The Analysis and Prevention of Failure in Railway Axles," *International Journal of Fatigue*, Vol. 20, No. 2, 1998, pp.135-144.

[10]   Maxwell, W. W., Dudly, B. R., Cleary A. B., Richards, J., and Shaw, J., "Measures to Counter Fatigue Failure in Railway Axles," *Journal of the Institution of Locomotive Engineers*, Vol. 58, No.2, 1968, pp.136-171.

[11]   Hirakawa, K., and Kubota, M., "On the Fatigue Design Method for High-Speed Railway Axles," *Proceedings of the Institution of Mechanical Engineers, Journal of Rail and Rapid Transit*, Vol. 215, Part F, 2001, pp.73-82.

[12]   Kondo, Y., and Bodai, M., "Study on Fretting Fatigue Crack Initiation Mechanism Based on Local Stress at Contact Edge," *Transaction of Japan Society of Mechanical Engineers*, Series A, Vol. 63, No. 608, 1997, pp.669-676.

[13]   Nishioka, K., and Hirakawa, K., "Fundamental Investigation of Fretting Fatigue, Part 5, Effect of Relative Slip Amplitude," *Bulletin of Japan Society of Mechanical Engineers*, Vol. 12, No. 52, 1969, pp. 692-697.

[14]   Nishioka, K., and Hirakawa, K., "Fundamental Investigation of Fretting Fatigue, Part 3, Some Phenomena and Mechanisms of Surface Cracks," *Bulletin of Japan Society of Mechanical Engineers*, Vol. 12, No. 51, 1969, pp. 397-407.

Author Index

S

Sakae, Chu, 434
Sato, Kenkichi, 76
Shiozawa, Kazuaki, 385
Shirai, S., 353
Suzuki, Kazutaka, 385
Swalla, D. R., 89

T

Taylor, Amy M. H., 291

V

Vincent, Léo, 17

W

Watanabe, Takashi, 159
Waterhouse, Robert B., 3

X

Xu, Jin-Quan, 33, 61

Y

Yoshimura, Toshihiko, 423

Subject Index